2010

09

Mathematical Modelling and Computer Simulation of Activated Sludge Systems

Mathematical Modelling and Computer Simulation of Activated Sludge Systems

Jacek Makinia

IWA
Publishing
London · New York

Published by **IWA Publishing**
Alliance House
12 Caxton Street
London SW1H 0QS, UK
Telephone: +44 (0)20 7654 5500
Fax: +44 (0)20 654 5555
Email: publications@iwap.co.uk
Web: www.iwapublishing.com

First published 2010
© 2010 IWA Publishing

Originated by The Manila Typesetting Company
Cover by designforpublishing.co.uk
Printed by Lightning Source

British Library Cataloguing in Publication Data
A CIP catalogue record for this book is available from the British Library

Library of Congress Cataloging- in-Publication Data
A catalog record for this book is available from the Library of Congress

ISBN: 97818433932385
ISBN 10: 1843392380

Contents

Chapter 1
Introduction

1.1 HISTORY OF THE ACTIVATED SLUDGE PROCESS

1.1.1 Initial period

The activated sludge process currently represents the most widespread technology for the secondary treatment of municipal wastewater and constitutes "*the heart*" of many wastewater treatment plants (WWTPs) (Lessard and Beck 1991). The scale of activated sludge plants ranges from package plants [for single houses] to huge plants serving large metropolitan areas with flows up to $5 \cdot 10^6$ m^3/d (Grady *et al.* 1999). With regard to the invention and initial development of the activated sludge process, various workers in both the USA and UK contributed useful results and ideas (Cooper and Downing 1998).

Even though the last two decades of the nineteenth century research efforts had concentrated on treatment by the promising biological filtration theories, experiments on the aeration of sewage had been carried out since the early 1880's (Cooper 2001). It is now generally accepted that Dr R. Angus Smith initiated in 1882 the earliest research in "*blowing air*" into sewage tanks to minimize undesirable odour problems associated with putrefying sewage (Lester and Birkett 1999; Metcalf and Eddy

2003). According to Martin (1927), these were the first tests leading to development of activated sludge. The aeration of wastewater was investigated subsequently (1884-1897) by several workers in the UK including Dupre and Dibdin, Hartland and Kaye-Parry and Fowler. At the same time (1891-1894), similar research were also conducted in the USA by Drown, Mason and Hine, Lowcock and Waring (Ardern and Lockett 1914; Mohlman 1917). All these attempts were derived from the idea that aeration *"per se"* could provide the desired oxidizing effect on sewage. Poor experimental results (little improvement in effluent quality) revealed, however, that this approach could not be considered *"a practicable adjunct in the process of sewage purification"* (Ardren and Lockett 1914).

In the following years, the most notable work in the aeration of sewage was that performed by Black and Phelps for the Metropolitan Sewerage Commission of New York (1910), and by Clark and his colleagues at the Lawrence, Massachusetts, Sewage Experiment Station of the Massachusetts State Board of Health in 1912 and 1913 (Babbitt 1922). Black and Phelps studied in 1910 the possibility of aerating sewage for the Metropolitan Sewerage Commission of New York (Black and Phelps 1914). The sewage was aerated for varying periods up to twenty four hours in tanks filled with closely spaced, wooden laths in order to achieve a higher surface area for desired slime accumulation. The effects of oxidation were practically insignificant in terms of ammonia removal, although some reduction in putrescibility was indicated by the incubation tests. Black and Phelps had recommended the process for a full-scale installation but eventually the idea was not adopted (Mohlman 1917). A similar unit with wooden laths had earlier been used as an anaerobic contact chamber and called Travis *"Colloider"* or *"Hydrolytic"* Tank (Alleman and Prakasam 1983).

According to Metcalf and Eddy (1922), the idea from which the activated sludge process for the treatment of sewage has been developed appears to originate from a series of experiments during 1912 and 1913 at the Lawrence Experimental Station of the Massachusetts State Board of Health. Clark, Gage and Adams conducted there a series of successful experiments on using aeration for preliminary treatment of sewage prior to filtration. The aeration of sewage was conducted in both bottles, in presence of algal growths on the walls, and in a tank with vertical slate walls placed one inch (25 mm) apart. The attention of the investigators was focused on the attached growth on the walls of the aeration vessels and they did not claim that the aeration would entirely obviate filtration (Mohlman 1917). At this point, however, it is important to note that the *"adherent growths and heavy deposit"* were not thrown out from the bottles but were retained to assist in treating the next dose of raw sewage (Fowler 1934). Vesilind (2003) noted that *"the researchers at the Lawrence Experiment Station did all the right research, but they did not understand the significance of their results"*.

In November 1912, Dr. Gilbert J. Fowler of the Manchester University was called to the USA along with other specialists to report upon the proposals of the Metropolitan Sewerage Commission of New York for disposing of the sewage of greater New York which amounted to the enormous volume of 1,000 million gallons per day (Fowler 1934). Under such conditions, the current methods including tricking filters or chemical treatment seemed impractical solutions. While considering the problem of New York pollution, Fowler also visited the Lawrence Experimental Station and saw the on-going experiments on the aeration of sewage in the presence of green organisms. The results of the experiments in Lawrence impressed Fowler so much that, shortly after his return to the UK, Fowler described them to two chemists, Edward Ardern and William T. Lockett, from the Rivers Committee of the Manchester Corporation, and suggested to his colleagues that similar experiments should be carried out at the Davyhulme Sewage Works of Manchester. However, the objective was a process which could be operated in an open tank, without the aid of filters (Fowler 1934).

The remarkable results which Ardern and Lockett had obtained during the course of their experiments (1913-1914) were presented to the society of the Chemical Industry at the Grand Hotel, Manchester on 3 April 1914 (Ardern and Lockett 1914). During the discussion their paper was called an "*epoch-making one*" and a "*bombshell fired into the camp*". They first continuously aerated sewage in glass bottles until complete nitrification was achieved. In comparison with the other investigations, the investigation in Manchester considered two novel aspects. Firstly, the bottles were protected from light by covering with brown paper to prevent the growth of algae. Secondly, the bottles were not emptied completely after each aeration period but the deposited solids were mixed with a new portion of raw sewage. This procedure was repeated several times. In the first run, aeration for about five weeks was required to achieve complete nitrification. The amount of solids deposited in the bottles was gradually increasing and the time required for complete nitrification was eventually reduced to twenty four hours. Once having accumulated a sufficient volume of the deposited solids, a series of tests were carried out to determine the effects of aeration of various samples of the Manchester sewage. In general, a proportion of one volume of the solids to four volumes of sewage was used and a well oxidized effluent was obtained within a period of 6-9 hours. The deposited solids resulting from the oxidation of sewage were indeed a suspension of viable microorganisms and named "*activated sludge*". Ardern and Lockett reported that:

> "*Activated sludge accumulated in the manner previously described is quite inoffensive, dark brown in colour and flocculent in character, and despite its low specific gravity separates from water or sewage at a rapid*

rate. After prolonged settlement the activated sludge however rarely contains less than 95 per cent of water. (...) Gelatine counts have shown a bacterial content of at least 30 million organisms per cubic centimetre. In addition, the sludge by reason of its nitrifying power must of necessity contain a large number of nitrifying organisms. It should also be noted that a fairly large number of protozoa are to be found (...)"

In conclusions, the authors were convinced that the new method, because of its simplicity, would find widespread applications:

"The method employed in producing a satisfactory purification of sewage is however of so simple a nature, that there would not appear to be any insuperable difficulties in translating the experiments described, on to a working scale."

In August, 1914, Dr. Edward Bartow, professor of chemistry and director and chief of the State Water Survey in Illinois, visited Fowler's group in Manchester and saw the work in progress. Upon his return to the USA Bartow and Floyd William Mohlman started their own bench- and pilot-scale experiments with activated sludge at the University of Illinois (November, 1914) (Mohlman 1917). Activated sludge was built up in the manner suggested by Ardern and Lockett (1914). The main finding was that during the aeration of sewage in contact with activated sludge, ammonia was oxidized to nitrate during 4-5 hours, whereas nitrite was evidently oxidized to nitrate almost as rapidly as it was formed. Furthermore, satisfactory activated sludge could be obtained with 6-hour aeration periods without complete nitrification needed from the beginning of the operation. In addition to the time of aeration, the study addressed several far-reaching issues, such as the required area for air diffusion, required amount of sludge for purification, quantity of sludge formed, composition of sludge including the content of nitrogen, dewatering of sludge and cost of the activated sludge process.

Indeed, within the period of a few years, development of the activated sludge process was proceeding very rapidly and an enormous amount of experimental work was initiated throughout the world. Carpenter and Horowitz (1915) reported that the literature on the subject of activated sludge was *"so recent, and so well known to the sanitary engineers"*. Porter's bibliography (Porter 1921) contains over 600 abstracts of the papers written between 1914 and 1920. Buswell (1923) noted that the bibliography listed over 80 experimental plants and 17 municipal activated sludge plants which had been completed or under construction. The first pilot and full-scale applications of activated sludge were based on a fill-and-draw operational mode (a precursor of the modern sequencing batch reactors), which was soon converted to continuous flow through aeration tanks, followed by sedimentation and biomass recycle (Grady *et al.* 1999). Babbitt (1922) described an activated sludge reactor as a rectangular tank with a depth of about 15 feet and a width of channel not to exceed 6 to 8 feet. According to the author, such proportions would allow better air and current distribution than larger tanks, and the level bottom should insure an even distribution of air.

The pioneering fill-and-draw laboratory studies of Ardern and Lockett were shortly thereafter successfully repeated in a pilot scale at Manchester's Davyhulme sewage treatment works. The first full-scale activated-sludge plant in England was put into operation in Worcester in 1916 (Cooper 2001). However, the major activated sludge plants, such as Mogden in London (which served 1.25 million people), Davyhulme in Manchester and Coleshill in Birmingham, were not built until mid-1930's. There were two principal reasons for this delay. Firstly, capital for investment was very limited in the UK after the First World War. Secondly, all the major cities had already invested in the biological filter technologies for sewage treatment in the period between 1890 and 1910 (Cooper 2001).

In the USA, by contrast, the development was more rapid and many of the activated sludge plants were the first form of sewage treatment ever used (Cooper 2001). Full-scale installations began to appear at about the same time as the experiments of Bartow and Mohlmann at the University of Illinois. In Milwaukee, a plant with the capacity of 1,600,000 gallons per day was erected in December, 1915. The plant was used for experimental purposes and was closed by 1922 Babbitt (1922). Platt (2004) noted that over a relatively short period of three years, Manchester (Fowler's group), Urbana (Bartow's group) and Milwaukee had become international leaders. They established the new method which became far superior to all previous methods of sewage treatment and disposal. By 1927 there were several full-scale systems in the US spread throughout the country (Table 1.1). Among them, Chicago North was largest with the capacity of 660,000 m^3/d. Stickney, another Chicago plant, went into operation in 1930 and was expanded in 1939. With the design capacity of over 4,500,000 m^3/d (1,200 MGD), Stickney is currently the largest activated sludge facility in the world. Due to litigations on patent infringements, a really widespread use of the activated sludge process in the US did not begin until the 1950s (Alleman and Prakasam 1983).

During the 1920's, the activated sludge process was gradually commenced in other countries. In 1924. Ontario, Canada had 7 municipal and 11 institutional activated sludge plants with a total capacity of 10,235,000 gallons per day (Wolman 1924). The first application in the continental Europe (outside the UK) took place in Denmark in the Soelleroed Municipality in 1922 (Cooper 2001). The first experimental plant in Germany was built and run in Essen by Karl Imhoff in 1924. It was a rectangular glass container with the volume of 0.5 m^3 which was divided into aerating and settling compartments in the ratio 6:1 (Miller 1927). Soon, the first full-scale plants in the country were built in Essen-Rellinghausen (1926) and in Stahnsdorf near Berlin (1929–1931) (Seeger 1999). The latter one was designed and operated as an experimental plant for the study of different sedimentation tank options. Miller (1930) reported brief descriptions of eight *"interesting"* plants in Germany. In 1927, an abattoir effluent at Apeldoorn (Holland) was treated using an activated sludge process equipped with a brush aerator developed by Kessener (Cooper 2001).

Table 1.1 List of early activated sludge installations in the UK and USA (Alleman and Prakasam 1983)

Year	Location	Flow rate, m³/d	Operational mode	Aeration system
	UK			
1914	Salford	303	Fill-and-draw	Diffused
		45	Continuous-flow	Diffused
1915	Davyhulme	378	Fill-and-draw	Diffused
1916	Worcester	7 570	Continuous-flow	Diffused
1917	Sheffield	3 028	Fill-and-draw	Mechanical
1917	Withington	946	Continuous-flow	Diffused
1917	Stamford	378	Continuous-flow	Diffused
1920	Tunstall	1 104	Continuous-flow	Mechanical
1920	Sheffield	1 340	Continuous-flow	Mechanical
1921	Davyhulme	2 509	Continuous-flow	Diffused
1921	Bury	1 363	Continuous-flow	Diffused
	USA			
1916	San Marcos, Texas	454	Continuous-flow	Diffused
1916	Milwaukee, Wisconsin	7 570	Continuous-flow	Diffused
1916	Cleveland, Ohio	3 785	Continuous-flow	Diffused
1917	Houston, Texas	20 817	Continuous-flow	Diffused
1917	Houston, Texas	18 925	Continuous-flow	Diffused
1922	Des Plaines, Illinois	20 817	Continuous-flow	Diffused
1922	Calumet, Indiana	5 677	Continuous-flow	Mechanical
1925	Milwaukee, Wisconsin	170 325	Continuous-flow	Diffused
1925	Indianapolis, Indiana	189 250	Continuous-flow	Diffused
1927	Chicago, Illoinis	662 375	Continuous-flow	Diffused

Yet before the World War II, the number of full-scale activated sludge plants could be counted in hundreds and was steadily increasing (Cooper and Downing 1998). For example, the number of U.S. activated sludge facilities raised to 203 in 1938 (Alleman and Prakasam 1983). The process reached other continents then Europe and America. For example, the Japanese city of Nagoya installed an activated sludge plant in December, 1924 and the daily volume of sewage treated was 15,600 ft³ (440 m³) (Miller 1927). While reviewing the initial twenty-five-year history of the activated sludge process for the Federation of Sewage Works Association in the USA, Mohlman noted that the process had been applied in five continents (cited in Cooper (2001):

> *"In 1938, the activated sludge process is in operation in hundreds of full-scale sewage treatment works and more than a billion gallons of sewage are treated every day. Activated sludge plants are now operated all over the world, extending from Helsinki, Finland to Bangalore, India; from Flin Flon, Manitoba, Canada to Glenelg, Australia; and from Golden Gate Park, San Francisco to Johannesburg, South Africa. Huge plants are in operation at London, New York, Chicago, Cleveland and Milwaukee."*

Almost 30 years later, Sawyer (1965) reviewed the developments and updates in the activated sludge process (*"as Dr. Mohlmann did similarly"*) and *"this new knowledge has shown the process to be extremely adaptable and, as a result, many modifications have been proposed to meet certain requirements or condition"*. In the period from the 1930s to the early 1950s, the major operational problems that motivated these modifications were sludge bulking, and oxygen transfer and utilization (McKinney, 1957). Lawrence and McCarthy (1970) summarized the ranges of process loading factors and SRTs for most common modifications of the conventional activated sludge process, such as extended aeration, tapered aeration, step aeration, contact stabilization, short term aeration (high loaded). More detailed descriptions of these systems can be found in Orhon and Artan (1994) (including the flow sheets) and Jeppsson (1996). It should be noted that not all modifications are recognized nowadays. For example, the so-called "Z" process, invented in 1944, faded into obscurity as it used asbestos fibers in the aeration tank to enhance floc sedimentation properties (Alleman and Prakasam 1983).

In the initial period of application of the activated sludge process, the discharge requirements were mainly to meet standards for suspended solids (SS) and 5-day Biochemical Oxygen Demand (BOD₅), e.g. the so-called *"Royal Commission 30:20 Standard"* in the UK. By the 1950s, effluent limits for ammonia concentration were started to be imposed for plants discharging into rivers providing water for public supply systems. Those limits were subsequently becoming more widespread and more stringent (Downing and Cooper 1998). In the 1970s, discharge requirements for nitrite and nitrate started to be enacted as well (Gujer 2010).

1.1.2 Biological nitrogen removal

Nitrification and denitrification were observed in the 19th century before the invention of the activated sludge process. Pasteur was first who suggested that the oxidation of ammonia to nitric acid taking place in nature was really due to microorganisms, and two French chemists, Schlösing and Muntz, actually proved that hypothesis (Fowler 1934). In their pioneering experiments, Ardern and Lockett (1914) reported the occurrence of complete nitrification after 10-18 hours when aerating normal Manchester sewage. The aeration time was reduced to 6 hours when treating wet weather sewage. In the early investigations of the activated sludge process considerable attention was paid to the degree of ammonia oxidation, but nitrification was not considered to be "*essential to the success of the [activated sludge] process*" (Buswell 1923). Until the early 1960's, nitrification was not reliable or predictable in activated sludge systems (Cooper 2001).

Skinner and Walker (1961) were among the first to investigate nitrification in continuous culture and obtained steady-state cultures of *Nitrosomonas europea* in an ammonia-limited chemostat. Boon and Laudelout (1962) studied the effects of different environmental conditions on the kinetics of the nitrite oxidation by Nitrobacter. A "*landmark*" work, as called by Barnard (2006), was performed by Downing *et al.* (1964) at the Water Pollution Research Laboratory (WPRL) in Stevenage (UK). The authors demonstrated that for consistent nitrification (i.e. preventing the washout of the slowly growing autotrophic microorganisms), the period of aeration of mixed liquor in the fully aerobic bioreactors would have to greater than the retention time in the bioreactor, t_m, defined as:

$$t_m = \frac{\Delta X}{\mu_N X} \tag{1.1}$$

where
ΔX - increase in concentration of activated sludge (X) during transit through an assumed plug-flow bioreactor
μ_N - specific growth rate of *Nitrosomonas spp* (assumed to be the organisms converting ammonia to nitrate), T^{-1}

The concept and results of the study of Downing *et al.* (1964) were later incorporated into design methods and mathematical models of activated sludge systems (Cooper 2001).

Denitrification was first studied in 1882 by Gayon and Dupetit who found that when a solution containing potassium nitrate together with sewage and a little urine was allowed to stand in absence of air the nitrate was reduced. Moreover, positive effects of the addition of organic compounds, such as carbohydrates, tartrates, were observed. It was concluded that denitrification was essentially the "*combustion*" of organic material by the nitrate oxygen and thus the process proceeded "*best*" in presence of a minimum air supply (Fowler 1934). In the early experiments with activated sludge, Buswell (1923) observed that the nitrate oxygen was utilized as a source of oxygen by the microorganisms in the sludge when low quantities of air were added while treating nitrate rich sewage. Fowler (1934) noted that the subject of denitrification has been investigated by numerous workers, notably Percy, Frankland and Beyerinck. Cooper and Downing (1998) found the studies of Kershaw and Finch (1936) and several years later by Edmondson and Goodrich (1943, 1947) as particularly significant in terms of the subsequent development of activated sludge systems. In both studies, the effluent from nitrifying biological filters was diverted into second-stage activated-sludge units where the oxidized nitrogen was effectively removed. Moreover, in the case of the latter study, sludge settleability in the activated-sludge unit significantly improved in comparison with the previously overloaded biological filter. Barnard (2006) reported another early study on denitrification while investigating sludge rising problems in final clarifiers (Sawyer and Bradney 1945).

Biological nitrification/denitrification as a process for nitrogen removal from wastewater gained attention during the early 1960s. Wuhrmann (1962, 1964) proposed a configuration of two tanks in series in a one-sludge system, known as the Wuhrman (or post-denitrification) process (Figure 1.1). A separate denitrification (anoxic) compartment was added after the aerobic compartment for denitrification using stored carbon in a high-rate process (Barnard 2006). As the process depended upon internal carbon sources, therefore its potential was limited. In fact, the original process configuration was not applied in full-scale without supplemental carbon addition (USEPA 1993). In Sweden, for example, WWTPs for carbon and chemical phosphorus removal (pre-precipitation) were commonly extended with post-denitrification where methanol was added for denitrification (Nyberg *et al.* 1992). A similar concept was also demonstrated in full-scale at the Blue Plains WWTP in the District of Columbia (Bailey *et al.* 1998). Ludzack and Ettinger (1962) introduced the "*semi-aerobic activated sludge process*", called the Ludzack-Ettinger process (Figure 1.1), to reflect simultaneous nitrification and denitrification observed in channel systems. The raw wastewater entering the "*semi-aerobic*" zone provided organic (readily biodegradable) carbon for denitrification. As the two zones ("*semi-aerobic*" and aerobic) were only partially separated, recycling of dissolved oxygen to the "*semi-aerobic*" compartment was induced, resulting in a low efficiency of nitrogen removal (35-50%) with weak sewage (Barnard 2006).

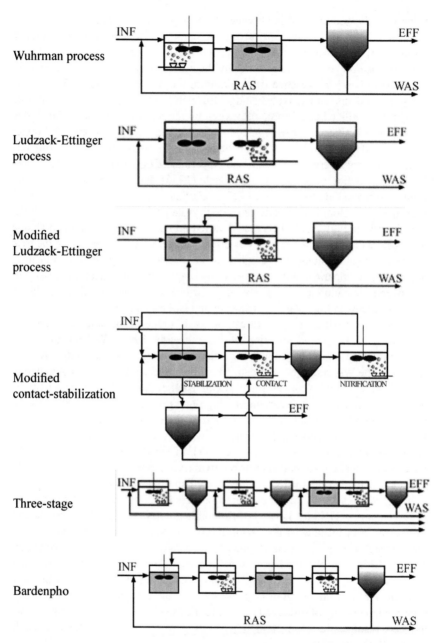

Figure 1.1 Process configurations of the activated sludge systems for nitrogen removal.

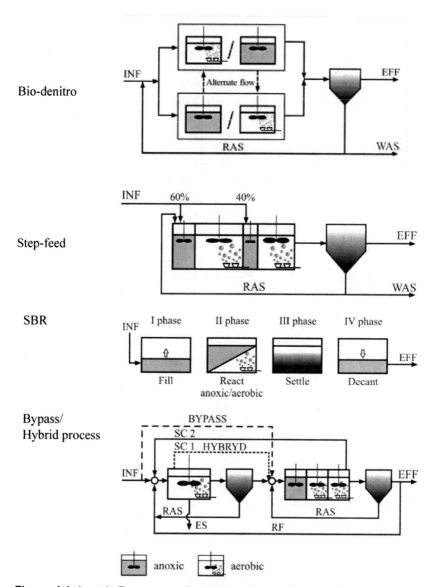

Bio-denitro

Step-feed

SBR

Bypass/
Hybrid process

anoxic   aerobic

Figure 1.1 (cont.). Process configurations of the activated sludge systems for nitrogen removal.

Fundamental studies for the development of modern biological nutrient removal activated sludge systems were initiated in South Africa at the beginning of the 1970's. Due to a water shortage and a rapidly growing population, the

South African water industry has protected its water resources very carefully and developed recycling processes (Cooper 2001). In some cases, however, eutrophication had already been a serious threat. For example, the receiving reservoirs for Johannesburg and Pretoria effluents had a "*pea-soup consistency*" (Barnard 1998). The first important contribution was made by Barnard (1973) who modified the Ludzack-Ettinger process by completely separating the anoxic and aerobic compartments and providing an additional recycle from the aerobic to the anoxic compartment. This process is known as modified Ludzack-Ettinger (MLE or pre-denitrification) process (Figure 1.1). It should be noted that Barnard developed this configuration by experimenting and modifying another flow sheet, called (modified) contact-stabilization (Figure 1.1), which had been proposed by Balakrishnan and Eckenfelder (1970) at the University of Texas. Because the work of Barnard was carried out within the framework of a water-reclamation project, where a very low nitrogen concentration in the effluent was required, the Wuhrmann concept was added to remove the remaining nitrate, resulting in the four-stage process which became the precursor of the Bardenpho process (Figure 1.1). This process was tested in pilot-scale (Q = 100 m^3/d) at the Daspoort (Pretoria) WWTP and later expanded to full-scale at 750,000 PE Goudkoppies (Johannesburg) WWTP in South Africa. Both plants were capable of achieving 92-95% removal of the total nitrogen.

In parallel to the process development in South Africa, interest in the removal of nitrogen in the UK led to the conversion of the Rye Meads activated-sludge plant, originally designed and built as a pre-denitrification system (1973), to a two-anoxic zone step-feed (cascade denitrification) process (Figure 1.1), in which 60% of the settled sewage was fed to the first anoxic zone and 40% to the second zone. The efficiency of nitrogen removal reached 77% and effluent nitrate concentrations dropped to less than 10 mgN/dm^3 (Cooper and Downing 1998). Full-scale applications of the step-feed concept, with up to three anoxic zones, were subsequently reported in other countries, such as Japan (Miyaji *et al.* 1980), Germany (Kayser *et al.* 1992; Schlegel 1992), USA (Fillos *et al.* 1996) and Canada (Barnard 1998). Recently, Johnson *et al.* (2005) compared the actual vs. theoretical benefits of implementing step-feed based on reviewing nine case studies in the USA (6 plants), Canada (1), New Zealand (1) and Singapore (1). The plants had three to six passes and were designed for nitrogen removal (7 plants) or combined nitrogen and EBPR (2 plants). The authors concluded that there were two primary reasons for considering a step-feed system including a reduced bioreactor volume for a defined capacity or performance (or increased process capacity given a fixed bioreactor volume) and more robust nitrification performance. In general, the step feeding flow configuration is considered an attractive process alternative for systems designed for nitrogen removal, primarily due to eliminating the need for internal recycling and using effectively organic carbon for denitrification. Moreover, in comparison with single-feed systems, step feeding enables to carry a higher solids inventory, i.e. a longer solids retention time (SRT) while reducing the solids loading in the feed to the secondary clarifiers. Important design parameters of the step-feed systems are volume ratios of anoxic and aerobic compartments, influent flow distribution and wastewater characteristics (especially C/N ratio) (Gorgun *et al.* 1996a).

Multi-stage (or separate-stage) systems have also been used to remove nitrogen based on the activated sludge process. The first two-stage system was put into operation at the Essen-Rellinghausen WWTP (Germany) in 1928, but the process was not adequately understood at the beginning (Imhoff 1955). The author also noted that the terms "*two-stage*" and "*multi-stage*" should only be used only when each stage includes its own aeration tank and its own secondary settling tank. A novel AB (Adsorption-Biooxidation) process was developed in the 1970s at the Aachen University of Technology (Bohnke 1977), when some German WWTPs had to work overloaded due to the lack of adequate land area for extension. Since the early 1980 the AB process has been commonly used in Germany. The inclusion of nitrogen removal is possible in this two-stage configuration when the second stage designed as a pre-denitrification system (the first high-loaded stage is primarily designed for carbon removal) (Rosenwinkel *et al.* 2007a).

In the USA, separate-stage systems using methanol as a carbon source for denitrification were widely considered in the 1970s, however, relatively few such systems were actually constructed (Sedlak 1991). At that time, a pilot plant at the Blue Plains WWTP (Washington DC) tested even a three-sludge system (Figure 1.1). Organic carbon was removed in the first stage, nitrification was enhanced in the second stage and methanol was added in the third stage for denitrification. Nitrogen removal to less than 3 mg N/dm^3 was reported (Barnard 1998). Another multi-stage process configuration developed in the USA was the above-mentioned modified contact-stabilization process (Balakrishnan and Eckenfelder 1970).

Recently (2000-2005), a flexible two-stage system was designed and implemented at the Vienna main WWTP. The novel configuration, presented in Figure 1.1, allows to run three different operational modes including Hybrid®- and Bypass- in addition to the conventional two-stage operation (Müller-Rechberger *et al.* 2001). The Hybrid®-mode can mainly be operated during dry weather conditions, when the influent flow rate does not exceed the maximum hydraulic loading of the first stage. If the influent flow rate to the plant exceeds this limit, the excess water bypasses the first stage and is directly sent to the second stage with the capacity greater by 50% compared to the first stage. In the Bypass-mode, a portion of the incoming flow bypasses the first stage under all weather conditions. Under normal operating conditions, the excess sludge is withdrawn from the first stage only. The excess sludge of the second stage (containing nitrifying bacteria) is pumped into the first stage and thus nitrification is enabled also there. The second stage is subdivided into a cascade of three compartments. The first compartment acts as a denitirifcation zone (15% of the volume), whereas compartments 2 and 3 are circulation tanks with aerated and anoxic zones for simultaneous nitrification and denitrification (Papp and Zelinka 2007).

In contrast to more and more sophisticated process configurations, a different line of modifications of the activated sludge process was related to changing the flow conditions from continuous to fill-and-draw. Even though the first pilot and full-scale activated sludge systems were based on the same basis, this operational mode did not receive attention for the next several decades. The sequencing batch reactor (SBR) is an activated sludge system operated on a fill-and-draw basis (Figure 1.1). With the advances in process control, the renewed interest in the SBR technology appeared in the early 1970s (Irvine and Davis 1971) and most intensive studies were conducted at the University of Notre Dame, Indiana (Irvine and Busch 1979). There is currently abundant recent literature reviewing the technology in terms of treatment performance, operation (including modes for nitrogen and phosphorus removal), and costs (e.g. Wilderer et al. 2001; Mace and Mata-Alvarez 2002; Keller 2005). The major advantages of the SBR can be attributed to two features: efficiency and flexibility. The process is very efficient in treatment municipal wastewater, different kinds of industrial effluents and landfill leachates. A very flexible operation is possible due to controlling the process temporarily (by adjusting the duration of each phase) rather than spatially.

The fill-and-draw operational mode with continuous feeding were used in popular oxidation ditches. These simple and low-cost treatment systems, developed in the Netherlands (Pasveer 1959), consist of one treatment unit without a clarifier. They were initially designed for small communities, however, the advantages led a wide-spread application including both large communities (up to several hundred thousand PE) and a variety of industrial wastewaters (Cooper and Downing 1998). The original ditch systems subsequently evolved to the continuously operated ditches with a separate clarifier. In Denmark, the oxidation ditch technology was modified in the 1970s as the phased isolation ditches which are continuous-flow activated sludge systems with intermittent operation (Bundgaard et al. 1989). The use of multiple ditches enables to operate the system with nitrogen removal, called Bio-Denitro (Figure 1.1). The operating mode consists of four phases with a total duration of 3-6 hours, during which both the flow direction and process conditions (aerobic/anoxic phases) in the bioreactors are controlled automatically. Bundgaard and Petersen (1991) reported that the Bio-Denitro process was implemented at approximately 35 Danish WWTPs and more than 20 were in the construction stage. The capacity of the plants varied from 1,000 to 450,000 PE and the effluent nitrogen concentrations were normally below 8 mg N/dm^3 (as low as 4 mg N/dm^3 under optimum conditions).

1.1.3 Enhanced biological phosphorus removal (EBPR)

The first observation of biological phosphorus removal beyond metabolic needs of the activated sludge was made in the mid-1950's by Greenburg *et al.* (1955). A few years later, Srinath *et al.* (1959) reported biological phosphorus removal in a plug-flow activated sludge plant. Levin and Sharpiro (1965) referred to this excess phosphorus removal as "*luxury uptake*" and proved that it was a biological mechanism. Based on that work, Levin (1970) developed a commercial (patented) process, named the Phostrip process (Figure 1.2) which combined enhanced biological phosphorus removal (EBPR) and chemical precipitation of phosphorus. At the time of Levin and Shapiro's work, high levels of phosphorus removal (reaching 90%) was also reported at several full-scale conventional activated sludge systems including San Antonio, Los Angeles and Baltimore (USEPA 1987a). A common feature of those plug-flow bioreactors was that phosphorus was released at the inlet zone where anaerobic conditions were very likely, whereas rapid phosphorus uptake occurred in the latter part of the bioreactors at elevated dissolved oxygen (DO) levels. Barnard (1976) postulated that an anaerobic zone, followed by an aerated zone, were necessary to obtain EBPR. This concept of a separate anaerobic basin ahead of the activated sludge bioreactor was generally termed Phoredox process, which stands for phosphorus and redox potential to emphasize the lower reduced conditions required in the anaerobic zone (Figure 1.2).

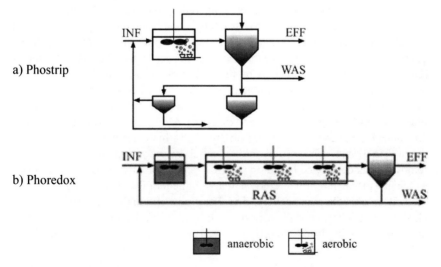

Figure 1.2 Activated sludge systems for phosphorus removal.

1.1.4 Integrated EBPR and nitrogen removal

The base for modern activated sludge systems for integrated phosphorus and nitrogen removal were investigations made by Barnard (1974, 1975, 1976) while studying the four-stage Bardenpho pilot bioreactor at the Daspoort WWTP (see above). Barnard observed a stable phosphorus reduction from 8 mg P/dm^3 to less than 0.2 mg P/dm^3, but only when a release of phosphorus (up to 30 mg P/dm^3) was taking place in the second anoxic zone. Barnard proposed the use of a separate anaerobic basin ahead of the four-stage Bardenpho system, thereby creating the modified (five-stage) Bardenpho process (Figure 1.3), or ahead of aerobic basins when nitrogen removal was not necessary (Figure 1.2). When only partial nitrogen removal was required, it could be reduced to three stages – anaerobic, anoxic and aerobic. This process is often referred to as the three-stage Bardenpho process or the A_2/O process (Figure 1.3). The concept of an initial anaerobic zone was first applied, in 1974, to the design of the Meyerton WWTP near Johannesburg. Experiments were also run at the Alexandra WWTP to verify the concept and anaerobic zones were also incorporated at the Goudkoppies WWTP which had already been under construction (Barnard 1998). Furthermore, Barnard (1976) postulated that the anaerobic zone should be prevented from any recycled nitrate loads from the clarifier or aerobic zone. This consideration later led to the development of other South African concepts, such as the University of Cape Town (UCT) process (Rabinowitz and Marais 1980), the modified UCT (MUCT) process (Siebritz et al. 1980) and the Johannesburg (JHB) process (Nicholls et al. 1987). These configurations were compared by Wentzel et al. (1992) and are illustrated in Figure 1.3. After constructing a number of other plants in South Africa followed and, these biological nitrogen and phosphorus-removal process configurations were patented in the United States in 1976 as the A/O flow sheets (Barnard 1998). A modification of the JHB process was implemented at the Westbank plant in British Columbia (Canada) (Figure 1.3). A constant base feed of primary effluent (60-90%) is added to the anaerobic zone, while the remainder plus storm flow are passed to the anoxic zones. All the VFAs from the fermenter are addend to the anaerobic zone (Barnard 1998). In the Netherlands, the BCFS® process (a Dutch acronym for biological-chemical phosphorus and nitrogen removal) was developed based on the UCT process configuration. The aim was to achieve efficient nutrient removal at low influent ratios of BOD/N and BOD/P as well as phosphorus recovery. The latter can optionally be applied by the chemical precipitation of phosphorus-rich water from the anaerobic zone using an in-line stripping installation (Figure 1.3) (van Loosdrecht et al. 1998; Meijer 2004).

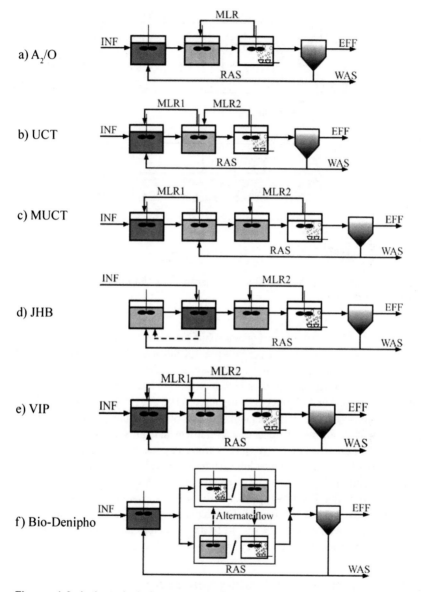

Figure 1.3 Activated sludge systems for integrated nitrogen and phosphorus removal.

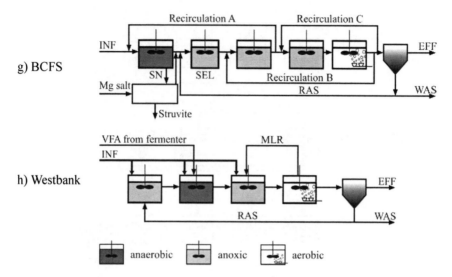

Figure 1.3 (cont.). Activated sludge systems for integrated nitrogen and phosphorus removal.

In the USA, EBPR facilities using anaerobic-aerobic zones were started up in the late 1970s at two sites in Florida. An inventory of full-scale EBPR facilities, carried out in 1984, revealed that 30 such facilities were identified as being in operation (11 facilities), under construction or design phase with 28 of these being either a Phostrip, modified Bardenpho or A/O process configuration (USEPA 1987a). In the mid-1980s, in response concerns about nutrient loading effects on water quality in the Chesapeak Bay, an extensive an overall reduction in nutrient loading (15-month) pilot plant program at the Lamberts Point WWTP in Norfolk, Virginia, was launched. The aim of that program was to investigate a biological nitrogen and phosphorus removal concept for upgrading secondary treatment. As a result, a new activated sludge process, called the Virginia Initiative Plant (VIP) process (Figure 1.3), was designed for the full-scale 150,000 m^3/d (40 MGD) facility (Daigger *et al.* 1988). The VIP was meant as a high-rate process with a total SRT of 5 to 10 days, whereas the similar UCT process configuration is generally designed for longer SRTs (13-25 days).

1.1.5 Nitrogen removal in sidestream processes

In WWTPs with sludge digestion, the sidestreams including return liquors (reject water) from dewatering processes and other internal flows can have detrimental effects to the mainstream treatment as early observed by Kappe (1957). The reject water flow rates are intermittent (depending on the schedule of dewatering) and typically constitute only 0.5-2% of the total influent flow rate. However, due to high ammonia concentrations (800-1200 mg N/dm^3), the recirculated reject water can contribute to 10-30% of the nitrogen load entering the biological stage (van Loosdrecht and Salem 2006; Fux *et al*. 2006; Gali *et al*. 2007; Rosenwinkel *et al*. 2007a; Smith and Oerther 2007). Furthermore, the reject water is warm (15-35 °C) with relatively low concentrations of alkalinity and biodegradable COD. This makes it characteristics completely different from raw municipal wastewater but removing ammonia from the sidestreams has a few benefits depending on the concept used. In the past decade, several side-stream treatment configurations have been developed for incorporation into the flow sheet of mainstream treatment while upgrading WWTPs (Figure 1.4). These configurations can essentially be grouped under two different concepts including treatment in the RAS lane (bioaugmentation) and separate treatment.

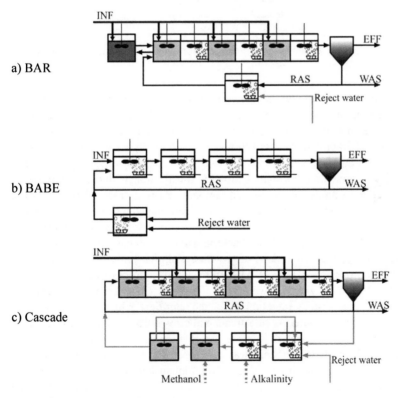

Figure 1.4 Process configurations for nitrogen removal in sidestream processes.

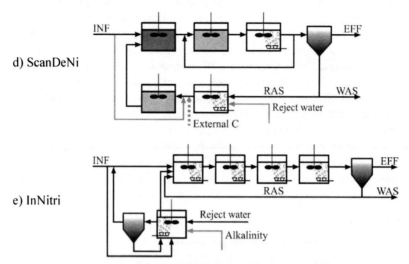

Figure 1.4 (cont.). Process configurations for nitrogen removal in sidestream processes.

The term "*bioaugmentation*" refers to processes aimed at the production of nitrifying bacteria in sidestream systems and their subsequent use for seeding in the mainstream bioreactors to support and/or enhance nitrification (Constantine *et al.* 2005). Parker and Wanner (2007) reviewed two types of bioaugmentation schemes, including external and "*in situ*", in activated sludge systems. The external bioaugmentation schemes promote nitrification within the mainstream process while being decoupled from its aerobic SRT (e.g. In-Nitri® process). The "*in situ*" schemes provide internal process enhancements which increase activity or enrich nitrifier population. In such cases, there is less concern about the loss of activity of the seeded nitrifying bacteria while transferring to the mainstream process, because their conditions of growth are similar to those prevalent in the mainstream process (e.g. BAR, BABE and Scan-Deni processes).

The In-Nitri®, or "*Inexpensive Nitrification*" process is basically an external activated sludge system (aeration tank and clarifier) which aims to oxidize ammonia from the dewatering sidestream and grow nitrifying bacteria at elevated temperatures. The generated bacteria are then used to seed the mainstream bioreactor. The In-Nitri® process was developed for plants in temper climates that did not have an adequate capacity to maintain long aerobic SRTs required for nitrification at low temperatures (Kos *et al.* 2000). As nitrifying bacteria are cultivated at the elevated temperatures in the external reactor, there were concerns that the bacteria would be subject to temperature shock conditions when transferred to the mainstream systems (Parker and Wanner 2007). Studies with SBRs at a lab scale showed that nitrification rates after transferring to the mainstream reactors were indeed lower (but no more than expected) and the changes proceeded gradually (Head and Oleszkiewicz 2004).

The Bioaugmentation R Process (R stands for regeneration zone in the Czech Republic or reaeration basin in the USA), called the BAR process, involves combining the RAS and reject water in a regeneration (reaeration) tank. The BAR concept was developed in the Czech Republic thanks to the full understanding of processes occurring in the regeneration (reaeration) zone of the so-called R-D-N (Regeneration-Denitrification-Nitrification) (Parker and Wanner 2007). This modification of activated sludge process was originally developed with two main goals including control of the excessive growth of filamentous bacteria and improvement of the conditions for the growth of nitrifying bacteria (Wanner et al. 1990). The positive effect of "in situ" bioaugmentation on the enhancement and stabilization of mainstream nitrifying population in activated sludge under various operating modes has been confirmed in several full-scale plants in the Czech Republic (Krhutkova et al. 2006). In the ScanDeNi® process (Rosen and Huijbregsen 2003; Rosenwinkel et al. 2007a), the aerobic tank in the RAS lane (receiving reject water) is followed by an anoxic tank where a carbon source is added (internal, typically a portion of the influent, or external). The ScanDeNi® process was developed in Sweden and first applied in 1998 at the Vasteras WWTP (150,000 PE), near Stockholm (Rosen and Huijbregsen 2003) The authors reported that the mainstream process would require 15-25% less volume or allow a 25-35% higher load.

The Biological Augmentation Batch Enhanced, or BABE®, process (Zilverentant 1999) is very similar to the BAR concept except that only a limited amount of return sludge from the mainstream process is passed through the sidestream reactor. In this way, the mainstream process can be operated at half of the minimum SRT. Furthermore, the BABE® process can also be extended by introducing extra denitrification space/phase in low loaded systems, on the expenses of the aerobic retention time, and hence reducing the effluent total N concentration (van Loosdrecht and Salem 2006). A full-scale application of the process took place at the Garmerwolde WWTP in the Netherlands. Operating results showed that the bioaumentation effect of the BABE® process improved the specific nitrification rate of the activated sludge by almost 60% (Salem et al. 2004).

Systems for separate treatment of reject water significantly reduce the nitrogen load to the main stream and improve overall nitrogen removal. The implementation of these systems usually requires new tanks or a comprehensive modification of available reactors. However, the overall investment costs are reduced as the separate treatment can be operated at high volumetric reaction rates. Nitrogen removal from reject water can be performed using the traditional nitrification-denitrification processes. However, the higher temperature and ammonia concentration allow a few alternatives for biological nitrogen removal including partial nitrification to NO_2-N (nitritation), followed by reduction of NO_2-N (denitritation), or autotrophic nitrogen removal via a novel Anammox process (see: Section 3.11). Separate full-scale sludge liquor treatment with various concepts for nitritation-denitritation has already been introduced at several WWTPs, primarily in the Netherlands, Austria, Switzerland and Germany (van Loosdrecht and Salem 2006; Fux *et al.* 2006; Rosenwinkel *et al.* 2007a). Nitrogen removal by nitritation-denitritation can be achieved in either SBR or continuous flow reactor (with or without biomass retention). The concept of the SHARON® process (Single reactor High activity Ammonia Removal over Nitrite) is based on higher growth rates of ammonia oxidizing bacteria (AOB) compared to nitrite oxidizing bacteria (NOB) in elevated temperatures (30-40 °C) (Hellinga *et al.* 1998). The process is performed in a chemostat with a retention time controlled (around 1 day) to maintain only the AOB (while the NOB are washed out). The reactor can be aerated continuously (van Dongen *et al.* 2001) or intermittently (Mulder *et al.* 2001). In the latter case, the anoxic phases could be used as a tool for controlling pH or obtaining full denitrification. Van Loosdrecht and Salem (2006) reported the full-scale application of this technology at six WWTPs in the Netherlands and New York. The SHARON® process can also be used to produce an appropriate feed composition (NO_2-N:NH_4-N ratio) to reactors performing the Annamox process. The first full-scale application of the combined technology took place at the Rotterdam-Dokhaven WWTP (the Netherlands) (van der Star *et al.* 2007).

1.1.6 Summary

During almost 100 years of their history, activated sludge systems have been expanding their capabilities from BOD removal to the integrated nitrogen and phosphorus removal. Moreover, several sidestream treatment technologies have been developed in the last decade to stabilize and enhance the efficiency of nitrogen removal. The growing number of interrelated biological processes has increased the level of complexity of modern activated sludge systems. The process development has been accompanied by improvements in the design methods, a better understanding of the process involved (including their capabilities and limitations), and ways to optimize operation (Stensel 2001). As a result, a large number of process configurations is available. The systems are also very efficient with respect to removing nutrients as the effluent concentrations below 0.1 mg P/dm^3 for phosphorus and 3 mg N/dm^3 for nitrogen can be achieved while treating typical municipal wastewater.

Empirical and semi-empirical methods can no longer provide optimal solutions with respect to the design, upgrade, and performance capabilities of the systems (Wentzel and Ekama 1997). In these areas, mathematical modelling and computer simulation are considered to be valuable tools. With the increasing practical experience, continuously improved activated sludge models have been used to quantify and predict process performance. This capability made them an important tool in the design and operation of WWTPs. Metcalf and Eddy (2003) emphasized that "*computer modeling provides the tool to incorporate the large number of components and reactions to evaluate activated sludge performance under both dynamic and steady-state conditions, and to easily design multiple staged reactors as well as a single-stage complete-mix reactor*". Moreover, Barnard (2006) noted that "*it can safely be said that most of the accumulated research in BNR that can be quantified today are incorporated into the models and in the computer simulations used and new findings are continuously added to perfect the simulations*".

1.2 DEVELOPMENT OF THE ACTIVATED SLUDGE MODELS

1.2.1 First period – empirical criteria

The period lasting from the process discovery until the early 1950s can be called "*empirical design, piloting and guesswork*" (Johnson 2009). The initial design methods of activated sludge tanks were simple and entirely empirical in nature. One of the first parameters used was a period of aeration. That period was dependent upon "*the strength of the sewage treated and the degree of purification required*" (Ardern and Lockett 1914) and "*the greater the degree of oxidation of the organic matter required, the longer must be the period of aeration*" (Metcalf and Eddy 1922). The recommended periods were 3-4 h and 8-10 h for dilute fresh domestic sewage and industrial wastewater, respectively. Later, Eckenfelder and O'Connor (1954) noted that "*the efficiency of the activated sludge process for the treatment of organic wastes is a function of the aeration time, the activated sludge solids concentration and the BOD loading*". Various loading factors were eventually developed involving those variables. Most commonly, this factor was expressed in terms of the daily load of 5-day biochemical oxygen demand (BOD_5) applied per mass of microbial solids present in the aeration tank (or volume of the tank). Garrett and Sawyer (1952) found the reported loading factors in the range of 0.25 to 0.5 kg BOD/(kg SS•d). At that time, aeration detention periods and mixed liquor suspended solids (MLSS) concentrations varied from 1 to 8 h and 1000 to 4000 mg/dm³, respectively (Eckenfelder and Porges 1957).

Other investigators interpreted these factors in terms of "*sludge age*" which in turn is related to the length of time the sludge has been undergoing aeration. Sludge age was generally considered as the reciprocal of the loading factor (Eckenfelder and Porges 1957). Gould (1953) defined the term "*sludge age*" as the ratio of the mixed liquor suspended solids to the daily load of suspended solids (SS) in the influent wastewater. Another interpretation of "*sludge age*" was proposed by Gellman and Heukelekian (1953) who considered it as the ratio of the MLSS mass in the tank to the mass of BOD removed per day. The modern definition of Solids Retention Time (SRT), synonymous to "*sludge age*" and mean cell residence time (MCRT), was firmly established by Lawrence and McCarthy (1970).

Eckenfelder and Porges (1957) proposed simple, empirical equations to estimate the amount of excess sludge produced and oxygen demand in the tank. These equations well summarize the approach used in the initial period of modelling activated sludge systems. The amount of excess activated sludge produced is proportional to the organic matter removed in the process, which could be estimated by the following material balance:

$$\text{Excess biological sludge} = aL_r - bS \qquad (1.2)$$

where

a — part of the BOD removed for synthesis, $M[M(BOD)]^{-1}$ (a = 0.50-0.75)
L_r — total BOD removed (assuming no storage), $M(BOD)L^{-3}$
b — rate of endogenous respiration, – (per cent per day)
S — sludge solids concentration, ML^{-3}

Furthermore, it was assumed that the presence of inert suspended solids would increase the total quantity of excess sludge for disposal. In order to take this contribution into account, Equation (1.2) was modified as follows:

$$\text{Excess sludge} = aL_r - bS + C \qquad (1.3)$$

where

C - inert suspended solids concentration, ML^{-3}

Finally, the total oxygen requirement was calculated based on the amount of organic matter removed and the concentration of sludge solids in the tank:

$$\text{Oxygen requirement} = (1-a)L_r + bS \qquad (1.4)$$

1.2.2 Second period – steady-state relationships of microbial growth and organic substrate utilization

The second phase in the development of activated sludge models can be characterized as the formal application of chemical reaction type kinetics to relate (at steady-state) microbial growth and organic substrate utilization under aerobic conditions (Wanner 1998). General aspects of kinetic principles, history of mechanistic microbial models and their ecological applications were comprehensively reviewed by Panikov (1995). The author noted that most of the initial kinetic studies originated from chemical kinetics and were focused on the solution of particular, well-defined problems, such as the microbial growth under steady-state and fully controlled environmental conditions. This resulted in the domination of models which had a narrow range of applicability and were incapable of predicting the diverse adaptive reactions under changeable (dynamic) environmental conditions. Moreover, most efforts were focused on studies with a limited range of specific, selected microorganisms whose growth properties could differ considerably from microbial populations that actually dominate in natural habits.

It is now generally accepted that the first kinetic principle for microbial growth was proposed by Penfold and Norris (1912) who studied the generation time of *Eberthella typhosa* as a function of peptone concentration (with and without glucose addition). Based on the experimental results, the authors provided a relationship between the specific growth rate constant (μ) and substrate concentration (S). They stated that this relationship is best described by a "*saturation-type*" curve where at high concentration of substrate, the microorganisms grow at a maximum rate (μ_{max}) regardless of the substrate concentration. These findings were only a confirmation of an earlier work of M'Kendrick and Pai (1911) as noted by Tanner and Wallace (1925).

Monod (1942) studied a relationship between growth rate and substrate (carbohydrate) concentration for strains of *Escherichia coli* and *Bacillus subtilis*. He demonstrated that there is a simple relationship between μ and S such that μ is proportional to S when this is low (growth-controlling) but reaching a limiting saturation value (μ_{max}) at high S according to the equation:

$$\mu = \mu_{max} \frac{S}{K_S + S} \tag{1.5}$$

Monod (1942) also developed a simple relationship between growth and utilization of substrate. In growth media containing a single organic substrate (e.g. glucose, ammonia and salts); the growth rate is a constant fraction, Y (known as the yield coefficient), of the substrate utilization rate:

$$\frac{dX}{dt} = -Y \frac{dS}{dt} \qquad (1.6)$$

Garrett and Sawyer (1952) stated that Equation 1.5 was "*merely another way of expressing the result obtained by Penfold and Norris*". The choice of a hyperbolic function (Equation 1.5) was entirely intuitive and empirical, except that a remark was made on the similarity with the adsorption isotherm. However, many years before the study of Monod, Michaelis and Menten (1913) derived the identical equation to describe the rate of an enzymatic reaction as a function of substrate concentration, which became one of the best known kinetic models in enzymology. Their equation was deducted from well-defined and clearly stated assumptions about the catalytic mechanism. Monod did not originally recognized any mathematical analogy to the Michaelis-Menten equation and this was only done in a later paper (Monod 1949). He stressed, however, that there was no relationship between the saturation constant, K_s, which characterized the affinity of microbial cells to substrate, and the Michaelis-Menten constant related to the enzyme dissociation. A detailed description of the development of Monod's theory was provided by Panikow (1995).

The Monod equation became the basis for all the widely accepted modern activated sludge models (see: Section 3.1), although it was also criticized in various respects. A particular subject of debate was its systematic deviations of μ at low substrate concentrations, where the actual growth rate lies above the prediction, and at high substrate concentrations, where $μ_{max}$ is approached too slowly (Kovarova-Kovar and Egli 1998). Furthermore, the original Monod equation did not consider the fact that cells may need substrate (or may synthesize product) even when they do not grow. In order to account for this phenomenon, the Monod equation was modified by introducing the maintenance or "*endogenous metabolism*" concept (first order expression with respect to the biomass concentration) as postulated by Herbert (1958). The combined equation, known as the Monod-Herbert model (Henze *et al.* 1987), became a basis for modelling the growth of different microbial groups in activated sludge.

In three decades following the studies of Monod, a variety of other mathematical expressions were put forward to describe this hyperbolic-type curve. In general, three approaches were used to apply refined equations for the growth kinetics of microbial cells (Kovarova-Kovar and Egli 1998):

- Incorporating additional constants into the original Monod model that provided corrections of substrate or product inhibition, endogenous metabolism (maintenance), substrate diffusion, or the dependence of μ_{max} on the biomass concentration;
- Proposing different kinetic concepts, resulting in both empirical and mechanistic models;
- Describing the influence of physicochemical factors on the Monod growth parameters.

The activated sludge process fundamentals came under close scientific study in the early 1950's (Goodman and Englande 1974). This can be attributed to the attempts of combining the kinetics of microbial growth and substrate utilization with biological reactor engineering principles. In 1950, simultaneously and independently, two identical approaches were used to describe mathematically a simple principle of the steady state attaining continuous culture in devices termed "*chemostat*" in the USA (Novick and Szilard 1950) and "*bactogen*" in France (Monod 1950). In the subsequent years, several investigators tried to apply the kinetics of microbial growth and substrate utilization along with mass balance equations to model performance of the activated sludge process. The greatest difficulties were identified with respect to a definition of the principal variables including substrate (S) and active biomass (X_v). In Monod's study (1942, 1949), these parameters were well defined, as single substrates were utilized by pure microbial cultures. In contrast, in an activated sludge system there are many different compounds that are all metabolized simultaneously by a mixed population of microorganisms.

The active biomass was measured in terms of the volatile suspended solids (VSS) concentration. Hoover and Proges (1952) developed an empirical formula for the composition of cell mass ($C_5H_7NO_2$) which became a widely accepted basis for different stoichiometric calculations. For example, Henze *et al.* (1987) used this formula to approximate the mass of nitrogen per mass of cell COD (0.086 g N/g COD). There has already been a century-long debate about whether a BOD-based method or a COD-based method is a more appropriate measure of organic substrate concentrations in biological wastewater treatment processes (Jenkins 2008). Initially, the BOD_5 concentration of the liquid phase of the mixed liquor was taken as the substrate concentration. Gaudy *et al.* (1964) noted, however, that the BOD measurement cannot be used for calculation of substrate recovery balances because the substrate, in the BOD test, is also partitioned between respiration and synthesis. On the other hand, the authors found that an energy balance comparing COD removed with the summation of oxygen uptake and the COD of the cells produced yielded average recoveries close to 100%.

Garrett and Sawyer (1952) and later Schulze (1956) applied the chemostat concept for control of a laboratory scale activated sludge reactor (Figure 1.5). Furthermore, Garrett and Sawyer (1952) also found that the growth of microorganisms in mixed cultures, such as activated sludge, was in accordance with the principles valid for pure microbial cultures. The authors related the growth rate and remaining substrate (BOD) concentration by a discontinuous function in which the growth rate was constant with high concentrations of the substrate, whereas the growth rate as directly proportional to the remaining substrate at low concentrations:

$$\mu = \begin{cases} \dfrac{\mu_{max}S}{K_B} & \text{if } S < K_B \\ \mu_{max} & \text{if } S \geq K_B \end{cases} \tag{1.7}$$

where

K_B - "*saturation*" constant, $M(BOD)L^{-3}$

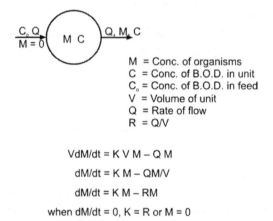

M = Conc. of organisms
C = Conc. of B.O.D. in unit
C_o = Conc. of B.O.D. in feed
V = Volume of unit
Q = Rate of flow
R = Q/V

VdM/dt = K V M – Q M
dM/dt = K M – QM/V
dM/dt = K M – RM
when dM/dt = 0, K = R or M = 0

Figure 1.5 Chemostat concept and model development by Garrett and Sawyer (1952) (the original notation kept).

Eckenfelder and O'Connor (1954) proposed a mathematical model for designing activated sludge systems including the calculation of volume of the aeration tank, air requirements, sludge production and nutritional requirements. That model was subsequently modified and expanded by Eckenfelder and co-workers during the period from 1960 to 1970 (Eckenfelder and O'Connor 1961; Eckenfelder 1961; Eckenfelder 1966; Eckenfelder 1970). These publications established a nomenclature and mathematical approach to design of the activated sludge process.

McKinney (1962) defined a fundamental model concept of a completely mixed reactor with the return activated sludge (RAS). The model was sequentially developed in a series of further publications (Burkhead and McKinney 1968; McKinney and Ooten 1969; McKinney 1970). For example, McKinney and Ooten (1969) proposed a structured approach for conversion of carbonaceous material including three volatile suspended solids (VSS) fractions: active, endogenous-inert and inert (from the influent).

Lawrence and McCarthy (1970) proposed a unified model describing common biological treatment processes (aerobic treatment of organic compounds, nitrification, anaerobic treatment of organic compounds) in three types of reactors including complete mixed – no solids recycle, completely mixed – solids recycle and plug flow – solids recycle. General relationships were developed for the SRT (termed a "*unifying parameter*") in relation to microbial growth, substrate utilization and process efficiency. The authors clearly concluded that:

> "The models presented herein are only a mathematical formalization of what has been observed to be the important parameters by designers, operators, and investigators in the past. Such formalization, hopefully, will furnish relationships with predictive value to serve not only in the design and control of existing treatment processes, but also will aid in the development of biological processes for other purposes (...) or for the more efficient removal of specific waste components (...)."

For many years, that model became the most widely accepted design model in the USA (Mines 1997). Details of its development can also be obtained in numerous references (e.g. Metcalf and Eddy 1991). Several other steady-state models of the activated sludge process were proposed in the USA (e.g. Busch and Kalinske 1956; Garrett 1958; Gaudy *et al.* 1960; Ramanathan and Gaudy 1971). Goodman and Englande (1974) compared the models of McKinney and Eckenfelder and concluded that they were essentially the same. Gaudy and Kinceannon (1977) demonstrated that the different design approaches developed by Lawrence and McCarthy (1970), and Ramanathan and Gaudy (1971) also yielded similar results to the models of McKinney and Eckenfelder.

Table 1.2 Most common models describing biochemical oxidation kinetics (Vavilin 1982)

No	Type of transfer function $S_e = f(S_0, X, T)$	Expression for treatment rate, $r(S_e)$ or $r(S)$	Comments
1.	$S_e = S_0 \bullet e^{-kXT}$	$r(S) = kS$	Eckenfelder's model for the design of plug-flow aeration tanks and trickling filters. As a rule, it correlates with the experiment (high-level treatment) poorly. It overestimates v^* at small S_e^* values.
2.	$S_e = S_0 \bullet e^{-(kXT)^n}$		Modification of model 1, proposed by Eckenfelder for the design of tricking filters. It improves the design of aeration tanks as well.
3.	$S_e = S_0 / [1 + kXT]$	$r(S_e) = kS_e$ $r(S) = kS^2/S_0$	Eckenfelder's model proposed for the design of completely mixing aeration tanks. It underestimates v^* at large S_e^* values and overestimates v^* at small S_e^* values.
4.	$S_e = S_0/[1 + (kXT)^n]$	$r(S_e) = \hat{k}S_e/X$	Modification of model 3, proposed by Eckenfelder for the design of tricking filters. It also improves the design of aeration tanks.
5.	$S_e = S_0/[1 + \hat{k}T]$		McKinney's model for the design of completely mixing aeration tanks. In the case of high-level treatment, its results are often no worse than those given by model 3. For McKinney's model, $k = kX = $ constant, and therefore it practically considers that at large X values the "activity" of microorganisms becomes lower.
6.	$S_e = S_0/(1 + r_{max}XT/S_0)$	$r(S_e) = r_{max}S_e/S_0$ $r(S) = r_{max}(S/S_0)^2$	Grau's model for the design of completely mixing aeration tanks. It is the best approximation of the generalized model 1 as compared to model 3.
7.	$S_e = S_0 e^{-r_{max}XT/S_0}$	$r(S) = r_{max}S/S_0$	Grau's model for describing a batch experiment. It correlates poorly with the experiment in the case of high-1evel treatment (small S_e^* values).

Table 1.2 (cont.). Most common models describing biochemical oxidation kinetics (Vavilin 1982)

No	Type of transfer function $S_e = f(S_0, X, T)$	Expression for treatment rate, $r(S_e)$ or $r(S)$	Comments
8.	$S_e = \dfrac{S_0 + S_h r_{max} XT/S_0}{1 + r_{max} XT/S_0}$	$r(S_e) = r_{max}\dfrac{S_e - S_h}{S_0}$	Adams and Eckenfelder's model for the design of completely mixing aeration tanks. It correlates better with model 1 than model 6.
9.	$S_e = S_0\left(\dfrac{(1+4kXT)^{1/2}-1}{2kXT}\right)$	$r(S_e) = kS_e^2/S_0$	Vavilin's model for describing high-level treatment in aeration tanks, tricking filters and rotating discs at S_e^* values far from "saturation."
10.	$S_e = P + (P^2 + k_s S_0)^{1/2}$ $P = (S_0 - k_s - r_{max} XT)/2$	$r(S_e) = r_{max}\dfrac{S_e}{k_s + S_e}$	Classical Monod's model. In this form, it is commonly applied to the deign of completely mixing aeration tanks. It may be also successful1y used for the design of plug-flow aeration tanks in the case of rough treatment and large S_e^* values near to "saturation."
11.	$S_e = S_0 + r_{max} XT - k_s \ln(S_0/S_e)$	$r(S) = r_{max}\dfrac{S}{k_s + S}$	Classical Monod's model is sometimes applied to the design of plug-flow reactors. It may be used for rough treatment design.
12.	$S_e = \hat{P} + (\hat{P}^2 + \bar{k}_s XS_0)^{1/2}$ $\hat{P} = S_0 - \hat{k}_s X - r_{max} XT/2$	$r(S_e) = r_{max}\dfrac{S_e}{\hat{k}_s X + S_e}$	Contois's model is a modification of Monod's model. It improves Monod's model in the case of transition to treatment at a higher level, since it considers the lowering of the "activity" of microorganisms with an increase in their concentration.
13.	$S_e = \left\{\dfrac{kXT}{2}\left[\left(\dfrac{4S_0}{(kXT)^2}+1\right)^{1/2}-1\right]\right\}^2$	$r(S_e) = kS_e^{1/2}$	Half-order model describes well the experiment in the case of rough treatment and large S_e^* values. It may be used for aeration tanks, trickling filters and rotating discs.
14.	$S_e = S_0/[1 + (n-1)kXT]^{1/(n-1)}$	$r(S_e) = kS(S/S_0)^{n-1}$	Fair's model the design of plug-flow reactors. It describes well high-level treatment at small S_e^* values. Fair's function follows from the generalized model (10) at $k_s^{n-p} S_0^p \gg S_e^n$ and $p = n-1$.

Vavilin (1982) proposed a unified approach to modelling aerobic biological treatment processes in different reactors (including aeration tanks). For this purpose, he reviewed models describing biochemical oxidation kinetics of single substrates and multicomponent pollutants utilized by either pure or mixed cultures (Table 1.2). By processing numerous experimental data it was shown that the high-level and rough aerobic treatment of multicomponent substrate, such as wastewater, could follow different principles. The classical Monod equation was fairly adequate only in the case of rough treatment, whereas models of an n order (n>1) with respect to the substrate were more appropriate in the case of high-level treatment. Moreover, the overall rate of treatment was dependent on the oxidation rates of individual components and their contribution to the overall rate. In raw wastewater, the contribution of *"easily oxidizable"* compounds was dominant among all the oxidized compounds, whereas the role of complex *"hard oxidizable"* compounds increased with the raising HRT and biomass concentration.

The steady-state models, developed based on the mass balances for microbial growth and substrate utilization, represented major advances when compared to older empirical equations (Sherrard 1984). The models helped in understanding the basic mechanisms of the activated sludge process, however, improvement in performance predictability under dynamic flow and load conditions was still needed. Busch (1984) noted that *"steady-state analysis is overly simplistic, and that transients overwhelm kinetic analysis based on steady-state assumptions, especially under oxygen-limited conditions."*

1.2.3 Third period – complex dynamic models

The third phase of development of activated sludge models can be characterized as the application of reactor engineering principles in combination with large matrixes of kinetic expressions and stoichiometric constants (Wanner 1998). A great complexity of the modern models resulted from the ability of defining the behaviour of influent wastewater components (i.e., organic, nitrogen and phosphorus fractions) and biomass components (i.e., heterotrophs, nitrifiers and PAOs with further subdivisions) and the reaction stoichiometry and kinetics (Jenkins 2008). The gradually increasing sets of extensive mathematical expressions with spatially and temporarily variable components could only be efficiently handled and solved only using computers.

The first dynamic models developed found only little practical applications and Henze *et al.* (2000) attributed this fact to several constraints including lack of trust in the models, a complicated way in which these models had to be presented in a written form, and limitations in computer power. The early use of digital computers (1960-1975) for simulation was primarily restricted to programming specialists possessing knowledge of numerical techniques and conventional languages, such as FORTRAN. Moreover, the users were not able to interact directly with the computer, but needed access to computer centres (Andrews 1993). In the 1970's, however, the increasing computer power coupled with falling prices for computers stimulated the development of dynamic models as these models could quickly be tested and validated (Jeppsson 1996). Since the beginning of 1980's, the low-cost personal computers (PCs) have become available. Moreover, the development of special, user-oriented simulation languages (e.g. SIMULINK or ACSL) allowed to concentrate on model development and interpretation of simulation results without knowing details of programming (Andrews 1993).

Dynamic models of the activated sludge process were formally formulated yet in the 1950s (see: Figure 1.5), but no tools were available at that time to solve even simple ordinary equations under dynamic conditions. Modelling activated sludge processes became a discipline in the mid-1960's (Henze et al. 2000). The authors implicitly referred to the study of Downing et al. (1964) who developed a simple Monod-type model for ammonia oxidation to nitrate (as a one-step conversion) by nitrifying bacteria in the activated sludge process. The model was tested against daily observations from laboratory scale systems treating domestic wastewater. That study not only greatly enhanced the understanding of nitrification in the activated sludge process, but also was an early attempt to introduce the concept of mass balances and kinetic relationships in the sanitary engineering literature (Gujer 2006). In the subsequent study, Knowles et al. (1965) showed "*the usefulness of electronic computers for the rapid integration of differential equations for bacterial growth, particularly those which describe sequential growth of interdependent bacteria and which can only be solved numerically*". The "*best fits*" of model predictions to experimental data were used to estimate the maximum specific growth rate constant for autotrophs ($\mu_{A,max}$) along with the ammonia saturation coefficient ($K_{NH,A}$) and the initial autotrophic biomass concentration ($X_{A,0}$).

The noticeable progress in the 1970s and 1980s with regard to dynamic models was briefly summarized by Lessard and Beck (1991). Even though the main attention was focused on carbonaceous oxidation, nitrification and denitrificaton, other mechanisms were also addressed including the behaviour of dissolved oxygen (DO), sludge bulking and biological phosphorus removal.

Some of the most fundamental work concerning the development of dynamic models for the activated sludge process was performed in the USA by Andrews and co-workers, as well as at the University of Cape Town (Republic of South Africa) by Marais' research group. In a review paper on early dynamic models and control strategies for wastewater treatment processes, Andrews (1974) noted that "d*ynamic modeling and computer simulation are useful tools in developing better procedures for process start-up, prediction and prevention of process failures, and improvement of process performance by consideration of dynamic* *behavior during both the design of a process and its associated control system"*. The author emphasized, however, that the development of dynamic models for wastewater treatment processes and the use of these models to improve control strategies were at that time in *"their infancy"*. As an example, an early dynamic model for the step feed activated sludge process was applied to simulate control strategies based on automatically controlling the points of addition of wastewater in terms of daily variations in the influent flow rate. Busby and Andrews (1975) presented a model, in which the biomass was structured into three parts (stored, active and inert) and a sorption expression was used to describe the rapid removal of substrate. The model was validated based on the literature data and applied to compare the dynamic performance of different configurations of activated sludge processes (conventional, extended aeration, high loaded, contact stabilization and step feed). Poduska and Andrews (1975) developed the first dynamic model for nitrification in the activated sludge process. Nitrification was modelled as a two-step process including ammonium and nitrite oxidation as the separate steps. The model was tested against the data obtained from a laboratory scale system and then used to investigate design criteria and operating techniques by which the efficiency of nitrification might be improved. Clifft and Andrews (1981) studied process oxygen requirements using a dynamic model that distinguished between particulate and soluble substrate metabolism. Such a fractionation was the key factor for accurate modelling the oxygen consumption in batch experiments with real wastewater. It was assumed that the soluble substrate can be directly synthesized or stored in the floc phase prior to synthesis, whereas the particulate substrate is removed from the liquid phase only by storage, which is followed by the synthesis for biomass growth. The paper of Clifft and Andrews (1981) was the last one to appear from Andrews' group (Grady 1989).

In the late 1970s, modelling activated sludge processes was reaching the most advanced level at the University of Cape Town (Henze *et al.* 2000). The first model of this research group (Marais and Ekama 1976) was proposed for the aerobic system including nitrification under constant flow and load conditions and constituted a development from a number of previous models for carbonaceous and nitrogenous material conversions (McKinney 1962; Downing *et al.* 1964; McKinney and Ooten 1969; Lawrence and McCarthy 1970). Marais and Ekama (1976) introduced several novel concepts that were later widely accepted in the area of activated sludge modelling:

- Biochemical Oxygen Demand (BOD) was rejected as a suitable parameter for measuring the concentration of organic material in wastewater and replaced by COD which linked electron equivalents in the organic substrate, biomass and utilized oxygen;

- Influent organic material was divided into three fractions: biodegradable, non-biodegradable particulate and non-biodegradable soluble;
- A similar fractionation was also adopted for the influent nitrogenous material (biodegradable, non-biodegradable particulate, non-biodegradable soluble, and free and saline ammonia);
- The nitrification process followed the approach of Downing *et al.* (1964) and was modelled as one-step conversion of ammonia to nitrate.

The steady-state model of Marais and Ekama (1976) gave predictions that deviated significantly from the observations made in experimental investigations by Ekama and Marais (1979). The authors found that aerobic completely mixed single reactor systems at short SRTs (1.5-3.0 d) fed under daily cyclic square wave loading conditions (12 hours - on, 12 hours - off), exhibited a precipitous drop in the OUR on termination of the feeding phase. In order to explain this step change of OUR, Dold *et al.* (1980) proposed that the influent biodegradable organic material can be further divided into two fractions: readily biodegradable and slowly biodegradable. The authors incorporated this bi-substrate concept in a dynamic kinetic model of an aerobic nitrification activated sludge process. The new model accurately predicted the observed precipitous drop in OUR and the behaviour of the COD, OUR and nitrogen in various aerobic systems under cyclic flow and load conditions. In addition, Dold *et al.* (1980) formulated the "*death-regeneration*" hypothesis for modelling the loss of active biomass (i.e. the reduction in active volatile mass with time). This approach described the phenomenon of active biomass loss in a substantially different way compared to the earlier steady state models adopting the endogenous respiration concept (see: Section 3.2). Van Haandel *et al.* (1981) extended the model of Dold *et al.* (1980) by incorporating the denitrification process. The extension required no change in the form of kinetic expressions describing the aerobic processes, except that the utilization rate of slowly biodegradable substrate under anoxic conditions needed to be reduced to about 1/3 that under aerobic conditions. Wentzel *et al.* (1992) summarized much of the previous research on activated sludge modelling conducted at the University of Cape Town (UCT). The UCT based research group model, consisting of the model of Dold *et al.* (1980) for aerobic systems with the extension of van Haandel *et al.* (1981) for denitrification, was later termed UCTOLD (Dold *et al.* 1991; Hu *et al.* 2003b).

The benefits to be gained by more common application of activated sludge modelling were recognized by the International Association on Water Pollution Research and Control (IAWPRC, the former name of IAWQ and IWA). In order to encourage to practitioners to use modelling more extensively, in 1983 the IAWPRC established a Task Group on Mathematical Modelling for Design and

Operation of Biological Wastewater Treatment. The task group had the following assignments (Henze *et al*. 1987):

- Review existing models in the literature;
- Reach a consensus concerning the simplest approach to modelling carbon oxidation, nitrification and denitrification;
- Develop a mechanistic model with a minimum of complexity that could be used for the design and operation of activated sludge systems performing carbon oxidation, nitrification and denitrification.

Grady (1989) reported that the task group met four times over a period of two and a half years for more than 15 days to accomplish its assignments and submitted its report to IAWPRC in 1986. A preliminary form of the model, called the Activated Sludge Model No. 1 (ASM1), was summarized in the technical literature (Grady *et al*. 1986). Dold and Marais (1986) conducted a comprehensive evaluation of the model and proposed some changes in the model structure, in particular relating to the way in which the fate of organic nitrogen was modelled. These suggestions were taken into account in the final form of ASM1 (Henze *et al*. 1987). The ASM1 has provided a consistent framework for the description of biological processes in suspended growth (activated sludge) systems, including carbon oxidation, nitrification, and denitrification.

The development of ASM1 constituted a significant contribution in the field of modelling biological processes of municipal wastewater treatment. The model consolidated a wide range of previous research, but a contribution of the UCT research group was identified of special significance by the task group. The description of organic matter and nitrogenous material conversions presented in ASM1 were adopted from the model UCTOLD. Apart from utilization of slowly biodegradable COD, which is conceptualized differently in UCTOLD and ASM1, both models are very similar and can predict the same processes., i.e. oxygen demand, sludge production, nitrification and denitrification (Hu *et al*. 2003b). Indeed, Wentzel *et al*. (1992) demonstrated that both models generated virtually identical predictions of OUR in a completely mixed aerobic reactor operated under daily square wave loading.

Extensive use of ASM1 in research and practice over the previous decade allowed to identify several deficiencies of the model. In order to correct them the International Association on Water Quality (IAWQ) task group (Gujer *et al*. 1999) presented a new model called the Activated Sludge Model No. 3 (ASM3). Although ASM1 and ASM3 are capable of simulating the same biological processes (i.e. carbon oxidation, nitrification and denitrification) and their complexity is similar, the conceptual differences exist between the two models. The major difference is that heterotrophic growth directly on external (primary) substrate as described in ASM1 is not considered in ASM3. In the latter model, it is assumed that all readily biodegradable substrate (S_S) is first taken up by heterotrophs and stored in the form of internal cell compounds (X_{STO}) prior to growth (Figure 1.6). Moreover, the "*death-regeneration*" hypothesis used in ASM1 is replaced by a more "*traditional*" concept of endogenous respiration. The term more "*traditional*" indicates here that a similar approach was already used earlier, e.g. in the model of Busby and Andrews (1975). Consequently, there is no production of X_S from decay of heterotrophs and autotrophs and thus the hydrolysis of slowly biodegradable substrate (X_S) plays a less significant role in ASM3 compared to ASM1. The other important features that differ ASM3 from ASM1 can be summarized as follows:

- Hydrolysis is independent of the electron donor and occurs at the same rate under aerobic and anoxic conditions;
- The loss rates of heterotrophic and autotrophic biomass are reduced under anoxic conditions;
- Two fractions of inert particulate material depending upon its origin, i.e. influent or biomass decay, are not differentiated;
- Lower anoxic yield coefficients are introduced;
- Ammonification and hydrolysis of biodegradable, particulate nitrogen are omitted;
- Soluble COD is a sum of only readily biodegradable substrate and soluble, non-biodegradable organic compounds.

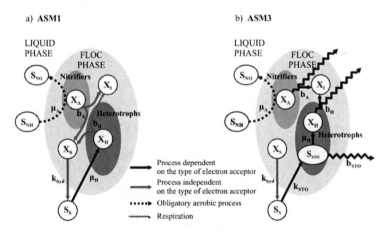

Figure 1.6 Comparison of the COD mass flows for ASM1 and ASM3 under aerobic/anoxic conditions.

Koch *et al.* (2000b) calibrated the model based on results of batch and full-scale measurements from various Swiss WWTPs. Subsequently, a new set of kinetic and stoichiometric coefficients was proposed for the reliable prediction of nitrification and denitrification. The influent concentration of readily biodegradable substrate was estimated from respiration measurements by curve fitting instead of using 0.45 μm filtration as recommended initially by the IAWQ task group (Gujer *et al.* 1999). This suggestion of Koch *et al.* (2000b) was incorporated in the revised version of the report on ASM3 (Henze *et al.* 2000), in which the following statement was added: *"filtration over 0.45 μm membrane filters cannot be used to differentiate model soluble from model particulate compounds in the influent"*. Koch *et al.* (2000b) concluded that ASM1 and ASM3 generated similar predictions of the dynamic behavior in common municipal WWTPs, whereas ASM3 performed better for WWTPs with substantial volumes of non-aerated zones or in situations where the storage of readily biodegradable substrate became a dominant process (e.g. WWTPs treating industrial wastewater with a high content of biodegradable COD).

Table 1.3 Principal organism groups included in biokinetic models for BNRAS systems, their functions, and the zones in which these functions are performed (Wentzel and Ekama 1997)

Organism group	Principal biological processes	Conditions
"ordinary" heterotrophs (unable to accumulate poly-P)	COD removal (organic degradation; oxygen uptake)	Aerobic
	Ammonification (organic N $\rightarrow$ NH$_4^+$)	Anaerobic/ Anoxic/ Aerobic
	Denitrification (organic degradation, NO$_3^-$ $\rightarrow$ NO$_2^-$ $\rightarrow$ N$_2$)	Anoxic
	Fermentation (fermentable RBCOD $\rightarrow$ VFA)	Anaerobic
PAOs (heterotrophs) (able to accumulate poly-P)	P release (VFA uptake; PHA storage)	Anaerobic
	P release (VFA uptake; PHA storage)	Anoxic
	P uptake (PHA degradation; denitrification)	Anoxic
	P uptake; P removal (PHA degradation; oxygen uptake)	Aerobic
Autotrophs (nitrifiers)	Nitrification (NH$_4^+$ $\rightarrow$ NO$_2^-$ $\rightarrow$ NO$_3^-$, oxygen uptake)	Aerobic

The authors of ASM3 also concluded that "*it is expected that future improvements of model structure may still be required*" (Gujer *et al.* 1999). For example, it has been hypothesized in ASM3 that all heterotrophic microorganisms are able to store substrate, however, microorganisms both with and without substrate storage capability could exist in reality (Hanada *et al.* 2002) or direct growth on external substrate is possible (Krishna and van Loosdrecht 1999). Alternative model concepts incorporating the storage phenomena are discussed in more detail in Section 3.3.

A restriction of ASM1 and ASM3 is that the process of EBPR is not included in these models. From the mid-80's to the mid-90's the EBPR process became very common in municipal WWTPs and at the same time the understanding of the principal phenomena of the EBPR process was increasing (Henze *et al.* 2000). Also significant efforts have been made to develop a mathematical model of the EBPR process. Wentzel and Ekama (1997) listed the following three principal microorganism groups that must be taken into account in models for activated sludge systems performing nitrification, denitrification and EBPR: (1) heterotrophic organisms unable to accumulate polyphosphate (poly-P), termed "*ordinary*" heterotrophs, (2) heterotrophic organisms able to accumulate poly-P, generically called PAOs, and (3) autotrophic organisms mediating nitrification, termed also autotrophs or nitrifiers. The functions of these groups under different oxic conditions are outlined in Table 1.3. A schematic diagram comparing the roles of "*ordinary*" heterotrophs and PAOs in a BNRAS system (UCT process configuration) is presented in Figure 1.7 (Barker and Dold 1997a). It should be noted that the approach adopted both by Wentzel and Ekama (1997) and Barker and Dold (1997a) does not include conversions of cell internal glycogen, which were recognized by several researchers of special importance in the PAO metabolism (see: Section 3.9.1).

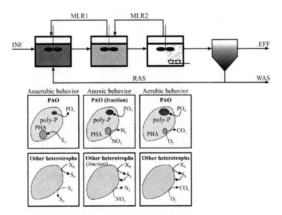

Figure 1.7 Schematic diagram outlining the role of PAOs and "*ordinary*" heterotrophs in a BNRAS (UCT) system according to the model of Barker and Dold (1997a).

In the recent twenty years, models that incorporated the EBPR process have become available. For example, Filipe and Daigger (1998) listed three models describing the anaerobic/aerobic behaviour of PAOs that were well recognized at the time of their study:

- Model of Wentzel *et al.* (1989b);
- Activated Sludge Model No. 2 (ASM2) (Henze *et al.* 1995a);
- Metabolic model of Smolders *et al.* (1995a), known also as Technical University of Delft bio-P model (TUD or TUDP).

Since then all these models have been revised and the latest modifications incorporate denitrifying PAOs as presented by Barker and Dold (1997a), Henze *et al.* (1999) and Murnleitner *et al.* (1997), respectively. The development of the most important complex activated sludge models is presented in Figure 1.8.

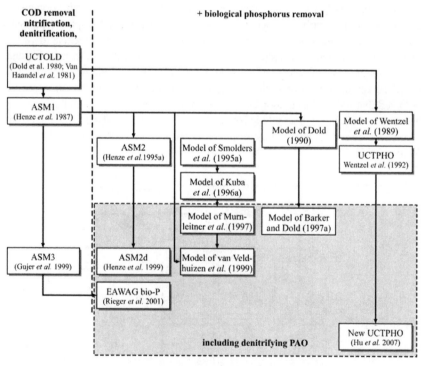

Figure 1.8 Development of the complex biokinetic models for BNRAS systems.

The model of Wentzel *et al.* (1989b) was also integrated with the UCTOLD model to form a general biokinetic model for BNRAS systems, termed UCTPHO (Wentzel *et al.* 1992). The UCTPHO model incorporates three main population groups mentioned earlier, however, without considering denitrifying PAOs. Recently, Hu *et al.* (2007) presented the extended UCTPHO model including the anoxic behavior of PAOs.

Eight years after the ASM1 release, a similar task group of the IAWQ presented an extended version of the model, called the Activated Sludge Model No. 2 (ASM2), which also considered EBPR (Henze *et al.* 1995a). The full ASM2 incorporates 19 processes and 19 components. Such a level of complexity was needed to describe many different process configurations which are used for EBPR. The authors noted in the report that "*the model is not the final answer. It should be used as a conceptual platform for further model development.*" (Henze *et al.* 1995). Working with ASM2 as a basis, Ekama and Wentzel (1999a) evaluated the difficulties and developments in modelling the EBPR processes. The authors concluded that recent developments in this area did not allow a significant improvement to be made with respect to ASM2 predictive capabilities. When ASM2 was completed, the role of denitrification with respect to the EBPR process was still unclear, so it was decided not to include that element (Henze *et al.* 2000). Mino *et al.* (1995a), Mauer and Gujer (1998) and Furumai *et al.* (1999) proposed to incorporate denitrifying PAOs in ASM2. This extension was taken into account in a revised version of the model, called ASM2d (Henze *et al.* 1999).

In parallel to the series of IWA ASMs, other comprehensive models of the BNRAS systems were developed. Dold (1990) combined (with some extensions and modifications) ASM1 and a model of Wentzel *et al.* (1989b) with respect to the EBPR process. Barker and Dold (1997a) presented a refined model of Dold incorporating denitrifying PAOs. The model (referred to as "*New General*") included some further refinements compared to the initial model of Dold. Barker and Dold (1997a) also discussed major differences between their model and ASM2, which can be summarized as follows:

- Although both models have the same number components, only some of them are essentially the same;
- A number of processes are modelled differently in the two models or are omitted in one of them;
- A number parameter names are used for different purposes in the two models.

The model of Barker and Dold (1997a) was evaluated against experimental data (both steady state and dynamic) from a number of aerobic, anoxic-aerobic and anaerobic-anoxic-aerobic activated sludge systems (Barker and Dold 1997b). With a single set of stoichiometic and kinetic coefficients (except for the $\mu_{A,max}$), the model was capable of reasonably predicting several key parameters, such as OURs in the aerated reactors as well as concentrations of soluble P, nitrate and volatile suspended solids (VSS). The authors concluded that "*improvements to the model no doubt will come about as the understanding of the complex interactions occurring within biological nutrient removal systems is expanded*" and identified several important aspects that would require further investigation (e.g. possible role of glycogen in the PAO metabolism, effect of the GAO presence, competition between denitrifying PAOs and "*ordinary*" denitrifiers, etc.).

Hu *et al.* (2003b) pointed out that the model of Barker and Dold (1997a) had not been validated for EBPR systems exhibiting significant anoxic phosphate uptake and the prediction capabilities of this model for such systems were unknown. Accordingly, the authors evaluated the model against experimental data from 30 different laboratory scale mixed culture systems including such configurations as modified Bardenpho, UCT, modified UCT (MUCT) and Johannesburg (JHB). Based on the obtained results, Hu *et al.* (2003b) identified a number of concepts included in the model of Barker and Dold (1997a) that would require more detailed investigation, in particular anaerobic hydrolysis of slowly biodegradable organic compounds, COD loss mechanisms, fraction of PAOs that can denitrify as well as reduced yields of PAO and "*ordinary*" heterotrophs under anoxic conditions.

The outlined ASM3 not only corrected the deficiencies of ASM1, but was also meant to be the core of a model that could be expanded with modules describing various processes of interest, such as EBPR, growth of filamentous organisms, pH calculations, etc. (Gujer *et al.* 1999). Rieger *et al.* (2001) developed a bio-P module, referred further to as the EAWAG bio-P module, and coupled it with a calibrated version of ASM3 (Koch *et al.* 2000b) to model also the EBPR process. The module contains eleven processes and four components (S_{PO4}, X_{PAO}, X_{PHA}, X_{PP}) which are identical to the ASM2d state variables. In comparison with ASM2d, there are several differences in the new model structure. The main differences can be summarized as follows (Rieger *et al.* 2001):

- Fermentation is neglected under assumption that there is no limitation of the phosphate release in typical municipal wastewater;
- The hydrolysis rate under anaerobic and anoxic conditions is not reduced in comparison with aerobic conditions;
- The concept of endogenous respiration is used for all groups of the microorganisms;

- The anaerobic decay is neglected for all groups of the microorganisms and the decay rates under anoxic conditions are reduced in comparison with aerobic conditions.

The newly developed model, consisting of ASM3 and the EAWAG bio–P module, was termed ASM3P (Gernaey et al. 2004; Meijer 2004). The ASM3P was tested against dynamic experimental data originating from several full-scale WWTPs in Switzerland (Rieger et al. 2001) and Germany (Wichern et al. 2003; Makinia et al. 2005a).

The EBPR models outlined above (ASM2d, "New General", ASM3P) assume a "grey box" approach for cell internal conversions of PAOs. It means that the model structures are directly related to the measured conversions in the bulk liquid, whereas the cell internal mechanisms are modelled using only polyphosphate and one COD storage compound (Meijer 2004). A novel metabolic approach to modelling the EBPR processes was developed under laboratory conditions in the Technical University of Delft (TUD). In a metabolic model, the conversions of components observed on the outside of the organisms are reduced to a number of internal characteristic reactions of the metabolism including, in the case of EBPR, storage of PHA, polyphosphate synthesis, or the production of ATP in the oxidative phosphorylation (Murnleitner et al. 1997). The metabolic approach is based on the degradation and formation of all relevant cell internal storage compounds (PHA, glycogen and polyphosphate) and offers a great promise because the stoichiometry for the model is derived entirely on fundamental biochemical principles (Filipe and Daigger 1998). Consequently, a reduced number of independent reaction rates is obtained, which also leads to a minimal number of necessary kinetic expressions and model parameters (Murnleitner et al. 1997).

The stoichiometry and kinetics of anaerobic and aerobic metabolism in the EBPR process were presented by Smolders et al. (1994a) and Smolders et al. (1994b), respectively. The model was developed based on the results of experiments carried out in a laboratory scale anaerobic/aerobic (A/O) SBR with a SRT of 8 days. The partial models were integrated by Smolders et al. (1995a) and the integrated model was validated with the results of two similar experiments at a SRT of 5 and 20 days (Smolders et al. 1995b). Kuba et al. (1996a) adopted the metabolic model of Smolders et al. (1995a) to anoxic conditions. The anoxic model was calibrated and validated based on the results of batch experiments carried out in a laboratory scale anaerobic-aerobic (A/A or A_2) SBR with different initial phosphate concentrations and over a range of SRTs (8 and 14 days). Murnleitner et al. (1997) compiled the results of previous research conducted in TUD and presented a revised (complete) version of the metabolic model. The revised model allowed to predict the previous SBR experiments with a single set of model parameters.

Van Veldhuizen *et al.* (1999a) developed a hybrid (metabolic-kinetic) model by integrating the complete metabolic model for the EBPR process with the ASM2 equations describing hydrolysis, heterotrophic processes and autotrophic processes. The new complex model was referred further to as "*Technical University of Delft Phosphorus*" (TUDP) model. The model was further tested against the data originating from full-scale WWTPs in the Netherlands (van Veldhuizen *et al.* 1999a; Brdjanovic *et al.* 2000; Meijer *et al.* 2001; Meijer *et al.* 2002b; Meijer 2004). In all of these cases, a reasonable correlation between measurements and model predictions was obtained without significant changes to the default values of model parameters. Moreover, the study of Meijer *et al.* (2001) confirmed that the TUDP model can be applied for predicting a full-scale WWTP performance without intensive calibration of the model kinetics, whereas the model stoichiometry can even be extrapolated to full-scale systems without calibration. It was also demonstrated that model predictions were highly sensitive to changes in operating data and thus the correct estimation of these data, especially the SRT, was crucial for an accurate model prediction. The authors preferred the adjustment of operating data rather than the calibration of kinetic parameters in order to fit model predictions to measured data.

The improvements of the metabolic model were also proposed in the studies of Filipe *et al.* (1998; 1999; 2001a) including the stoichiometry (Filipe and Daigger 1998) and the description of acetate uptake kinetics (Filipe and Daigger 1999; Filipe *et al.* 2001a). However, none of these modifications was incorporated in the latest version of the TUDP model (Meijer 2004). Meijer (2004) summarized the previous research conducted in TUD on the development of the metabolic model with a specific goal to identify bottlenecks in modelling full-scale WWTPs. The TUDP model structure was further revised by describing hydrolysis as a function of the total heterotrophic biomass being a sum of PAOs and "*ordinary*" heterotrophs.

In summary, the development of the IWA-type Activated Sludge Models (ASM1, ASM2, ASM2d and ASM3) constituted the most significant contribution in the field of modelling biological processes of municipal wastewater treatment in the past 20 years. These models have received a widespread acceptance, first in the research community, and later among practitioners, which was confirmed by the number of citations in the Scopus database, i.e. ASM1 - cited 593 times, ASM2 - cited 272 times, ASM2d - cited 163 times and ASM3 - cited 210 times (in September 2008). The uniform structure of these models constituted a convenient base for further development of model concepts and incorporation of additional processes or state variables. In order to facilitate this, most of the existing commercial simulation programs (simulators) have became more flexible by allowing the users to introduce and examine their own ideas in a simple way (i.e. through Petersen matrix editors). During the recent years, new developments have also been proposed to a standardized organization of the simulation study and increase the reliability of wastewater treatment design and analysis based on simulations with well-calibrated models. An increasing interest in modelling activated sludge systems in the recent years has been reflected by the fast growing number of relevant publications per year (Figure 1.9).

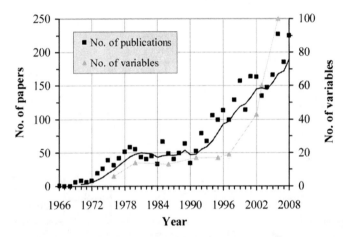

Figure 1.9 Number of publications on activated sludge modelling referenced in the SCOPUS database between 1967 and 2008 (keywords used for searching were "*activated sludge*" and "*model*"; trend lines are fitted with a moving average with the period of 5 years) (the secondary y-axis shows the model complexity in terms of the number of model variables).

1.3 BASIC DEFINITIONS IN MATHEMATICAL MODELLING AND COMPUTER SIMULATION

Originally, the term "*simulate*" meant to imitate or feign something (Roberts *et al.* 1983). More recent definitions of simulation are more precise and refer to using a model to predict the performance of a system under different conditions (Patry and Chapman 1989), exploring the effects of changing conditions on the behaviour of a real system (USEPA 1993), or designing a model of a system and conducting experiments with that model (Smith 1999).

Thus a simulation involves a well defined object of interest in the real world (system) and its description (model) used to understand and predict certain behavioural aspects of the system, or evaluate operational strategies for the system. The system and its model are capable of generating behaviour and are a source of data. The process of generating behaviour of the system and gathering data is called experimentation. According to the above definition of Smith (1999), simulation can be seen as virtual experimentation on the model. In other words, the terms modelling and simulation refer to the respective purposes of conceptual understanding and actual approximation (Aris 1999). The modelling and simulation concept of a real system is presented in Figure 1.10.

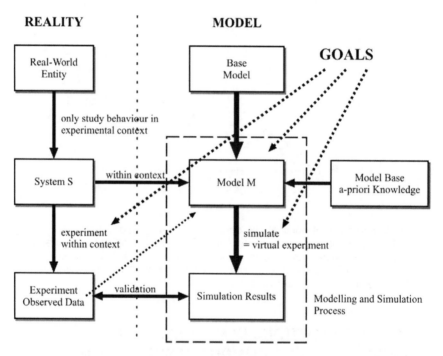

Figure 1.10 Modelling and simulation concept of a real system (Vangheluve *et al.* 2002).

1.3.1 System

Any real world entity or object can exhibit widely varying behaviour depending on the context in which it is studied. An object, which is well defined in terms of specific aspects of its behaviour under specific conditions, is called a "*system*" and it can be both physical (i.e. obeying conservation and constraint laws) and non-physical (informational, such as software) (Vangheluve *et al.* 2002). In other words, a system is a set of various interrelated components (elements) within boundaries and the components interact with one another in an organized fashion toward a common, specified objective (Figure 1.11). Furthermore, a system is a complex entity of higher order than any and all properties borne by its components and the functional objective. Such a situation is called "*synergy*" (from the Greek *syn-ergos* = working together) which simply means that "*a system is more than the sum of its parts*". The components may be quite diverse, consisting of persons, organizations, procedures, software, equipment and/or facilities. Most systems are sufficiently complex that their components are grouped into *subsystems,* which must function in a coordinated way for the system to accomplish its objectives. On the other hand, every system exists in the context of a broader *supersystem,* i.e. a collection of related systems (NASA 1995).

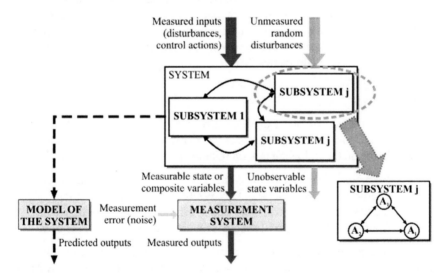

Figure 1.11 A schematic diagram representing a complex system with its sub-systems, inputs, outputs and a model.

1.3.2 Experimentation

Reality is always observed and modelled within the experimental frame. The concept of experimental frame refers to a limited set of circumstances under which a system is to be observed or subjected to experimentation, i.e. the act of carrying out an experiment. In its most basic form, an experimental frame consists of a set of the conditions under which the system behaviour is to be observed, and two sets of variables, the system inputs and outputs, through which the system interacts with the surrounding environment (Kops *et al.* 1999). The inputs and outputs which may comprise such variables as material resources, energy and information. The inputs are generated by the environment that influence the behaviour of the system. The measured inputs are either disturbances or control actions, whereas unmeasured inputs are random disturbances. Observations of the system behaviour can be made by outputs which are determined by the system and influence the behaviour of the environment. The observations yield measurements (Figure 1.11).

An experiment is the process of extracting data from a system through applying a set of external conditions to the inputs of a system (i.e. the accessible inputs), observing the reaction of the system, and recording the behaviour of the outputs (i.e. the accessible outputs). By performing experiments, knowledge about a system is gathered. At the beginning, the knowledge is unstructured. By understanding the cause and effect relationships and by placing observation in both a temporal and spatial order, the knowledge gathered during the experiment becomes organized (Hydromantis 2006).

1.3.3 Model

Simulation involves a kind of model, which is a description of reality, used to understand and predict certain aspects of reality (Meijer 2004). Any model should be built based on the following criteria (Gottman and Kumar Roy 1990):

- Parsimony : a model should contain the fewest number of parameters;
- Interpretation : a model should be easily and meaningfully interpretable;
- Significant effects : terms included in the model should be significant;
- Goodness of fit : a model should provide a sufficient goodness of fit.

It should be realized that a perfect model is never built and it is always a simplification of reality. This is especially true with respect to natural systems containing living organisms (Jeppsson 1996). The extent of simplification will be determined by the intended application of the model (Meijer 2004). Since models should be built for a specific purpose, the value of every model is based entirely upon the degree to which it solves someone's real world problem (Smith 1998). Consequently, the acceptance of models should be guided by "*usefulness*" rather than "*truth*" or "*perfection*". The golden rules of modeling state that "*no model is perfect, some are useful"* and *"a model should be as simple as possible, and only as complex as needed"*. The latter means that the ability of a model to simultaneously consider many factors must be balanced against the value of gaining a basic understanding of a system. In general, the simplest model should be preferred as it provides an adequate description of a given system. Simplicity often leads to clarity in thinking and evaluation. Moreover, it allows to avoid errors caused by a lack of understanding of the fundamental cause-and-effect relationships in the system (USEPA 1993).

> *"Truth is ever to be found in simplicity, and not in the multiplicity and confusion of things. As the world, which to the naked eye exhibits the greatest variety of objects, appears very simple in its internal constitution when surveyed by a philosophical understanding, and so much the simpler by how much the better it is understood."* **(Isaac Newton)**

The development of a new model can be meant as an ongoing, iterative and sometimes long-lasting process of conceptual progress and actual understanding (Figure 1.12). Preliminary ideas/hypotheses about the system are brought together in a first model. The model is then evaluated and interpreted by comparing its predictions with observed data from especially designed experiments or undirected experience. Initially formulated models can give inaccurate or even completely false predictions. In such a case, the concepts incorporated in the first model may be modified and the cycle of formulation, experimentation, solution and comparison is repeated again. The entire process is continued until adequate agreement is obtained between predictions and measurements. It should be realized that it can take a long time (sometimes years) a research model reaches the stage, if ever (!), of being ready for practical applications Andrews (1992).

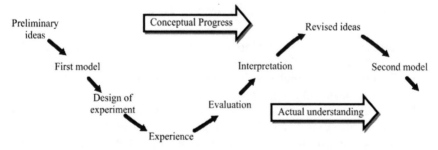

Figure 1.12 An ongoing, iterative process of model development (Aris 1994).

A simulation model can be a mental conception (a conceptual model), a physical model, a mathematical model or a combination of all of these (Figure 1.13). While capturing essential aspects of a given system, a conceptual model represents an understanding of the cause-effect relationships between components of that system. The relationships can be described qualitatively or/and quantitatively, and presented in the form of functional flow block diagrams. Conceptual models form the basis for most of the design and operational decisions made in environmental engineering (USEPA 1993). Some simulations can involve physical models, e.g. pilot plants, but these models are usually relatively expensive to build and unwieldy to move. Therefore, mathematical models are often preferred (Roberts *et al*. 1983), even though they cannot (and even should not) completely replace physical models (Bingley and Upton 2000; Hao *et al*. 2001a).

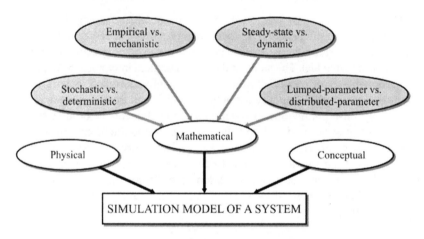

Figure 1.13 Classification of simulation models of a system.

Mathematical models are used to quantitatively describe certain aspects of a system, such as effectiveness, performance or technical attributes, and cost. Apart from this restriction (quantitative description), mathematical models have many features in common with conceptual models and, in fact, are in one sense their subset (USEPA 1993). The heart of any mathematical model is an equation or a set of equations relating meaningful inputs, outputs and characteristics (parameters) of a given system and ideally, the relationships express causality, not just correlation (NASA 1995). The term "*mathematical model*" is sometimes attributed to large sets of complex equations but if a studied system is simple enough, its model may be solved analytically using mathematical methods (such as algebra, calculus or probability theory) to obtain exact information on questions of interests (Law and Kelton 2000). For example, a simple expression such as $y = mx + b$ is an analytical solution representing the system output (y) related to the system input (x) by the system parameters (m,b).

However, most problems of interest in the real world are usually too complex that a simple mathematical model can not be constructed and it is not possible to solve the equation(s) analytically. In such cases, the problems can be evaluated using a computer simulation which is defined as the use of a computer software to predict the performance of a real system under different conditions. By changing variables, predictions may be made about the behaviour of the system. In a computer simulation, a model is evaluated numerically, and data are gathered in order to estimate the desired true characteristics of the model (Law and Kelton 2000). In terms of the manner in which the variables change, simulations are usually referred to as either discrete event or continuous. The former term refers to the fact that state variables change instantaneously at distinct points in time, whereas in a continuous simulation, variables change continuously, usually through a temporarily variable function (Smith 1999). Simulation is performed by a simulator (or simulation platform) which is defined as a user friendly software package that encodes a conceptual model for a particular application (McHaney 1991). The simulator essentially consists of two elements - an internal representation and a solver. The former is a representation of the model which can be understood by the solver, whereas the solver "*solves*" the model, i.e. generates behaviour (Kops *et al.* 1999).

It is essential that simulations are evaluated in terms of their accuracy and appropriateness for solving specific problems. When a simulator produces nonsense predictions, it is not always clear whether this is due to errors in the conceptual model, programming errors, or even the use of faulty data. Whether the simulation model itself and its implementation in the software package accurately represent the real system can be checked in a two-stage process including verification and validation. Verification is the process of determining that the simulator performs the operations as they have been described in the conceptual model. It comprises debugging the model code and ensuring that the simulated responses appear to be feasible (Jeppsson 1996). The term validation means a model ability to mimic system behaviour. The validation step is essential to be confident of any simulation model and involves checking that the model, correctly implemented, generates responses consistent with the data obtained from the real system. Three different levels of model validity can be distinguished (Kops *et al.* 1999):

- Replicative: the model is able to reproduce the input/output behaviour of the system;

- Predictive: the model is able to be synchronised with the system into a state, from which unique prediction of future behaviour is possible;
- Structural: the model can be shown to uniquely represent the internal (structural) workings of the system.

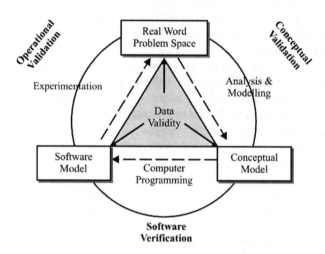

Figure 1.14 Process of model conceptualization, verification and validation (Smith 1999).

Validation is the ultimate check on the success of model building but "*no recipe exists for doing this*" (Bratley *et al.* 1987). A failed attempt at validation usually suggests modifications to the model concept and/or repeating the verification/validation analysis. Model conceptualization, verification, and validation often are in a dynamic, feedback loop as shown in Figure 1.14. Validation should also include validation of the data used to develop the conceptual model as well as the data used for testing. It is usually time-consuming and costly to obtain accurate and appropriate data, and initial attempts to validate a model may fail if insufficient effort has been applied to this problem (Bratley *et al.* 1987).

In addition to verification and validation, Smith (1999) introduced a term accreditation as an official determination that the simulation is acceptable for some specified purposes. No simulation provides a universal solution to all problems in a domain, and accreditation defines the set of problems for which the simulation is appropriate and useful.

In engineering applications, many types of mathematical models can be applied and these models can be categorized in several different ways depending upon specific attributes considered, e.g. deterministic vs. stochastic, mechanistic vs.

empirical, steady-state vs. dynamic, lumped-parameter vs. distributed-parameter (Figure 1.13). These terms allow to enter a dialogue between model builders and model users about the type of model used in a particular analysis, but no hierarchy is implied in the above list of the categories (NASA 1995).

- **deterministic vs. stochastic models.** Deterministic models incorporate direct links between inputs, outputs and system parameters through rate equations to seek the future course of the system behavior at some fixed point of time (Grady 1989). In other words, the solution to the model has a fixed number of definite answers for any given set of conditions (Andrews 1992). The deterministic models are based on first engineering principles, i.e. the rate equations are developed from general balance equations applied to mass and other conservatives (Gernaey *et al.* 2004). In contrast, the principle of uncertainty (a distribution of all possible outcomes) is introduced in stochastic models, sometimes called probabilistic models, and statistical techniques are used to express the model in a mathematical form (Andrews 1992).

- **mechanistic vs. empirical models.** Mechanistic models, also called "*white-box*" or internal models, are developed from the fundamental engineering and scientific knowledge of a process occurring within a system being modeled (Patry and Chapman 1989). Prosser (1990) noted that mechanistic models represent hypotheses that may be tested by comparing predicted system behaviour with experimental data. The aim of this kind of models is to predict behaviour from assumptions regarding controlling mechanisms when many of these assumptions are introduced merely to simplify the theoretical description. According to the USEPA (1993), the mechanistic models offer the best potential for providing a realistic representation of the system as well as the best opportunity to accurately predict system response over a broad range of operating conditions, including extrapolation. The mechanistic models are usually deterministic (Grady 1989). In the case, when details of the mechanisms involved in a system are either scarce, non-existent or irrelevant, empirical ("*black-box*", external or input/output) models can be applied. Such models identify and quantify relationships between process variables without reflecting knowledge of the process and typical examples include different autoregressive time series models, artificial neural networks (ANN) and multivariate statistical methods (MVS) (Gernaey *et al.* 2004). These authors emphasized that the empirical models complement and support the knowledge about the wastewater treatment process in situations where the mechanistic models do not accurately describe a process (e.g. sedimentation), insufficient data are available for calibration of the mechanistic model or rain events occur in the influent data. The empirical models are, however, highly system specific and thus not easily adaptable to new conditions (Grady 1989). The "*black-box*" models are incapable of describing the full dynamics of the activated sludge process but they may be useful for on-line control of some well-defined parts of the process (Jeppsson 1996). The advantages of "*white-box*" and "*black-box*" modelling can be combined in a hybrid scheme, called "*grey-box*", in which models are based on first engineering principles with specific elements, e.g. derived empirically from process data (Gernaey *et al.* 2004). For example, influent flow rate predictions can be represented by the sum of deterministic sinusoidal daily variation with a random (stochastic) component.

- **dynamic vs. steady-state (static) models.** Steady-state models predict the equilibrium behaviour of a system without insight into its temporal variations. Dynamic models consist differential equation(s) and predict the time-varying performance of a process (Party and Chapman 1989), are especially useful for modelling highly dynamic systems, such as WWTPs. More than three decades ago, Andrews (1974) emphasized the need for consideration of the dynamic behaviour of wastewater treatment processes due to large temporal variations encountered during the operation of WWTPs. These variations can be classified as "*regular*" variations, internally generated (self-inflicted) disturbances and unexpected disturbances. Typically, the "*regular*" variations in hydraulic loading, pollutant loading and influent characteristics reveal almost regular daily variations, on top of which are weekly and annual cyclic variations (Olsson and Newell 1999). The flow rate variations are affected by the size of the community and the length of the wastewater collection system (Metcalf and Eddy 1991). The influent pollutant loadings change from one municipal wastewater to another due to the following factors: socioeconomic factors, water usage, degree of inflow and infiltration, use of garbage grinders, presence of any industrial wastewater, characteristics of the sewer system (e.g. size, retention time, storage capacity, etc.) and presence of a phosphate detergent ban (Barker and Dold 1997a). Initially, it was assumed that specific wastewater exhibits a rather stable composition of various fractions even when overall concentrations in the wastewater vary (Henze 1992). This conclusion was illustrated by several examples from various WWTPs with respect to suspended solids, soluble Chemical Oxygen Demand (COD_{sol}) and enzyme activities. Later studies (e.g. Kristensen *et al.* 1998; Melcer, 1999) have indicated, however, that considerable changes in the wastewater characteristics can be observed. During the plant operation, internally self-inflicted disturbances are generated, e.g. by returning filter-wash water and supernatant from sludge dewatering. The unexpected disturbances may include rain storm events, snow melts, industrial discharges, toxic release, equipment failures, etc. (Nielsen 2001).

- **lumped-parameter vs. distributed-parameter.** If the dynamic behaviour of a system is modelled without providing spatial details, then such a model is called a lumped-parameter (or state-space) model and consists of a set of ordinary differential equations. Using this approach, the properties of the system are assumed to be uniform over a given volume which is represented by a continuously stirred tank reactor (CSTR). A distributed-parameter model, described by a set of partial differential equations, aims to predict both the temporal and spatial behaviour of the system, e.g. the distribution of a pollutant concentration along the length of an activated sludge reactor (Patry and Chapman 1989). Using multiple CSTR elements, it is possible to predict spatial variations with a lumped-parameter model by (Beck 1989) (see: Section 2.4.2)

1.3.4 Advantages and disadvantages of mathematical modelling and computer simulation

The development of a mathematical model and subsequent computer simulations with the model have many advantages compared to experimenting with a real system (McHaney 1991):

- Experimentation conducted without disruptions to existing systems (testing of new ideas may be difficult, costly or otherwise impossible in systems that already exist);
- Testing a concept prior to installation, which may reveal unforeseen design flaws and improve the design concept;
- Detection of unforeseen problems or bugs, which may exist in the system's design (debugging time and rework costs can be avoided) or operation (improvements to system operation may be discovered);
- Gaining in system knowledge, which might be dispersed at the beginning;
- Much greater speed in analysis (simulation permits "*time compression*" to fractions of seconds or minutes representing minutes, hours, days, or even years of system time. (This feature is especially important in wastewater treatment systems where the rates of biological processes are relatively slow and physical experimentation may require weeks or even months (Andrews 1992));
- Forcing system definition in order to produce a valid working model of a system;
- Enhancing creativity which can be exercised without the risk of failure.

All the seven advantages of using simulation for analysis of "*what-if*" scenarios result in the reduction of risk and increasing the certainty about the expected operation of a new system or about the effects of changes to an existing system. Unexpected problems can be exposed before undertaking costly and time-consuming investments, with simulation results sometimes contradicting intuition.

Although computer simulation can be a powerful method of analysis, certain limitations and disadvantages must be acknowledged:

- It is neither cheap nor easy to apply this tool correctly and effectively (Bratley *et al.* 1987). Moreover, is not generally set up to produce quick answers to questions. In many cases, data collection, model development and implementation, analysis, and report generation will be costly and require considerable amounts of time.
- Simulation results can be no better than the model (and data) on which they are based on (USEPA 1993). Since a simulation model encodes concepts that are difficult to completely define, it is easy to create a model that is not a reasonable representation of the real system. Another limitation is the availability of accurate and appropriate data for describing the behaviour of the system (Smith 1999). Incorrect or incomplete models and/or poor data can result in simulations generating large quantities of worthless, inaccurate or even completely misleading results.
- Due to approximations made while creating the model, it is known in advance that the real system and its model do not have identical output distributions (Bratley *et al.* 1987). Therefore, it should be realized that a simulation yields only approximate results, i.e. measurements of general trends, rather than exact data for specific problems (Smith 1999).
- The attempt to use computer simulation to find an optimum solution to a problem might rapidly degenerate into a trial-and-error process.
- In the case of wastewater treatment, computer simulation is a much cleaner job than physical experimenting. This factor is dangerous and can result in neglecting the validation of the simulation model (Andrews 1992).

Chapter 2
Model building

2.1 COMPONENTS OF A COMPLETE MODEL OF AN ACTIVATED SLUDGE SYSTEM

An activated sludge system is a complex physical-chemical-biological system with numerous internal interactions between state variables and dynamic behaviour of input variables and local conditions. The starting point for any model development is the description of the submodels that make up a complete model of the activated sludge system (Figure 2.1). The submodels covered in this chapter include a hydraulic configuration model, influent wastewater characterization model, bioreactor model and sedimentation tank (clarifier) model. The hydraulic configuration model describes tank volumes and the liquid flow rates between tanks and is discussed in terms of different process configurations of biological nutrient removal (BNR) systems. The influent characterization model calculates the influent concentrations of state variables (COD, nitrogen and phosphorus fractions) in terms of the physical state and biodegradability. The bioreactor model couples a hydrodynamic mixing (flow pattern) model with a biokinetic model (describing biochemical conversions in the activated sludge) or other source terms that describe environmental conditions, e.g. process temperature and oxygen transfer. The flow patterns in a bioreactor are described at the extremes as plug flow or completely mixed. A value of the dispersion number (or inverse Peclet number) indicates which of the two patterns is approached. The process temperature can be calculated from a heat balance over the bioreactor accounting for such components as solar radiation, atmospheric radiation, conduction and convection, evaporation, aeration, mechanical energy from mixing and biological processes. The clarifier models are described in terms of various degree of complexity ranging from simple ideal point settlers to two-dimensional (2-D) hydrodynamic models.

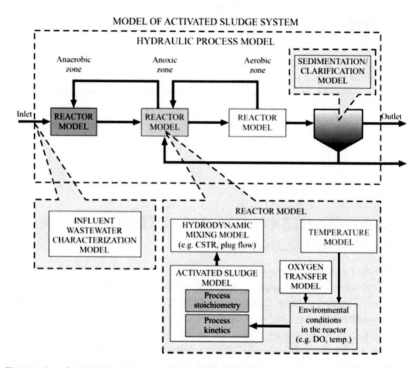

Figure 2.1 Schematic representation of a complete model of an activated sludge system (modified from Melcer (2003) and Meijer (2004)).

2.2 HYDRAULIC CONFIGURATION MODEL

The entire activated sludge system consists of two or three main elements including a mainstream bioreactor, a clarifier and, in some cases, a sidestream sub-system for reject water treatment. The mainstream and sidestream bioreactors may be divided into different compartments (reactors) depending upon the actual electron conditions in the given compartment (anaerobic, anoxic and aerobic).

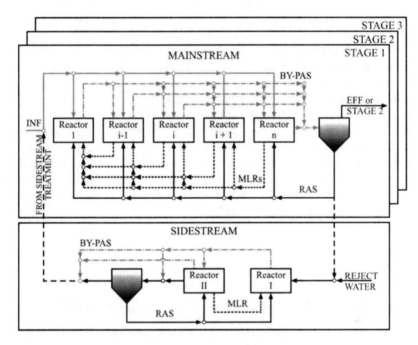

Figure 2.2 General hydraulic configuration model including the mainstream and sidestream treatment systems.

The hydraulic configuration model describes an activated sludge system in terms of the physical characteristics (such as reactor volumes and clarifier dimensions) and the liquid flows. The latter comprises a mixed liquor flow from one reactor to the following one, mixed liquor recirculations (MLRs), return activated sludge (RAS) recirculations and possibly by-passes to a clarifier (Figure 2.2). Most common process configurations of biological nutrient removal (BNR) systems have been presented in Figures 1.1-1.4.

2.3 INFLUENT WASTEWATER CHARACTERIZATION MODEL

Sherrard (1984) noted that *"wastewater composition plays a small part in mathematical analyses made, but may have a major impact on predicted results"*. A similar conclusion was derived by Barker and Dold (1997a): *"if the model is to provide reasonable predictions of system behavior, adequate knowledge of wastewater characteristics is extremely important"*.

Wastewater components can be characterized using physical or chemical criteria. Physical characterization, important primarily for estimating the potential efficiency of the primary clarifiers (Water Research Commission 1984), involves separation of the pollutants in wastewater based on the filter pore size (Figure 2.2). The conventional fractionation into soluble (dissolved) vs. particulate (suspended) components is usually performed using a 0.45 μm pore size filter, whereas an additional colloidal component is included in the *"true"* fractionation. A sampling program on several WWTPs in the Netherlands showed that chemical precipitation can alternatively be used instead of 0.1 μm filtration when determining the content of soluble fraction. The difference in soluble COD for both methods was only approximately 1% (Roeleveld and van Loosdrecht 2002).

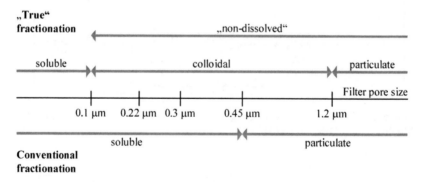

Figure 2.2 Physical COD and nitrogen fractionation in terms of the filter pore size (Czerwionka *et al.* 2009).

The average concentrations of various components in the Danish municipal wastewater are presented in Table 2.1. The dominating fractions were suspended and soluble for COD and nitrogen respectively. Very similar proportions of the components was found for much stronger wastewater (COD = 500–800 mg COD/dm³, $N_{tot.}$ = 60–80 mg N/dm³) entering four large WWTPs in northern Poland (Figure 2.3).

Table 2.1 The average composition of Danish municipal wastewater based on physical fractionation (Henze and Harremoes 1992)

Quality parameter Fraction	COD g COD/m^3	BOD$_5$ g BOD$_5$/m^3	N$_{tot.}$ g N/m^3	P$_{tot.}$ g P/m^3
Suspended	200	85	4	1
Colloidal	65	35	6	1
Soluble	85	40	20	6
Total	350	160	30	8

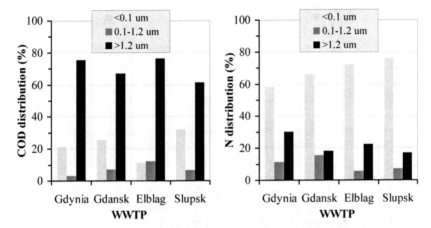

Figure 2.3 Distributions of COD (a) and nitrogen (b) in the settled wastewater in four large WWTP in northern Poland (adapted from Pagilla *et al.* 2008).

Chemical characterization of wastewater comprises the identification and measurement of organic (carbonaceous) and nitrogenous constituents including biomass as well as inorganic constituents, such as alkalinity, acidity, pH, phosphorus, etc. (Water Research Commission 1984). Organic constituents in wastewater can be further categorized based on the characterization of specific groups of chemical compounds, such as lipids, proteins, carbohydrates, etc. or fractionation of the organic constituents according to their rate of biodegradation (Wanner 1994). The latter approach (organic fractionation) is primarily related to modelling the activated sludge process and has its origin in the multi-substrate models of Dold *et al.* (1980) and Henze *et al.* (1987). Figure 2.4 illustrates sample relationships between physical fractions, specific organic compounds and model components determined for the typical Danish municipal (Henze *et al.* 1994).

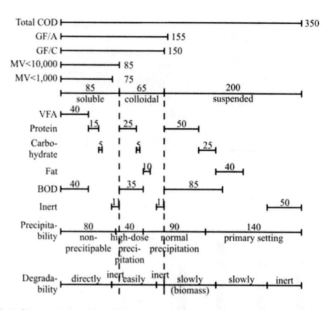

Figure 2.4 Characterization of organic matter in Danish municipal wastewater (Henze *et al.* 1994) (related to the average composition listed in Table 2.1).

The major difference between the organic fractions is the rate with which they are utilized by microorganisms (Figure 2.5). Expressed in "*oxygen demand*" units the organic compounds can be divided into three major categories: readily biodegradable, slowly biodegradable, and non-biodegradable (inert) organic compounds (Henze *et al.* 1987; Henze 1992; Wanner 1994; Barker and Dold 1997a; Orhon and Cogkor 1997). In the case of performing preliminary simulation studies, it would be very useful to be able to translate the traditional data available in the records of WWTPs into a form that can be adapted for the complex biokinetic models (Grady *et al.* 1999). A rough complete organic fractionation of the influent wastewater can be performed using only 2-3 input parameters, i.e. COD (and/or BOD$_5$) and volatile suspended solids (VSS), and a few stoichiometric coefficients (e.g. Hydromantis 2006). Single conventional analytical parameters such as BOD and COD, routinely used to quantify collectively the concentrations of organic compounds in wastewater, cannot directly differentiate between the individual categories (Orhon and Cokgor 1997). The advantage of using COD is that it provides a consistent basis for the description of the activated sludge process including relationships between substrate, biomass and electron acceptor (DO and nitrate) (Barker and Dold 1997a).

Table 2.2 Reported wastewater fractionation in various countries

Reference	WWTP	COD g COD/m^3	S_I	S_S	X_S	X_I	$X_{b,H}$
Roeleveld and van Loosdrecht (2002)	21 Dutch WWTPs	241–827	3–10	9–42	10–48	23–50	
Meijer et al. (2001)	Hardenberg (Holland)	604	7	33	31	29	–
Brdjanovic et al. (2000)	Haarlem (Holland)	331	7	32	42	19	–
Van Veldhuizen et al. (1999a)	Holten (Holland)	516	5	41	37	17	–
Petersen et al. (2002)	Zele (Belgium)	44	12	16	22	50	–
Ginestet et al. (2002)	7 French WWTPs	663	1–9	<1–10	24–49	29–55	6–17
Choubert et al. (2009)	22 French WWTPs	500±180	4	20	59	17	–
Penya-Roja et al. (2002)	Valencia (Spain)	240	16	51	24	9	–
Ferrer et al. (2004a)	2 Spanish WWTPs	419–939	4–9	35–50	33–41	8–20	–
Kappeler and Gujer (1992)	2 Swiss WWTPs	250–325	12–20	8–11	53–55	9–10	7–15
Rieger et al. (2001)	2 Swiss WWTPs	250–380	4	10	53	20	13
Carucci et al. (1999)	Rome-Est (Italy)	290	6	15	56	8	15
Andreottola et al. (2003)	Media Pusteria (Italy)	534	5	29	33	27	6
Wichern et al. (2001, 2003)	6 German WWTPs	Not provided	3–11	10–19	45–62	10–24	9–19
Makinia (2006)	2 Polish WWTPs	546–609	4–7	21–40	34–51	20–24	–
Xu and Hultman (1996)	Kungsangen (Denmark)	183	15	27	33	17	8
Sahlstedt et al. (2003)	Helsinki (Finland)	373	169	25	35	21	–
Orhon et al. (1994b)	Istanbul (Turkey)	431	2	10	81	7	–
Gokcay and Sin (2004)	Ankara (Turkey)	249	20	32	40	8	–
Makinia and Wells (2000)	Rock Creek (USA)	180–300	7–9	22–26	53–54	12–15	–
Latimer et al. (2007)	11 US WWTPs	Not provided	3–11	11–28	46–70	8–35	

The ultimate BOD (BOD_U) can be converted to biodegradable COD (BCOD) provided that growth of biomass is taken into account during the BOD test. Furthermore, it can be assumed that 35-40% of the VSS in domestic wastewater is non-biodegradable (Grady *et al*. 1999).

Henze (1992) presented an overview of the early methods for determining wastewater fractions. These procedures have been continuously developed and evaluated (e.g. Mamais *et al*. 1993; Spanjers and Vanrolleghem 1995; Orhon and Cokgor 1997; Kristensen *et al*. 1998; Spanjers *et al*. 1998; Vanrolleghem *et al*. 1999; Petersen *et al*. 2002; Roeleveld and van Loosdrecht 2002). More details about a practical approach to wastewater characterization can be found in Section 4.3.2. The reported fractionation in various countries is presented in Table 2.2.

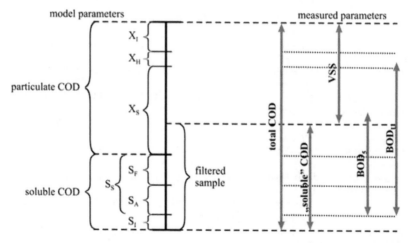

Figure 2.5 COD fractionation in wastewater and various analytical techniques for measuring parts of total COD.

Readily biodegradable compounds, S_S. The readily biodegradable fraction of COD consists of soluble compounds with low molecular weights and usually constitutes 10-15% of total COD in raw wastewater (Henze 1992). Higher contributions of the S_S fraction (up to 50% of total COD) were reported in several later studies (Table 2.2). These compounds can be immediately metabolized by microorganisms, i.e. transported to the cells and oxidized or converted to storage products or biomass. Typical examples in raw wastewater are alcohols (methanol, ethanol), VFAs (especially acetic acid), glucose and other monosacharides, and lower amino acids (Wanner 1994). Only readily biodegradable compounds are considered to be the substrate in heterotrophic growth, either under aerobic or anoxic conditions (Henze et al. 1987). For modelling EBPR, a further subdivision of S_S is needed into fermentable readily biodegradable compounds (S_F) and fermentation products (S_A), covering a wide range of compounds, but assumed to be only acetate (Henze et al. 1995a).

Non-biodegradable (inert) organic compounds (S_I and X_I). These compounds are assumed to be biologically inert and can be present in raw wastewater in both soluble and suspended forms (Henze et al. 1987). The soluble non-biodegradable compounds (S_I) pass unchanged through any biological treatment process. The particulate non-biodegradable organic compounds (X_I) are incorporated in the activated sludge flocs and removed from the system with excess biomass (Wanner 1994). The significance of particulate inert products for activated sludge modelling was recognized in the late 1950's when Symons and McKinney (1958) found that the total oxidation of activated sludge was not possible and a small amount of non-oxidizable polysaccharide material remained (Orhon et al. 1999). The concentration of inert particulates in the settled wastewater typically ranges from 15 to 40 mg/dm^3 (Stensel 1992) and they constitute 35-40% of organic particulates in domestic wastewater (Grady et al. 1999). In terms of COD, the contribution of inert particulates usually varies from 8 to 18% of total COD in raw wastewater (Xu and Hultman 1996), however, the X_I/total COD ratios in a broader range (7-50%) were reported in other studies (Table 2.2).

In addition to the inerts present in the influent, non-biodegradable organic compounds are generated during biological treatment processes and hence the chemical composition of this fraction in the effluent may significantly differ from that in the influent. The soluble inerts are produced as metabolism by-products during decay or hydrolysis (Orhon et al. 1989; Boero et al. 1991). Their amount may exceed the amount of soluble inerts in the influent (Henze 1992). The particulate inerts are a fraction of the net biomass decay (Henze et al. 1987). Kappeler and Gujer (1992) found that the coefficient for the production of inert particulate COD from endogenous respiration equals to approximately 0.2 g COD/g COD consumed. For models using the death-regeneration concept of Dold et al. (1980), this fraction is actually less than 20% (Henze et al. 1987). Alternatively, a simplified approach can be used by means of a fictive influent concentration which includes the generated inerts as well as the true inerts in the influent (Henze 1992).

Slowly biodegradable compounds, X_S. The slowly biodegradable fraction of COD usually constitutes 40-60% of total COD in raw wastewater (Henze 1992). This fraction was originally defined as particulate organics in the model of Dold *et al.* (1980), but later on it became evident that a wide range of compounds (soluble, colloidal and larger organic particles of complex structure) could be classified as slowly biodegradable (Orhon and Cokgor 1997). The common feature of slowly biodegradable compounds is that they cannot be immediately metabolized by microorganisms. Therefore, these compounds are first hydrolyzed, or degraded by other complicated reactions treated as hydrolysis, inside the microorganism flocs by means of extracellular enzymes (Henze *et al.* 1987). The hydrolysis products, chemically similar to readily biodegradable compounds, can then be transported into the cells for intracellular metabolism (Wanner 1994). Due to a possible significant variation in the composition of slowly biodegradable compounds it may be difficult to characterize them only by a single hydrolysis rate (Orhon and Cokgor 1997). In such cases, this fraction can further be subdivided into rapidly hydrolysable COD (S_H) and slowly hydrolysable COD (X_S) (Henze 1992). A common assumption is that the high molecular weight soluble compounds are termed rapidly hydrolysable while the organic suspended solids are termed slowly hydrolysable (Wanner 1994). Actually, however, some soluble compounds hydrolyze slowly whereas some suspended compounds hydrolyze rapidly (Henze 1992). The hydrolysis rate is lower then the utilization rate of readily biodegradable compounds which makes hydrolysis the limiting step in heterotrophic growth only on the slowly biodegradable substrate (Henze *et al.* 1987). The hydrolysis rate variations are also associated with the electron acceptor present (see: Section 3.5).

Viable (active) heterotrophic biomass, X_H. The heterotrophic biomass may account for 10-80% and 10-20%, respectively, in terms of the volatile suspended solids and total organic matter in raw wastewater (Henze 1986; Kappeler and Gujer 1992). Orhon *et al.* (2002) found that the active heterotrophic biomass constituted a significant fraction (40-50%) of the settled COD. Most of the biomass in the primary effluent may have its origin in excess sludge recycled to the influent of the plant (Kappeler and Gujer 1992). For wastewater characterization, the fraction of biomass in wastewater may not taken into consideration, primarily due to the lack of reliable and easy methods for measurement. Without considering the presence of biomass, its fraction is lumped into the slowly biodegradable organic fraction. This assumption does not affect the general modelling significantly when the yield coefficient is increased by approximately 10% (Henze 1992). Roeleveld and van Loosdrecht (2002) explained further why it is not strictly necessary to consider biomass fractions in the influent. For heterotrophs, the growth rate is so high that wash-out never occurs in practice provided that the initial concentration used for simulations assumes the presence of these organisms. With regard to autotrophs ($X_{b,A}$) and PAO ($X_{b,PAO}$), these biomass fractions are usually neglected in the influent as they are very small in comparison with the total COD. However, due to a low growth rate, the autotrophs and PAOs have to be considered in the influent to high-loaded systems to avoid the possibility of washing out.

Nitrogen components. The primary source of nitrogen in raw municipal wastewater is human excretion of which approximately 75% is excreted as urea while the rest is organic nitrogen (Henze 1992). In the municipal wastewater entering a treatment facility, over 90% of nitrogen is present in the form of ammonia or unstable organic compounds (e.g. amino acids, protein and nucleotides), which are converted to ammonia during treatment (Gray 1989). The nitrogen fractions can be estimated according to the same rules as the organic fractions, i.e., biodegradable and non-biodegradable (Henze *et al.* 1987). The first group is composed of ammonia (free ammonia and ammonium ions) and nitrogenous organic compounds - soluble and particulate. The content of non-biodegradable nitrogen is not significant, i.e. approximately 2–3% of the total nitrogen mass (Henze 1992). Nitrogen fractionation in wastewater (Figure 2.6) is usually based on the TKN concentration, but when appreciable nitrate/nitrite concentrations are encountered this parameter must also be included as an additional fraction (Barker and Dold 1997a). Nitrogen contents of various organic fractions (typical values and ranges) are presented in Table 2.3.

The nitrogen characterisation should at least get sufficient attention to predict appropriately the nitrogen content of the sludge in the bioreactor (Roeleveld and van Loosdrecht 2002). This issue was addressed in the studies of Meijer *et al.* (2001) and Makinia *et al.* (2005a) and resulted in significant adjustments of the N content of X_S, $i_{N,XS}$, and X_I, $i_{N,XI}$. With lower effluent criteria for nitrogen, the contribution of effluent organic nitrogen (EON) becomes very important. For BNR systems designed to maximize nitrification/denitrification and effluent solids removal, the effluent TN concentration may range from 2.0 to 4.0 g N/m^3 with about 40% as EON; mainly as colloidal and soluble fractions (Stensel *et al.* 2008). Present knowledge on the characteristics and behaviour of both fractions is limited and insufficient to estimate BNR process effluent concentrations as a function of plant design and influent characteristics. Elevated values in the effluent might potentially be a result of industrial discharges or high strength wastes and the simplest approach in such cases is to adjust the N content of S_I (Roeleveld and van Loosdrecht 2002).

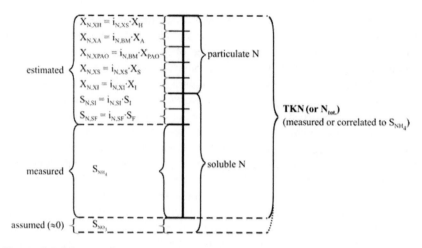

Figure 2.6 Nitrogen fractionation in wastewater.

Table 2.3 Nitrogen contents of various organic fractions in wastewater

Definition	Symbol	Unit	Typical value[1]	Range[2]
N content of S_I	$i_{N,SI}$	g N/g COD	0.01	0.01–0.02
N content of S_A	$i_{N,SA}$	g N/g COD	0.00	0.00
N content of S_F	$i_{N,SF}$	g N/g COD	0.03	0.02–0.04
N content of X_I	$i_{N,XI}$	g N/g COD	0.03	0.01–0.06
N content of X_S	$i_{N,XS}$	g N/g COD	0.04	0.02–0.06
N content of biomass (X_H, X_{PAO}, X_A)	$i_{N,BM}$	g N/g COD	0.07	

[1] Values from Henze *et al.* (1995a)
[2] Values from Roeleveld and van Loosdrecht (2002)

Phosphorus components. Grady *et al.* (1999) classified phosphorus forms present in domestic wastewater as orthophosphate, inorganic condensed phosphates (pyrophosphate, tripolyphosphate and trimetaphosphate) as well as organic phosphate (e.g. sugar phosphates, phospholipids and nucleotides). Condensed phosphates and organic phosphate are converted to orthophosphate through microbial activity. For the modelling purposes, characterization of the phosphorus content in wastewater can be based on the total phosphorus concentration (Barker and Dold 1997a). Phosphorus fractionation in wastewater and phosphorus contents of various organic fractions are presented in Figure 2.7 and Table 2.4, respectively. Concerning a detailed characterization of the phosphorus fractions, limited data exist in literature. Barker and Dold (1997a) suggested that concentrations of PO_4-P (considered as soluble reactive phosphorus) can be calculated as 0.85-0.90 of total P concentrations. Henze *et al.* (1999) assumed a constant composition of all organic fractions (i.e. constant P to COD ratio). Brdjanovic *et al.* (2000) calculated concentrations of PO_4-P by subtracting the content of P in all these fractions from the measured concentrations of total P. This approach was compared with routine measurements from two large WWTPs in northern Poland (Makinia *et al.* 2002). The PO_4-P/$P_{tot.}$ ratio in the primary effluent varied within the range of 0.55-0.77 (actual) vs. 0.56-0.71 (calculated) at the Gdansk WWTP and 0.63-0.84 (actual) vs. 0.67-0.80 (calculated) at the Gdynia WWTP, respectively.

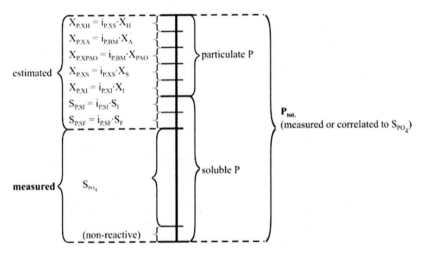

Figure 2.7 Phosphorus fractionation in wastewater.

Table 2.4 Phosphorus contents of various organic fractions in wastewater

Definition	Symbol	Unit	Typical value[1]	Range[2]
P content of S_I	$i_{P,SI}$	g P/g COD	0.00	0.002–0.008
P content of S_A	$i_{P,SA}$	g P/g COD	0.00	0.00
P content of S_F	$i_{P,SF}$	g P/g COD	0.01	0.010–0.015
P content of X_I	$i_{P,XI}$	g P/g COD	0.01	0.005–0.010
P content of X_S	$i_{P,XS}$	g P/g COD	0.01	0.010–0.015
P content of biomass (X_H, X_{PAO}, X_A)	$i_{P,BM}$	g P/g COD	0.02	

[1] Values from Henze *et al.* (1995a)
[2] Values from Roeleveld and van Loosdrecht (2002)

2.4 BIOREACTOR MODEL

For modelling a particular compartment of the activated sludge bioreactor, a biokinetic model (source term) needs to be combined with the appropriate flow terms forming a hydrodynamic mixing model. Other source terms that describe environmental conditions, e.g. oxygen transfer and process temperature, may also be considered. An oxygen transfer model is essential for predicting the behaviour of dissolved oxygen (DO) in aerobic compartments and estimating sufficient amounts of oxygen for aerobic biochemical reactions. A temperature model is important not only for modelling the link between temperature and biological kinetics (especially nitrification), but also to understand heat losses and gains through a treatment plant.

2.4.1 Biokinetic model

The most common approach to describing biochemical conversions in activated sludge systems is based on biokinetic (or activated sludge) models. These models are mechanistic in nature (see: Section 1.3.3) and are supposed to provide hypothetical explanations for complex interactions occurring within the systems. Sometimes, chemical reactions (e.g. phosphorus precipitation) may be incorporated as an additional module.

In order to model biochemical reactions of various components in the activated sludge systems, these reactions have to be written in the form of mass balance equations. These equations are usually derived in molar units, and thus the molar units have to be converted to appropriate mass units. The entire procedure of converting molar-based stoichiometric equations into the mass-based equations was described in detail by Irvine *et al.* (1980) and Grady *et al.* (1999). In the final stage of the procedure, a single mass-based stoichiometric equation can generally be written as:

$$A_1 + v_2\,A_2 + ... + v_{m'}\,A_{m'} \longrightarrow v_{m'+1}\,A_{m'+1} + v_{m'+2}\,A_{m'+2} + ... + v_m\,A_m \quad (2.1)$$

where

v_1 to v_m - Stoichiometric coefficients
A_1 to $A_{m'}$ - Reactants
$A_{m'+1}$ to A_m - Products

Because the sum of the stoichiometric coefficients in any mass-based stoichiometric equation equals to 0, Equation 2.1 may be rewritten as:

$$(1)\,A_1 + (v_2)\,A_2 + ... + (v_{m'})\,A_{m'} + v_{m'+1}\,A_{m'+1} + v_{m'+2}\,A_{m'+2} + ... + v_m\,A_m = 0$$

$$(2.2)$$

The stoichiometry of the biokinetic models of BNRAS systems is implicitly based on four materials (conservatives) subjected to continuity considerations including ThOD (conservative form of COD), nitrogen, phosphorus and in most cases ionic charges. A continuity equation applies to each conservative k, which can be written as:

$$\sum_{i=1}^{m} v_i \, \iota_{i,k} = 0 \tag{2.3}$$

where
$\iota_{i,k}$ - Conversion factor to convert the units of component A_i to the units of the conservative k, to which continuity is to be applied, MM^{-1}

Since there is a relationship between the masses of the components, it follows that there is also a relationship between the rates at which they are utilized or formed. If r_i represents the rate of degradation (or formation) of component i in reaction j, it follows that:

$$\frac{\rho_{1,j}}{(1)} = \frac{\rho_{2,j}}{(v_2)} = ... = \frac{\rho_{m',j}}{(v_{m'})} = \frac{\rho_{m'+1,j}}{(v_{m'+1})} = \frac{\rho_{m,j}}{(v_m)} = \rho_j \tag{2.4}$$

where
ρ_j - Generalized process (reaction) rate, $ML^{-3}T^{-1}$

When the reaction rate has been determined for one component, then the reaction rates in the same units are known for all other components.

Multiple transformation processes taking place simultaneously in activated sludge systems can be presented as a set of mass-based stoichiometric equations, in which subscript i denotes component A_i (i = 1 to m) participating in process (reaction) j (j = 1 to n):

$$(1) A_1 + (v_{2,1}) A_2 + ... + (v_{m',1}) A_{m'} + v_{m'+1,1} A_{m'+1} + v_{m'+2,1} A_{m'+2} + ... + v_{m,1} A_m = 0 \quad \rho_1$$
$$(v_{1,2}) A_1 + (1) A_2 + ... + (v_{m',2}) A_{m'} + v_{m'+1,2} A_{m'+1} + v_{m'+2,2} A_{m'+2} + ... + v_{m,2} A_m = 0 \quad \rho_2$$
$$...$$
$$(v_{1,n}) A_1 + (v_{2,n}) A_2 + ... + (v_{m',n}) A_{m'} + v_{m'+1,n} A_{m'+1} + v_{m'+2,n} A_{m'+2} + ... + (1) A_m = 0 \quad \rho_n$$

$$\tag{2.5}$$

where
$v_{i,j}$ - Normalized mass-based stoichiometric coefficient for A_i in reaction j, MM^{-1}

Within the stoichiometric matrix one stoichiometric coefficient, $v_{k,j}$, of which process j may be chosen as dimensionless with the value of $+1$ or -1.

The overall net conversion (utilization/formation) rate, r_i, for component A_i will be obtained by considering the sum of the rates for all parallel reactions in which A_i participates:

$$r_i = \sum_{j=1}^{n} v_{i,j} \rho_j \qquad (2.6)$$

where

r_i - Overall conversion (utilization/formation) rate of component i, $ML^{-3}T^{-1}$

All conservatives k are subject to continuity equations for all processes j:

$$\sum_{i=1}^{m} v_{i,j} \iota_{i,k} = 0 \qquad (2.7)$$

The conversion factors $\iota_{i,k}$ can be incorporated in a composition matrix $I_{i,k}$ as proposed by Gujer and Larsen (1995). Each continuity equation allows to predict the value of one stoichiometric coefficient, provided that the other coefficients are known. These equations are usually used to calculate the stoichiometric coefficients of oxygen from ThOD, ammonia from nitrogen, inorganic soluble phosphorus from phosphorus, alkalinity from ionic charge and total suspended solids (TSS) from total solids continuity, respectively.

In order to facilitate the description of complex biokinetic models, Henze *et al.* (1987) proposed a matrix format based on the work of Petersen (1965). Such a notation ensures a transparent presentation of the model structure, i.e. the choices of model constituents and the parameters occurring in the expressions describing the interactions between model components (Beck 1989). Their symbols are listed across the top of the table heading and the considered processes are listed down the left side of the table. The elements within the matrix comprise the stoichiometric coefficients, $v_{i,j}$, whereas the process kinetic rates ρ_j, are placed down the right column adjacent to the stoichiometric matrix. An example of the Petersen matrix for a simple biokinetic model is presented in Table 2.5. The original Monod-Herbert model has been extended with two components, i.e. ammonia nitrogen (S_{NH}) and inorganic soluble phosphorus (S_{PO4}) to demonstrate the link between the stoichiometric and composition matrices.

Table 2.5 Process kinetics and stoichiometry for heterotrophic growth in an aerobic environment according to the Monod-Herbert model with the conservation matrix (modified from Henze *et al.* 1987)

Continuity →

Process j \ Component i	1 X_b	2 S	3 S_O	4 S_{NH}	5 S_{PO4}	Process rate, p_j [ML^{-3}T^{-1}]
1 Growth	1	$-\dfrac{1}{Y}$	$-\dfrac{1-Y}{Y}$	$v_{4,1}$	$v_{5,1}$	$\mu_{max}\dfrac{S}{K_S+S}X_b$
2 Decay	-1		-1	$v_{4,2}$	$v_{5,2}$	bX_b
Observed conversion rates [ML^{-3}T^{-1}]	$r_i = \sum_j r_{i,j} = \sum_j v_{i,j}\rho_j$					

(Mass balance ↓)

Stoichometric parameters: Y	Biomass [M(COD)L^{-3}]	Substrate [M(COD)L^{-3}]	Oxygen (negative COD) [M(–COD)L^{-3}]	Ammonia N [M(N) L^{-3}]	Soluble P [M(P)L^{-3}]	Kinetic parameters: μ_{max}, K_S, b

Conservation matrix						
Nitrogen	$i_{N,X}$	$i_{N,S}$		1		
Phosphorus	$i_{P,X}$	$i_{P,S}$			1	

Without introducing the matrix notation, a presentation of the overall degradation (or formation) rates in complex biokinetic models would be much more difficult to read and understand. Gujer (2006) discussed this problem using the model of Dold *et al.* (1980) as an example. The complete model was described by 63 numbered equations, which now could be presented in a single table. For comparison, the reaction rates for the individual components of the extended Monod-Herbert model are presented in Table 2.6. Moreover, this table also contains the stoichiometric coefficients for selected components, i.e. S_{NH} and S_{PO4}, which are calculated from Equation 2.7 assuming the continuity for nitrogen and phosphorus, respectively.

As described in Section 1.2.3, a variety of biokinetic models is available to predict the behaviour of activated sludge systems so Gujer (2006) even called for "*a model developing moratorium*". Henze *et al.* (2000) pointed out that different models might turn out to generate equivalent predictions in spite of conceptual differences between them. Indeed, Makinia *et al.* (2006b) demonstrated that two models including ASM2d (Henze *et al.* 1999) and ASM3P (Rieger *et al.* 2001) can predict with similar accuracy the long-term behaviour of nitrogen compounds (NH$_4$-N and NO$_3$-N) in a BNR pilot plant. The relationships between ASM2d and ASM3P predictions for NH_4-N and NO_3-N were highly correlated ($r^2 = 0.83$-0.99) with the slopes remaining close to 1.0.

Table 2.6 A list of the resulting overall degradation/formation rates of the model presented in Table 2.1 and the selected stoichiometric coefficients calculated from Equation 2.11

Mass balance equation	Stoichiometric coefficient
$r_X = \mu_{max} \dfrac{S}{K_S + S} X_b - bX_b$	$v_{4,1} = -i_{N,X} + \dfrac{1}{Y} i_{N,S}$
$r_S = -\dfrac{\mu_{max}}{Y} \dfrac{S}{K_S + S} X_b$	$v_{4,2} = i_{N,X}$
$r_{SO} = -\dfrac{1-Y}{Y} \mu_{max} \dfrac{S}{K_S + S} X - bX$	$v_{5,1} = -i_{P,X} + \dfrac{1}{Y} i_{P,S}$
$r_{SNH} = v_{4,1}\mu_{max} \dfrac{S}{K_S + S} X - v_{4,2}bX$	$v_{5,2} = i_{P,X}$
$r_{SPO4} = v_{5,1}\mu_{max} \dfrac{S}{K_S + S} X - v_{5,2}bX$	

The comparison of concepts used for modelling specific processes in the most common biokinetic models can be found elsewhere (Hu *et al.* 2003b; Makinia 2006). Table 2.7 contains a comparison of five models in terms of the number of incorporated processes, state variables, kinetic and stoichiometric coefficients, composition factors and temperature correction factors. Components in ASM2d and their equivalents in the other four models are listed in Table 2.8. In the following chapter, approaches to modelling specific biochemical processes occurring in activated sludge systems are discussed in terms of possible modifications and extensions of the existing models.

Table 2.7 Complexity of the most common, complex activated sludge models

Element of the model structure	Model				
	ASM2d	ASM3P	TUDP	New General	New UCTPHO
Processes	21	23	22	36	35
State variables	19	17	17	19	16
Kinetic coefficients	45	43	50	30	28
Stoichiometric coefficients	9	12	18	17	14
Composition (conversion) factors	13	15	18	16	12
Temperature correction factors	11	11	15	18	10

Table 2.8 Components in ASM2d and their equivalents in the other common activated sludge models

Component	ASM2d	ASM3P	TUDP	New General	New UCT-PHO
Soluble components					
Dissolved oxygen	S_O	+	+	+	+
Fermentable, readily biodegradable substrate	S_F	part of S_S	+	+	+
Fermentation products (acetate)	S_A	part of S_S	+	+	+
Inert soluble organic matter	S_I	+	+	+	+
Ammonia nitrogen	S_{NH}	+	+	+	+
Nitrate (+nitrite) nitrogen	S_{NO}	+	+	+	+
Dinitrogen	S_{N2}	+			
Inorganic soluble phosphorus (Orthophosphate)	S_{PO4}	+	+	+	+
Alkalinity	S_{ALK}	+	+		
Particulate components					
Inert particulate organic material	X_I	+	+	+	+
Slowly biodegradable substrate	X_S	+	+	+	
Heterotrophic organisms	X_H	+	+	+	+
Autotrophic organisms	X_A	+	+	+	+
Phosphate accumulating organisms (PAOs)	X_{PAO}	+	+	+	+
Cell internal storage product of PAOs	X_{PHA}	+	+	+	+
Stored polyphosphate (releasable)	X_{PP}	+	+	+ (X_{PP-LO})	+
Metal hydroxides	X_{MeOH}				
Metal phosphate	X_{MeP}				
Additional components in the other models					
Readily biodegradable substrate		$S_S = S_A + S_F$			
Organics stored by "*ordinary*" heterotrophs		X_{STO}			
Glycogen			X_{GLY}		
Endogenous mass				X_E	X_E
Fixed stored polyphosphate				X_{PP-HI}	
Particulate biodegradable organic N				X_{ND}	
Soluble biodegradable organic N				S_{ND}	
Non–biodegradable soluble N				S_{NI}	
Adsorbed organics					X_{ADS}
Enmeshed organics					X_{ENM}

2.4.2 Hydrodynamic mixing model

2.4.2.1 Types of reactors

Many important parameters are influenced by the flow patterns in the activated sludge reactor including organic matter (BOD or COD) removal and settling properties of the activated sludge (Horan 1990). The extreme (ideal) patterns of flow in a reactor are described, based on the residence time distribution (RTD) curves, as plug flow or completely mixed. In an ideal plug-flow reactor (PFR), there is no axial (longitudinal) mixing and each element of the flowing fluid moves along the axis of the reactor in the same order relative to all other elements. Consequently, a response to a conservative (inert, non-reactive) tracer injected at the reactor inlet will appear in the reactor outlet after the time corresponding to a theoretical hydraulic detention time ($\tau = V/Q$). If the tracer is injected as an impulse or continuously, the response will appear in the outlet as an impulse or step function, respectively (Figure 2.8a). In a completely mixed reactor, commonly termed a continuous stirred tank reactor (CSTR), the contents are instantaneously mixed and homogenous at every point. Under such conditions, the effluent composition is identical the reactor content. In the CSTR, the tracer injections emerge in the effluent tracer concentrations as an exponential decay (impulse injection) and cumulative function (continuous injection) (Figure 2.8b). For actual-flow reactors (AFRs), also termed arbitrary, non-ideal or dispersive flow reactors, the results must lie between these two extreme cases (Figure 2.8c). The imperfectness of the flow pattern, reflected by such features as short circuiting zones and/or regions of stagnant fluid (dead zones), may due to the following factors (Metcalf and Eddy 2003):

- The scale of mixing phenomenon,
- Geometry (i.e. aspect ratio) and size of the reactor,
- Power input per unit volume (i.e. mechanic or pneumatic),
- Type and location of the inlets and outlets,
- Inflow velocity and its fluctuations,
- Density and temperature differences between the inflow and the contents of the reactor.

Flow fields in AFRs are very complex and difficult to model completely. Therefore, in traditional wastewater treatment practice, reactors have generally been designed on the basis of the ideal flow configurations (PFR or CSTR). A value of the inverse Peclet (termed also dispersion or Bolenstein) number, defined as a ratio of the mass transport due to dispersive to advective flow, indicates which of the two patterns is approached:

$$\frac{1}{Pe} = \frac{E_L}{uL} = \frac{\text{dispersive flow}}{\text{advective flow}} \text{ in the axial direction} \qquad (2.8)$$

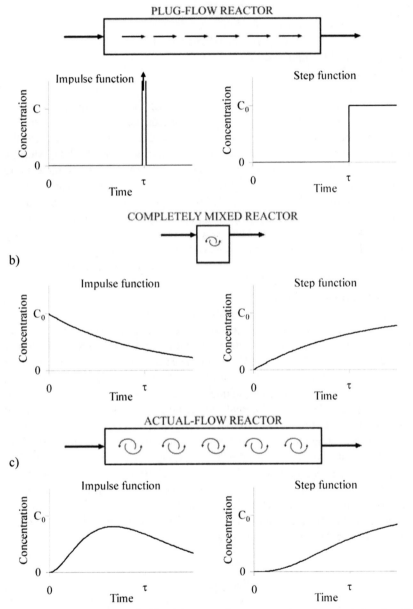

Figure 2.8 Responses of different types of reactors to impulse and step function inputs of an inert tracer.

where
Pe - Peclet number
E_L - Longitudinal (axial) dispersion coefficient, L^2T^{-1}
u   - Mean velocity along reactor, LT^{-1}
L   - Reactor length, L

When the dispersion number $\to 0$ (no dispersion), the flow pattern of a reactor approaches that of a PFR, while at the dispersion number $\to \infty$ (maximum dispersion), the flow pattern approaches that of a CSTR. In practice, when the dispersion number is greater than 0.5-4 (Khudenko and Shpirt 1986; Murphy and Timpany 1967; USEPA 1993), completely mixing can be assumed. Long and narrow tanks, for which the dispersion number is smaller than 0.05-0.2 (Eckenfelder et al. 1985; Khudenko and Shpirt 1986; USEPA 1993) are considered an approximation of plug flow. However, typical dispersion numbers in activated sludge reactors in practice range between 0.1 and 4 (San 1994; Metcalf and Eddy 2003) which suggests that the existing deviations from an ideal flow pattern have to be taken into consideration. Moreover, the effect of longitudinal mixing on sludge settleability is essential when the dispersion number is greater than 0.06 (Chambers and Jones 1988).

Several complex models are available to describe these deviations, of which the tank-in-series model and the advection-dispersion equation (ADE) have found widespread application (Horan 1990; Metcalf and Eddy 2003). Using the former approach, an AFR can be modelled by a simple cascade of identical CSTRs (Figure 2.9a) or a more complicated network of CSTRs (Figure 2.9b–d). In the latter case, the possible degrees of freedom include not only the number of tanks, but also their respective volumes and internal connections between the tanks (DeClercq et al. 1999). For example, Petersen et al. (2002) constructed a complex model of an activated-sludge system to fit simulation results to experimental data from a tracer test. That model consisted of the aeration tank (24 tanks-in-series), the channel from the aeration tank to the secondary clarifiers (2 tanks-in-series), an ideal point-settler, a "buffer tank", and the recycle channel from the secondary clarifiers to the aeration tank (5 tanks-in-series).

A general mass balance equation for i tank for a cascade of N CSTRs of equal volumes (Figure 2.9a) is given by:

$$\frac{dC}{dt}\frac{V_{tot}}{N} = C_{i-1}Q - C_iQ \qquad (2.9)$$

where
C   - Inert tracer concentration, ML^{-3}
V_{tot} - Volume of entire reactor, L^3
N   - Number of CSTRs composing reactor
Q   - Flow rate through reactor, $L^3 T^{-1}$

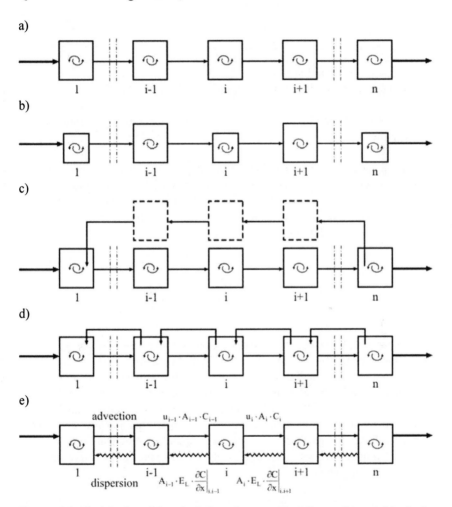

Figure 2.9 Models describing deviations from an ideal flow pattern: (a) tanks-in-series (equal volumes), (b) tanks-in-series (non-equal volumes), (c) tanks-in-series with virtual reactors in the recirculation lane, (d) tanks-in-series with recirculations, (e) dispersive flow reactors.

Thus, the response curve to a pulse input can be determined for such a system (Murphy and Timpany 1967):

$$\frac{C_e}{C_0} = \frac{N}{(N-1)!}\left(\frac{Nt}{\bar{t}}\right)^{N-1} \exp\left(\frac{-Nt}{\bar{t}}\right) \tag{2.10}$$

where
C_e - Effluent concentration of inert tracer, ML^{-3}
C_0 - Weight of tracer added divided by volume of reactor, ML^{-3}
t - Time, T
$\bar{t}$ - Theoretical hydraulic retention time or mean residence time for entire reactor volume, T

Typical aeration reactors have shown mixing patterns to be equivalent to 3-12 tanks (Ottengraf and Rietema 1969; Sundstrom and Klei 1979, Grady and Lim 1980; Fall and Loaiza-Navia 2007). The actual number of equivalent tanks in the tanks-in-series model can be found from the following relationship (Fujie et al. 1983; Chambers and Jones 1988):

$$N = \frac{Pe}{2} \tag{2.11}$$

Another alternative for the description of flow conditions is the advection-dispersion equation (ADE). In comparison with the tank-in-series model, which reasonably describes small deviations from plug flow at a large number of tanks, the ADE is more flexible and can simulate all types of behaviours between the ideal limits of no mixing (PFR) and complete mixing (CSTR). Murphy and Timpany (1967) were first who found that this partial differential equation provided a better representation of the response curve than the equal, or non-equal tanks-in-series models (ordinary differential equations) when the variance of the curve was used as the criteria of comparison.

2.4.2.2 Longitudinal advection-dispersion model

Considering the ADE, a basic partial differential equation can be developed from a mass balance on an element cross section (A) with both bulk flow with velocity

(u) and longitudinal (axial) dispersion (E_L) of an inert tracer with concentration C (Figure 2.9e):

$$V_i \frac{C_i^{n+1} - C_i^n}{\Delta t} = A_i E_L \frac{C_{i+1}^n - C_i^n}{\Delta x} - A_{i-1} E_L \frac{C_i^n - C_{i-1}^n}{\Delta x} + u_{i-1} A_{i-1} C_{i-1}^n - u_i A_i C_i^n$$

(2.12)

where
V - Volume of reactor element, L^3
A - Cross section of reactor element, L^2

Assuming the limit $\Delta x \rightarrow 0$, Equation 2.12 becomes a partial differential equation which presented (after transformations) as:

$$\frac{\partial C}{\partial t} + \frac{1}{A} \frac{\partial(uAC)}{\partial x} = \frac{1}{A} \frac{\partial}{\partial x}\left(AE_L \frac{\partial C}{\partial x}\right)$$

(2.13)

The advection term, related to the velocity distribution in the reactor, depends on the flow rate in the reactor. The dispersion term is described by the E_L coefficient in the analogous way to the diffusion term in Fick's law of diffusion. Assuming constant u, A and E_L in the reactor, Equation 2.13 can be reduced to the following form:

$$\frac{\partial C}{\partial t} + u \frac{\partial C}{\partial x} = E_L \frac{\partial^2 C}{\partial x^2}$$

(2.14)

Using dimensionless parameters: $\eta = x/L$, $\tau = t/\bar{t}$ and $C_t = C/C_0$, Equation 2.14 can be presented as:

$$\frac{\partial C_t}{\partial \tau} + \frac{\partial C_t}{\partial \eta} = \frac{E_L}{uL} \frac{\partial^2 C_t}{\partial \eta^2}$$

(2.15)

The exact solution to Equation 2.15 in a closed vessel (assuming the boundary conditions of $E_L=0$ at inlet and outlet) was given by Thomas and McKee (1944):

$$\frac{C_e}{C_0} = 2 \sum_{n=1}^{\infty} \frac{\mu_n(U \sin\mu_n + \mu_n \cos\mu_n)}{U^2 + 2U + \mu_n^2} \exp\left[U - \left(\frac{U^2 + \mu_n^2}{2U}\right)\tau\right]$$

(2.16)

where

$$U = \frac{uL}{2E_L} \tag{2.17}$$

$$\mu_n = \cot^{-1}\left[\frac{1}{2}\left(\frac{\mu_n}{U} - \frac{U}{\mu_n}\right)\right] \tag{2.18}$$

The appropriate boundary conditions for the system were proposed by Danckwerts (1953):

- at the reactor inlet (x=0):

$$E_L \frac{dC}{dx} = u(C - C_{in}) \tag{2.19}$$

- at the reactor outlet (x = L):

$$\frac{dC}{dx} = 0 \tag{2.20}$$

These conditions state that, at the entrance of the reactor, the amount of tracer in the influent stream and available by dispersion and flow are equal, and that at the reactor exit, no further change in concentration of the effluent streams occurs. Equation 2.14 with the boundary conditions given by Equations 2.19 and 2.20 is difficult to solve analytically so the solution must generally be obtained numerically. However, for small deviations from plug flow an approximate analytical solution is available (Grady and Lim 1980). In such a case, the non-reactive tracer concentration at the outlet (x=L) is given by:

$$\frac{C_e}{C_0} = \frac{1}{2}\left[1 - \text{erf}\left(\frac{1-\tau}{2\sqrt{\tau\left(\frac{E_L}{uL}\right)}}\right)\right] \tag{2.21}$$

where $\text{erf}(x) = \frac{2}{\sqrt{\pi}} \int_0^x \exp(-y^2)\,dy$ and is a tabulated function.

2.4.2.3 Combining ADE with source terms (biokinetic models)

Based on a basic differential material balance for a soluble substrate S at any cross-section of the bioreactor, the differential equation can be developed to describe concentration S for n^{th} order reaction with a rate constant k as presented by Levenspiel (1972):

- under dynamic conditions

$$\frac{\partial S}{\partial t} + u\frac{\partial S}{\partial x} = E_L\frac{\partial^2 S}{\partial x^2} - kS^n \tag{2.22}$$

- at steady state:

$$u\frac{\partial S}{\partial x} = E_L\frac{\partial^2 S}{\partial x^2} - kS^n \tag{2.23}$$

where
S - Soluble substrate concentration, ML^{-3}
k - Reaction rate constant, (dimension variable)
n - Reaction order

In general, a numerical method must be used to solve Equation 2.23 with its boundary conditions (Grady and Lim 1980). However, for the first-order reaction, $r_i = -kS$, the equation was solved analytically by Wehrner and Wilhelm (1956) for concentration S as a function of relative position η, along the reactor length L:

$$\frac{S}{S_{in}} = \frac{2\exp\left(\frac{\eta}{2}\frac{uL}{E_L}\right)\left\{(1+a)\exp\left[\frac{a}{2}\frac{uL}{E_L}(1-\eta)\right] - (1-a)\exp\left[\frac{a}{2}\frac{uL}{E_L}(\eta-1)\right]\right\}}{(1+a)^2\exp\left(\frac{a}{2}\frac{uL}{E_L}\right) - (1-a)^2\exp\left(-\frac{a}{2}\frac{uL}{E_L}\right)} \tag{2.24}$$

$$a = \sqrt{1 + 4k\,\bar{t}\left(\frac{E_L}{uL}\right)} \tag{2.25}$$

where
S_{in} - Influent concentration of soluble substrate, ML^{-3}

For the effluent concentration, S_e, Equation 2.24 can be reduced to the following form:

$$\frac{S_e}{S_{in}} = \frac{4a \exp\left(\frac{1}{2}\frac{uL}{E_L}\right)}{(1+a)^2 \exp\left(\frac{a}{2}\frac{uL}{E_L}\right) - (1-a)^2 \exp\left(-\frac{a}{2}\frac{uL}{E_L}\right)} \qquad (2.26)$$

where
S_e - Effluent concentration of soluble substrate, ML^{-3}

Equation 2.26 may be further simplified by neglecting the second term in the denominator which is relatively small (Horan 1990). Another exception to the use of numerical methods for solving Equation 2.23 is when the dispersion number is small (Levenspiel 1972):

$$\frac{S_e}{S_{PF}} = 1 + n\left(\frac{E_L}{uL}\right)\frac{\Psi}{1+(n-1)\Psi}\ln\left(\frac{S_{in}}{S_{PF}}\right) \qquad (2.27)$$

$$\Psi = kS_{in}^{n-1}\,\bar{t} \qquad (2.28)$$

where
S_{PF} - Effluent concentration of soluble substrate in a plug flow reactor, ML^{-3}

Until now, there have been a very few attempts (Ottengraf and Rietema 1969; Turian et al. 1975; Olsson and Andrews 1978; San 1994) that combined the 1-D ADE with a Monod type equation as the reaction term to describe biochemical processes occurring in the activated sludge reactors.

Stamou (1994) developed a mathematical model of a completely aerobic oxidation ditch. The model involved the 1-D ADE combined with basic biological processes (source terms) comprising biomass growth, substrate utilization and dissolved oxygen consumption:

$$\frac{\partial\varphi}{\partial t} + \frac{\partial(u\varphi)}{\partial x} = \frac{\partial}{\partial x}\left(E_L\frac{\partial\varphi}{\partial x}\right) + r_\varphi \qquad (2.29)$$

where
φ - Component of the simplified ASM1, $M(COD)L^{-3}$

The source terms (r_φ) were adopted from the simplified ASM1 (Henze *et al.* 1987) as written in Table 2.5.

Zima *et al.* (2008) compared predictions obtained using the 1-D ADE and conventional "*tanks-in-series*" model (6 tanks) with the following source term representing anitrification rate incorporated as a source term:

$$r_n = r_{n,max,20} X_{vss} \frac{S_{NH}}{K_{NH,A} + S_{NH}} \frac{S_O}{K_{O,A} + S_O} \theta_A^{(T-20)} \tag{2.30}$$

where

r_n - Specific nitrification rate, $M(N)M^{-1}T^{-1}$

$r_{n,max,20}$ - Maximum specific nitrification rate at $T = 20\ ^oC$, $M(N)M^{-1}T^{-1}$

X_{VSS} - Activated sludge concentration (organic fraction); ML^{-3}

The results of simulations revealed that the hydraulic model plays a minor role compared to the biochemical transformations in predicting the longitudinal ammonia concentration profiles (Figure 2.10). The average differences between the two model predictions ranged from 0.01 mg N/dm^3 in compartment 4 to 0.37 mg N/dm^3 in compartment 6 (reactor outlet).

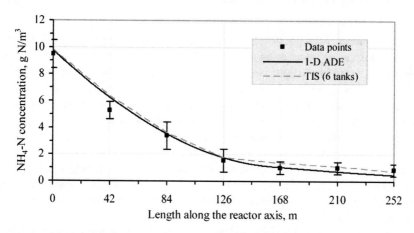

Figure 2.10 Measured vs. predicted longitudinal ammonia concentration profiles in the selected points of the activated sludge bioreactor at the Wschod WWTP in Gdansk (Poland) (Zima *et al.* 2008).

The 1-D ADE was also combined with a full complex biokinetic model (ASM1) to accurately predict performance of a pilot-scale oxidation ditch (Stamou *et al.* 1999) and a full-scale pre-denitrification system (Makinia and Wells 2000). Despite still limited experiences with using 1-D partial differential equations, the flow conditions in activated sludge reactors have already been studied in three dimensions. Alex *et al.* (2002) applied a 3-D model (conceptualized as a number of ideally stirred control volumes with advection and diffusion between those volumes) to investigate the undesirable phenomena, such as short circuiting, dead or stagnant zones or sludge settling within the studied tank. Guimet *et al.* (2004) used a 3-D ADE with a source term for dissolved oxygen to check the potential occurrence of dead oxygen zones resulting from an inappropriate location of the air supply inlet in the aeration tank. Oda *et al.* (2006) developed the most advanced CFD technique to optimize an intermittent agitation in anoxic zones by coarse bubbles. This technique was based on the 3-D multiphase Navier-Stokes equations combined with the ASM2d (Henze *et al.* 1999), settling velocity model of suspended solid, and model of oxygen mass transfer in the studied reactor. Gujer (2006) noted, however, that the application of partial differential equations provides spatial resolution of predicted concentrations but drastically increases a computational complexity.

2.4.3 Oxygen transfer model

2.4.3.1 Introduction

In the majority of wastewater treatment processes, aeration accounts for the largest portion of plant energy consumption, ranging from 45 to 75 % of the operating costs (Rosso 2005). Aeration in the activated sludge bioreactors serves two functions. First of all, the adequate amount of dissolved oxygen (DO) is essential for maintaining aerobic biochemical reactions occurring in the aerobic compartments of bioreactors. Oxygen transport from the aeration devices to microorganism cells is complex and proceeds in several stages: bringing wastewater into contact with oxygen, transferring oxygen across the gas-liquid interface to dissolve in the liquid, and transferring the DO through the liquid to the microorganisms (Winkler 1981). Secondly, aeration should insure adequate mixing in order to prevent the settling of solids and is frequently accompanied by mechanical agitation (Cooney and Wang 1971).

The transfer across the gas-liquid interface occurs only when a transfer driving force exists, which is defined as the difference in activity between the oxygen in the gas phase and the oxygen dissolved in the liquid (Winkler 1981). In the gas phase, the driving force is a partial pressure gradient, whereas in the liquid phase it is a concentration gradient (Eckenfelder 1989). Among three common physical mass transfer models, i.e. two-film (Lewis and Whitman 1924), penetration (Higbie 1935), and surface renewal (Danckwerts 1951), only the first one has widely been used to describe the absorption of oxygen by liquid in wastewater treatment processes (Metcalf and Eddy 1991). Therefore, only this model will be further outlined in the following discussion. Both the penetration and surface renewal models are more theoretical and account for the liquid agitation, thus embodying the flow regime parameters into the mass transfer calculation (Rosso 2005). Metcalf and Eddy (2003) noted that in more than 95% of the situations encountered, the results obtained from all the three models are essentially the same and it is still unclear which approach is correct. A detailed comparison of all the models can be found elsewhere (Schroeder 1977; Mueller *et al.* 2002).

The two-film theory concept considers physical mass transport across a two-film layer that consists of a gas film and a liquid film. At steady-state, the total flux of oxygen through the gas film (F_G) must be the same as the flux through the liquid film (F_L). Both fluxes can be expressed as the product of a transfer driving force, a total interfacial area, and a transfer coefficient:

$$F_G = K_G A_t (p_G - p_L) \qquad (2.31)$$

$$F_L = K_L A_t (S_{O,sat} - S_O) \qquad (2.32)$$

where

F_G - Total mass flux of oxygen through the gas film, $ML^{-2}T^{-1}$

K_G - Gas-phase oxygen transfer coefficient, $L^{-3}T$

A_t - Total interfacial area, L^2

p_G - Partial pressure of oxygen in the bulk of the gas phase, $ML^{-1}T^{-2}$

p_L - Partial pressure of oxygen at the gas-liquid interface, $ML^{-1}T^{-2}$

F_L - Total mass flux of oxygen through the liquid film, $ML^{-2}T^{-1}$

K_L - Liquid-phase oxygen transfer coefficient, $L^{-1}T^{-1}$

$S_{O,sat}$ - DO (saturation) concentration at the gas-liquid interface, ML^{-3}

S_O - DO concentration in the liquid phase, ML^{-3}

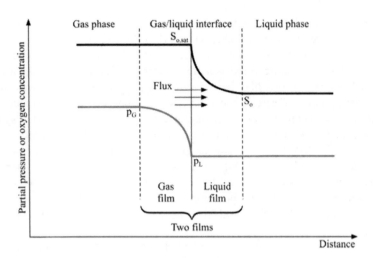

Figure 2.11 Schematic representation of the two-film concept for oxygen transfer (absorption) from a gaseous phase to a liquid phase.

Since the solubility of oxygen in water is very low, the liquid-film resistance controls the rate of mass transfer (Schroeder 1977; Winkler 1981; Eckenfelder 1989). Therefore, the oxygen transfer rate, defined as the rate of oxygen transfer per unit time and unit volume of the liquid, may be given by the following equation:

$$OTR_C = K_L \frac{A_t}{V}(S_{O,sat} - S_O) = K_L a(S_{O,sat} - S_O) \qquad (2.33)$$

where
OTR$_C$ - Oxygen transfer rate in clean water, $ML^{-3}T^{-1}$
V   - Volume, L^3
$K_L a$ - Overall mass transfer coefficient, T^{-1}

In practice, the total interfacial area (A_t) is difficult to determine due to turbulent flow or irregular geometries (Montgomery 1985). Therefore, it is convenient to measure the product of the mass transfer coefficient and the specific interfacial area, known as the overall mass transfer coefficient. Other names have also been used for $K_L a$, e.g. "*volumetric mass-transfer coefficient*", "*oxygenation efficiency*", "*absorption coefficient*", "*sorption-rate coefficient*", and "*mass-transfer product*" (Winkler 1981).

Dold and Fairlamb (2001) noted that care should be taken when quoting the overall mass transfer coefficient in diffused air systems where bubbles are distributed from diffusers at the bottom of the reactor. In many cases, that coefficient has been recorded incorrectly as "K_La" without recognizing a clear definition of the volume basis in the term A_t/V (Equation 2.33). In well-aerated reactors, the liquid volume (V_l) can be significantly smaller compared to the total volume of the bubble-liquid dispersion (V_{tot}), whereas the volume basis has not always been clearly reported. Depending on whether the overall mass transfer coefficient is quoted in terms of the total gas-liquid volume (V_{tot}) or the liquid volume (V_l) only, two distinctions in nomenclature should be made:

$$K_La = K_L \frac{A_t}{V_{tot}} \tag{2.34}$$

$$K_La_L = K_L \frac{A_t}{V_l} \tag{2.35}$$

where
V_{tot} - Total volume of the bubble-liquid dispersion, L^3
K_La_L - Mass transfer coefficient in terms of the liquid volume, T^{-1}
V_l - Liquid volume, L^3

By combining Equations 2.34 and 2.35, and noting that $V_{tot}=V_b+V_l$, the following relationship can be derived:

$$K_La = K_La_L \left(1 - \frac{V_b}{V_{tot}}\right) = K_La_L(1-\varepsilon_b) \tag{2.36}$$

where
V_b - Gas (bubble) volume, L^3
ε_b - Volume fraction of bubbles in the air-water dispersion (V_b/V_{tot})

If ε_b is small then $K_La_L \sim K_La$, otherwise K_La can be calculated from Equation 2.36 provided that K_La_L and ε_b have been determined previously. Dold and Fairlamb (2001) discussed in detail the calculation procedure in relation to the dynamic batch experimental procedure of estimating K_La (see: Section 4.3.5).

A modification of Equation 2.33 to account for microbial respiration results in the following equation used to describe oxygen transfer (Mines and Sherrard 1987):

$$OTR_{AS} = K_La(S_{O,sat} - S_O) - OUR_T \tag{2.37}$$

where

OTR_{AS} - Oxygen transfer rate in the activated sludge process, $M(O_2)L^{-3}T^{-1}$

OUR_T - Biochemical reaction (respiration) rate, $M(O_2)L^{-3}T^{-1}$

In order to solve Equation 2.37, in addition to the biochemical reaction (respiration) rate (OUR_T), also the parameters dealing with oxygen transfer from aeration devices (K_La, $S_{O,sat}$) have to be known. These parameters are described below in more detail.

If the oxygen uptake rate exceeds the oxygen transfer predicted by Equation 2.37, then an additional oxygen transfer pathway must exist. Tsao (1968), Mines and Sherrard (1987), and Clifft and Barnett (1988) attributed this phenomenon to a direct transfer at the bubble-microorganism interface by highly active dispersed bacteria entrapped in the bubble interfacial film (interfacial transfer). In order to account for this phenomenon, Equation 2.37 may be further modified as follows (Mines and Sherrard 1987, Clifft and Barnett 1988):

$$OTR_{AS} = K_La(S_{O,sat} - S_O) - OUR_T + OTR_{if} \qquad (2.38)$$

where

OTR_{if} - Interfacial oxygen transfer rate, $ML^{-3}T^{-1}$

Another possible mechanisms for increasing oxygen transfer rate include an alternate direct oxygen transfer from the gaseous phase (Albertson and DiGregorio 1975) or adsorption of the microbial cells onto the bubbles in the bulk liquid and the overlapping of the films (Bennet and Kempe 1964).

2.4.3.2   Overall mass transfer coefficient, K_La

According to the two-layer theory, K_La is a constant for a given aeration system. It was reported, however, that in the activated sludge process K_La may vary with the MLSS concentration (Albertson and DiGregorio 1975), solids retention time (Mines and Sherrard 1987; Rosso et al. 2008) or oxygen uptake rate (Albertson and DiGregorio 1975; Mines and Sherrard 1987; Clifft and Barnett 1988). The presence of preceding unaerated (anaerobic or anoxic) zones also increases the K_La values (Rosso et al. 2008). The effects of several physical and chemical parameters associated with the given aeration system have been discussed in many handbooks (e.g. Schroeder 1977; Winkler 1981; Eckenfelder 1989; Metcalf and Eddy 1991) and a classical paper of Stenstrom and Gilbert (1981). These parameters including temperature, liquid depth, and wastewater characteristics and process conditions are described below.

Temperature. Oxygen transfer will increase with increasing temperature, which can be presented by one of the following equations (Schroeder 1977; Stenstrom and Gilbert 1981):

$$\frac{K_L a(T_1)}{K_L a(T_2)} = \sqrt{\frac{T_1 \mu_2}{T_2 \mu_1}} \tag{2.39}$$

$$K_L a(T) = K_L a(20°C) + \theta(T - 20) \tag{2.40}$$

$$K_L a(T) = K_L a(20°C)\theta^{(T-20)} \tag{2.41}$$

where
μ_1, μ_2 - Dynamic viscosity of liquid, $ML^{-1}T^{-1}$
θ - Temperature correction factor

A comparison of Equations 2.39 and 2.40 revealed that these two expressions gave similar results below 30 °C (Schroeder 1977). Equation 2.41 is most commonly used to correct the effects of temperature on oxygen transfer. Some factors, such as changes in viscosity, surface tension, and diffusivity of oxygen are lumped together. This empirical approach produced a great variety of correction factors ranging from 1.008 to 1.047 with a typical value equal to 1.024. It appears that a θ factor of 1.024 should be used unless it is known that a different θ factor is more suitable (Stenstrom and Gilbert 1981).

Liquid depth. The effect of liquid depth on $K_L a$ will depend on the method of aeration. For most types of diffused air systems, the following relationship can be applied:

$$\frac{K_L a(H_1)}{K_L a(H_2)} = \left(\frac{H_1}{H_2}\right)^{n_1} \tag{2.42}$$

where
H_1, H_2 - Liquid depth, L
n_1 - Empirical constant ($n_1 = 0.7$ for most systems)

Wastewater characteristics and process conditions. The combination of various effects can be taken into account using a relative oxygen transfer coefficient (factor), α, which is defined as a ratio of wastewater to clean water mass transfer:

$$\alpha = \frac{K_L a(\text{wastewater})}{K_L a(\text{clean water})} \tag{2.43}$$

where
α - Relative oxygen transfer coefficient

The most important process parameter to affect aeration efficiency is the SRT which is directly related to the biomass concentration, and dictates oxygen requirements (Gillot and Heduit 2007; Rosso et al. 2008). The α factor is higher at higher SRTs. Using literature data, Gillot and Heduit (2007) showed that its value was <0.5 and >0.8 for SRTs shorter than 5 d and longer than 25 d, respectively. However, their own measurements from 14 nitrifying plants were not able to explain the discrepancy obtained in the α factor values (0.44-0.98). Based on the experimental data from more than 50 case studies, Stenstrom and Rosso (2008) demonstrated that the average value of α factor raised from 0.43 to 0.54, for systems operated at the SRT = 5 d and 15 d, respectively. Conventional activated sludge systems, typically operated at lower SRTs, have lower biomass concentrations, and less chance for the dissolved substrate to be quickly sorbed by the biomass. Systems operated at higher SRTs, including BNR systems, have the advantage of higher biomass concentration. The results of 28 measurements carried out at various activated sludge systems revealed that the average α factor was 0.37, 0.48 and 0.59, respectively, for conventional (low SRT), nitrifying and nitrogen removal (nitrification-denitrification) systems (Rosso et al. 2008). The change in α factor between the nitrifying and nitrogen removal systems resulted from the impact of readily biodegradable COD removal in the preceding unaerated (anoxic) zones. These authors also noted the effect of diffuser condition (new or recently cleaned diffusers vs. used and old diffusers). Observations of approx. 100 field tests showed that the α factor decreased with time (log fit) and the greatest rate of decrease occurred in the first 24 months of operation (Stenstrom and Rosso 2008).

The presence of certain substances (e.g., surface active agents or dissolved salts) may have both adverse and advantageous effects on oxygen transfer (Wagner and Popel 1996). For example, surface active agents or surfactants, which are typically discharged as oils, soaps and detergents, because of their amphiphilic nature, accumulate at the air-water interface of rising bubbles. The surfactant accumulation increases the rigidity of the interface which results in reduction of internal gas circulation and overall transfer rate (Rosso and Stenstrom 2006). Due to a greater diffusion rate and greater accumulation at the bubble surface, surfactants with high diffusivity suppresses the transfer rate more dramatically compared to surfactants with lower diffusivity. The surfactant effect can be partially offset by increasing the flow regime (i.e., coarse bubbles) (Rosso et al. 2008).

The α factor also varies with many process conditions including intensity of mixing or turbulence, suspended solids concentration, basin geometry, and method of aeration. Especially the method of aeration affects the value of α factor significantly (Stenstrom and Gilbert 1981). Typical α values for diffused air systems range between 0.4 and 0.8, whereas for surface aeration systems they range between 0.6 and 1.2 (Metcalf and Eddy 2003).

In a plug-flow reactor, some compounds (e.g., surfactants) may be expected to have a gradually reducing effect on oxygen transfer from the inlet to the outlet (Thomas *et al.* 1989). The α factor may be as low as 0.3 at the reactor inlet, and up to 0.8 at the reactor outlet (Horan 1990). The experimental data from a WWTP with two independent activated sludge lanes (Figure 2.12) revealed that in a low SRT conventional lane, the α factor was increasing rapidly at a relatively constant rate along the length of bioreactor. In the case of a pre-denitrification system (parallel lane), the average α factor was greater and the gradient was gradually reduced reaching almost a constant value at the end of the bioreactor (Rosso *et al.* 2008). Popel and Wagner (1994) recommended the α value of approximately 0.55 for the design of fine bubble aeration systems.

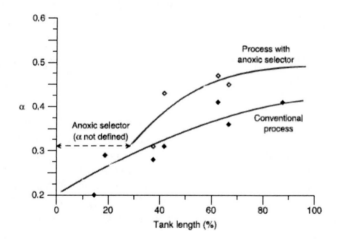

Figure 2.12 Effect of tank length on the α factor in a conventional vs. nitrogen removal system (Rosso *et al.* 2008).

The value of K_La is most often related to changes in the intensity of aeration, measured as the total air flow rate, Q_A, to the aeration basin (Table 2.6). In the early stage of research, Eckenfelder and O'Connor (1954) proposed a general power function of Q_A to determine K_La. The authors also noted that for various diffused aeration systems, the exponent b_1 is directly proportional to the bubble size approaching unity as a limit. The same type of function was used by Chen *et al.* (1980). Holmberg (1986) reduced that relationship to a simple linear function. Goto and Andrews (1985) extended the linear relationship by adding a non-zero (negative) intercept (-b_1). Reinius and Hultgren (1988) also used a similar equation, but their results produced a positive intercept (+b_1). Bocken *et al.* (1989) compared the accuracy of equations proposed by Holmberg (1986) and Goto and Andrews (1985). The latter relationship was found to overcome offsets (particularly in oxygen uptake rates) resulting from an incorrectly assumed relationship. Holmberg *et al.* (1989) suggested that a more accurate model assumption would be the use of exponent $b_1 = 0.5$ in the power function used by Chen *et al.*

Table 2.6 Expressions relating $K_L a$ to the total air flow rate to the aeration basin

Reference	Expression
Eckenfelder and O'Connor (1954), Chen *et al.* (1980)	$K_L a = m_1 Q_A^{b_1}$
Holmberg (1986)	$K_L a = m_1 Q_A$
Goto and Andrews (1985)	$K_L a = m_1 Q_A - b_1$
Reinius and Hultgren (1988)	$K_L a = m_1 Q_A + b_1$
Holmberg (1989)	$K_L a = m_1 \sqrt{Q_A}$

Dold and Fairlamb (2001) noted that the $K_L a$ coefficient for a given air flow per diffuser obviously increases with increasing diffuser density due to a greater total air flow rate, however, this relationship is not directly proportional. The exponent parameter m_1 in the equation of Chen *et al.* (Table 2.6) is assumed to raise with density while b_1 remains constant for a specific diffuser type. Based on analysis of data for a number of different diffuser types and sizes, the authors developed the following equation relating the parameter m_1 to the diffuser density (in the equation of Chen *et al.*):

$$m_1 = m_2 DD^{0.25} + b_2 \qquad (2.44)$$

where
DD   - Diffuser coverage (100·area diffusers / area reactor), %
m_2, b_2 - Empirical constants, T^{-1} (e.g. 2.5656 d^{-1}, 0.0432 d^{-1})

Dold and Fairlamb (2001) also emphasized that despite the empirical origin of the exponent parameter value of 0.25 in Equation 2.44, there is some underlying theoretical basis for this value. Kawase and Moo-Young (1990) indicated through a theoretical analysis of aerated systems with separate mixing that the m_l parameter value should be related to the power dissipation rate raised to the 0.25 power. Furthermore, in fine bubble diffuser systems, the power dissipation rate should in turn be related to the diffuser density for a given air flow rate per diffuser.

2.4.3.3 Saturation concentration of dissolved oxygen in mixed liquor, $S_{O,sat}$

The equilibrium concentration of oxygen in contact with liquid, $S_{O,sat}$, is defined by Henry's law as (Eckenfelder 1989):

$$S_{O,sat} = \frac{p_G}{H_e} \qquad (2.45)$$

where
H_e - Henry's constant, L^2T^{-2}

Henry's constant, H_e, increases with an increase of temperature and dissolved solids concentration, thereby reducing $S_{O,sat}$. The combined impact of these factors on $S_{O,sat}$ can be expressed as follows:

- Winkler (1981):

$$S_{O,sat} = \frac{468}{31.6 + T} - \frac{0.0036TDS}{21.2 + T} \qquad (2.46)$$

- Weber (1972):

$$S_{O,sat} = \frac{475 - 0.00265TDS}{33.5 + T} \qquad (2.47)$$

where
T - Temperature, ^oC
TDS - Total dissolved solids concentration, ML^{-3}

Standard handbooks provide tables where the saturation solubility of oxygen in water at various temperatures and total dissolved solids (TDS) concentrations are summarized. The expression for the oxygen saturation concentration in clean water at 1 atm pressure can be determined from the following equations:

- Standard Methods (1992)

$$S_{O,sat(cw)} = \exp\left\{\begin{array}{l}\left[-139.344+\left(\dfrac{1.575701\cdot10^5}{T}\right)-\left(\dfrac{6.642308\cdot10^7}{T^2}\right)+\left(\dfrac{1.2438\cdot10^{10}}{T^3}\right)\right.\\[3mm]\left.-\left(\dfrac{8.621949\cdot10^{11}}{T^4}\right)-Chl\left[3.1929\cdot10^{-2}-\left(\dfrac{19.428}{T}\right)+\left(\dfrac{3.1929\cdot10^3}{T^2}\right)\right]\right]\end{array}\right\}$$

(2.48)

where

Chl   - Chloride concentration, ML^{-3}

- ASCE (1997)

$$S_{O,sat} = 14.65 - 0.41T + 7.99\cdot10^{-3}T^2 - 7.78\cdot10^{-5}T^3 \qquad (2.49)$$

The saturation solubility of oxygen changes with wastewater characteristics due to the presence of particulates, salts and surface-active substances in wastewater. To account for this effect, a β factor is introduced. It has been defined as the ratio of the DO saturation concentration in wastewater and in clean water, or tap water, as follows:

$$\beta = \frac{S_{O,sat}(\text{wastewater})}{S_{O,sat}(\text{clean water})} \qquad (2.50)$$

where

β - Ratio of the DO saturation concentration in wastewater and in clean water

Values of β vary from about 0.7 to 0.98 (Metcalf and Eddy 2003). The β factor for domestic wastewater is generally about 0.95, but it can vary over a much broader range for industrial wastewaters (Stenstrom and Gilbert 1981).

Values of the saturation concentration in mixed liquor can be estimated in terms of the type of aeration system. For surface aeration, devices it can be taken from handbook values for DO with appropriate corrections for pressure, temperature, and TDS. For submerged aeration systems, the $S_{O,sat}$ values may be estimated based on clean water saturation studies performed under similar geometric and gas flow conditions (Mueller and Boyle 1988).

In most diffused aeration systems, the gas is introduced to the liquid at a depth 3.0 to 4.5 m below surface of the liquid. The oxygen saturation concentration $S_{O,sat}$ increases in proportion to relative pressure. Hence, the depth-dependent saturation can be formulated as a function of the depth (H-h_b) as follows (Popel and Wagner 1994):

$$S_{O,sat}(z) = S_{O,sat}[1 + f_1(H - h_b)] \qquad (2.51)$$

where

$S_{O,sat}(z)$ - DO saturation concentration at depth z, $M(O_2)L^{-3}$
f_1 - Conversion factor, L^{-1} ($= 0.0968$ m^{-1})
H   - Reactor depth, L
h_b - Depth variable measured from reactor bottom, L

The saturation concentration decreases, on the other hand, because when a bubble is rising in the tank to the surface, the partial pressure in the air bubble is reduced as oxygen is absorbed (Eckenfelder 1989). It may be written as follows (Popel and Wagner 1994):

$$S_{O,sat}(z) = S_{O,sat}[1 - OTE(h_b)] \qquad (2.52)$$

where
$OTE(h_b)$ - Oxygen transfer efficiency as a function of h_b

In practice, the average saturation concentration of DO, $\bar{S}_{O,sat}$, corresponding to the aeration basin mid-depth, is used (Eckenfelder 1989). Ewing et al. (1979) and Eckenfelder (1989) proposed the following formula to calculate the $\bar{S}_{O,sat}$ value in diffused aeration systems:

$$\bar{S}_{O,sat} = S_{O,sat} \frac{1}{2}\left(\frac{P_b}{P_a} + \frac{O_t}{0.21} \right) \qquad (2.53)$$

where

$\bar{S}_{O,sat}$ - Average DO saturation concentration corresponding to reactor mid-depth, $M(O_2)L^{-3}$
P_b - Absolute pressure at the depth of air release, $ML^{-1}T^{-2}$
P_a - Atmospheric pressure, $ML^{-1}T^{-2}$
O_t - Exit oxygen molar fraction

The exit oxygen molar fraction is given by (Ewing et al. 1979):

$$O_t = \frac{0.21(1 - OTE)}{0.79 + 0.21(1 - OTE)} \qquad (2.54)$$

2.4.4 Process temperature model

2.4.4.1 Effects of temperature in activated sludge systems

Temperature is an important factor influencing microbial kinetics, physiochemical parameters (DO saturation and diffusion) and settling velocity. Controlling temperature is not normally possible in activated sludge systems. However, incorporating factors influencing temperature in the design process, such as aeration system, surface to volume ratio, and tank geometry can reduce the range of temperature extremes and improve the overall process performance (Lippi *et al.* 2009). Therefore, an accurate temperature model allows for determining the effects of various design and upgrade options on prediction of the equilibrium process temperature.

The processes especially sensitive to changes in temperature below 10°C are nitrification, and to a lesser extent – denitrification. For example, Scherfig *et al.* (1996) noted that using the temperature dependence on the kinetic coefficients presented by USEPA (1993), a drop of 2 °C from 9 °C to 7 °C requires an increase of 20% and 16% in the required volumes for nitrification and denitrification, respectively. Also la Cour Jansen *et al.* (1992) reported that the cold and changeable winter temperatures make up the most serious operational problems in the nitrifying wastewater treatment plants in Denmark. In higher temperatures (from 10°C to 17°C) the process of nitrification in the activated sludge process can be very sensitive to small changes in temperature resulting in elevated nitrite concentrations in activated sludge plant effluents. There is a critical temperature below which the rate of formation of nitrate is less than the rate of formation of nitrite (Randall and Buth 1984). Ruano *et al.* (2007) concluded that temperature effects are very influential on model predictions, although this issue is rarely incorporated in simulation studies.

The temperature dependence of rates and equilibria of chemical and biochemical reactions has been most commonly represented by the van't Hoff-Arrhenius (or modified Arrhenius) equation, which relates the rate and equilibrium constants k to the absolute temperature T and activation energy, E_a:

$$\frac{d(\ln k)}{dT} = \frac{E_a}{RT^2} \tag{2.55}$$

where
k   - Reaction rate (or equilibrium) constant, T^{-1}
E_a - Activation energy, J/mol
T   - Temperature, K (273.15+°C)
R   - Ideal gas constant, 8.314 J/(mol·K)

Table 2.7 Sample values of the temperature characteristic term, θ, for various kinetic coefficients in activated sludge processes

Parameter	Dold et al. (1980)	Metcalf and Eddy (1991)	Hamilton et al. (1992)	Brezonik (1994)	Henze et al. (1995a)	Barker and Dold (1997a)	Grady et al. (1999)
μ_H on S_S	1.2	1.00–1.08	1.072	1.047–1.075	1.07	1.029	
μ_H on X_{STO}	1.029						
b_H			1.120			1.029	
K_S	1.0					1.000	
K_{STO}	1.1						
μ_{AOB}	1.123	1.10		Nitrosomonas 1.055-1.10 Nitrobacter 1.06	1.12	1.123	Nitrosomonas 1.098–1.118 Nitrobacter 1.068–1.112
μ_{NOB}							
$K_{NH,AOB}$	1.029						Nitrosomonas 1.125 Nitrobacter 1.157
$K_{NO2,NOB}$						1.123	
b_A	1.029		1.120			1.029	
k_a			1.072	1.08		1.029	
k_{hyd}			1.116		1.04	1.05	

Equation 2.55 may be integrated between the temperature limits (T and T_0) yielding:

$$\ln \frac{k_T}{k_{T_0}} = \frac{E_a (T - T_0)}{RTT_0} \qquad (2.56)$$

Because most of activated sludge systems are operated in a relatively narrow temperature range, in practice, the term $E_a/(RTT_0)$ may be considered to be constant. Thus,

$$k_T = k_{T_0} e^{C_T(T-T_0)} \qquad (2.57)$$

The term e^{C_T} in Equation 2.57 is expressed as a temperature coefficient (temperature correction factor or Arrhenius coefficient), θ, which gives

$$k_T = k_{T_0} \theta^{(T-T_0)} \qquad (2.58)$$

The accuracy of k_T is strongly dependent on the accuracy of θ. In order to avoid an error in k_T in excess of 10%, the temperature must be accurate to within 1 °C to 1.5 °C for values of θ between 1.07 and 1.10 (Argaman and Adams 1977). Table 2.7 contains a list of θ values for various biochemical reactions in the activated sludge process.

Sayigh and Malina (1978) reported that the modified Arrhenius equation (2.58) does not accurately apply when evaluating temperature effects on treating domestic wastewater by the completely mixed continuous flow activated sludge process in the range of temperature 4-31 °C. The authors found that the coefficient θ depends on variability in substrate characteristics and the gross bacterial culture adaptation. Furthermore, the latter depends on two other factors, such as shifts in the population and the acclimation of specific bacteria within the mixed culture. In contrast, many earlier and later studies revealed that the biochemical reactions in mixed microbial cultures, such as activated sludge, would follow the modified Arrhenius relationship over the temperature range from 5 °C to 20 °C (Makinia 1998). This is valid because the species dominating growth for 0 °C to 25 °C are likely to be fairly uniform. A potential gap occurs between 20 °C and 25 °C where a decrease in overall reaction rate may occur in a mixed culture because of a shift in the dominant species from psychotrofilic to mesophilic (Randall *et al.* 1982). Also Horan (1990) noted that significant changes in the composition of microbial population occur outside the range 5-25 °C which invalidate the van't Hoff-Arrhenius relationship.

Zwietering *et al.* (1991) compared several expressions for bacterial growth as a function of temperature (Table 2.8), either available from literature or newly developed, which are capable of predicting both the increasing and decreasing microbial growth rates with increasing temperature (unlike the van't Hoff-Arrhenius equation, which only describes the increase). A model which has only been seldom used is the model of Hinshelwood (1946), although it is a simple model with a biological basis including both the enzyme reaction and the high-temperature denaturation. Schoolfield *et al.* (1981) proposed a nonlinear Arrhenius type of model on a biological basis, describing the specific growth rate as a function of temperature over the whole biokinetic temperature range. Further empirical models were proposed by Ratkowsky *et al.* (1982, 1983), i.e., the square root model, describing the specific growth rate up to 15 °C, and the expanded square root model, describing the growth rate over the whole biokinetic temperature range. Zwietering *et al.* (1991) modified the latter equation of Ratkowsky *et al.* to decline the growth rate above T_{max}.

Recently, Hiatt and Grady (2008) used the expanded square root equation of Ratkowsky *et al.* (1983) to describe the effects of temperature on microbial growth of heterotrophs and autotrophs (both ammonia and nitrite oxidizing bacteria). The authors concluded that the use of the equation of Ratkowsky *et al.* (1983) represents "*an advance over the previous models and removes the need for setting upper limits on the valid temperature range*". In that equation, the parameter b_2 is the regression coefficient of the square root of growth rate constant for temperatures below the optimal temperature, whereas c_2 is an additional parameter to enable the model to fit the data for temperatures above the optimal temperature.

Figure 2.13 illustrates the observed growth rates of *Lactobacillus plantarum* and their predictions obtained with the expanded square root equation of Ratkowsky *et al.* (Zwietering *et al.* 1991). The estimated values of model parameters (b_2, c_2, T_{min}, T_{max}) are listed in Table 2.9. Hellinga *et al.* (1999) found a similar temperature dependency of the maximum respiration rate of ammonium oxidizing bacteria (AOB) in a sidestream treatment system (SHARON process) operated at temperatures between 15 and 55 °C at pH 7. Below 40 °C, biomass activity increased with temperature rising, but over 40 °C, biomass activity rapidly decreased due to degradation. This confirms a limitation of the van't Hoff-Arhenius equation only to certain temperature ranges.

Table 2.8 Formulae describing the temperature dependency of bacterial growth rates (based on Zwietering *et al.* 1991)

Equation/Reference	Formula
Equation of Hinselwood (1946)	$$\mu(T) = k_{T,1}e^{\left(\frac{-E_{a,1}}{RT}\right)} - k_{T,2}e^{\left(\frac{-E_{a,2}}{RT}\right)}$$ $k_{T,1}$ – frequency factor for the enzyme reaction, T^{-1} $k_{T,2}$ – frequency factor for the high-temperature denaturation, T^{-1} $E_{a,1}$ – activation energy for the enzyme reaction, J/mol $E_{a,2}$ – activation energy for the high temperature denaturation, J/mol
Equation of Schoolfield *et al.* (1981)	$$\mu(T) = \frac{k_{T,a}e^{\left(\frac{-E_a}{RT}\right)}}{1 + k_{T,l}e^{\left(\frac{-E_l}{RT}\right)} + k_{T,h}e^{\left(\frac{-E_h}{RT}\right)}}$$ $k_{T,a}$ – frequency factor for the controlling enzyme reaction, T^{-1} $k_{T,l}$ – frequency factor for high-temperature inactivation $k_{T,h}$ – frequency factor for low-temperature inactivation $E_{a,a}$ – activation energy for the controlling enzyme reaction, J/mol $E_{a,l}$ - activation energy for the controlling enzyme reaction, J/mol $E_{a,h}$ - activation energy for the controlling enzyme reaction, J/mol
Square root equation of Ratkowsky *et al.* (1982)	$$\mu(T) = \left[b_1(T - T_{min})\right]^2$$
Expanded square root equation of Ratkowsky *et al.* (1983)	$$\mu(T) = \left[b_2(T - T_{min})(1 - e^{c_2(T - T_{max})})\right]^2$$
Modified Ratkowsky equation (Zwietering *et al.* 1991)	$$\mu(T) = \left[b_3(T - T_{min})\right]^2 (1 - e^{c_3(T - T_{max})})$$ T_{min} – the minimum temperature at which growth is observed (°C) T_{max} - the maximum temperature at which growth is observed (°C) b_1, b_2, b_3 – empirical parameters, 1/°C $d^{-0.5}$ c_1, c_2, c_3 – empirical parameters, 1/°C

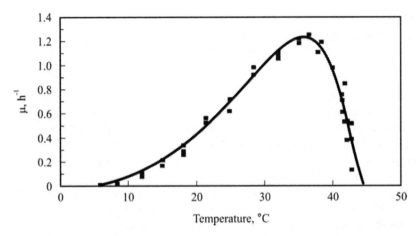

Figure 2.13 Growth rates of *Lactobacillus plantarum* modelled using the expanded square root equation of Ratkowsky *et al.* with the parameter values listed in Table 2.8 (Zwietering *et al.* 1991).

Table 2.9 Results of parameter estimation for the expanded square root equation of Ratkowsky *et al.* (Zwietering *et al.* 1991)

Parameter	Estimated value	95% confidence interval
b_2	0.0377	0.0321 to 0.0433
T_{min}	2.82	−0.223 to 5.86
c_2	0.25	0.173 to 0.326
T_{max}	44.9	44.2 to 45.5

2.4.4.2 Historical background of temperature modelling in activated sludge reactors

The earliest temperature model was developed by Eckenfelder (1966). This was a simple empirical equation for estimating the equilibrium temperature in aerated lagoons employing surface aeration equipment:

$$\frac{\bar{t}}{H} = \frac{T_{in} - T}{f_H(T - T_a)} \tag{2.59}$$

where

T_{in} - Temperature of wastewater in inlet to reactor, deg

f_H - Proportionality factor containing all of the heat transfer characteristics, LT^{-1}

T_a - Air temperature, deg

The coefficient f_H combines all of the heat transfer characteristics including the heat transfer coefficients, the surface area increase from aeration equipment, and wind and humidity effects. An approximate value of f_H was 27 m/d (90 ft/day) for most aerated lagoons. Another simple empirical equations for aerated lagoons were proposed by van der Graf (1976) and Grady et al. (1999).

Ford et al. (1972) developed an empirical method for estimating temperature changes in activated sludge reactors equipped with mechanical surface aerators. This technique included calculation of heat losses from aerator spray as a function of the differential enthalphy of the air flow into the basin.

Novotny and Krenkel (1973) presented a comprehensive steady-state model for calculating the equilibrium temperature in aeration basins based on theoretical energy balances using a cooling pond approach. The model included four energy balance terms: short-wave (solar) radiation, long-wave (atmospheric) radiation, evaporation, and convection. Argaman and Adams (1977) extended Novotny and Krenkel's model by adding the heat gains from mechanical energy input and biochemical reactions, and the heat loss through the basin walls. Talati and Stenstrom (1990) integrated the best parts of the previous models and improved the accuracy of temperature prediction to ±1.2 °C and reduced the amount of site-specific information needed. Sedory and Stenstrom (1995) used the equations from the model by Talati and Stenstrom (with slight changes with respect to solar radiation and biochemical process energy), and developed a dynamic model to predict activated sludge basin temperature. In concurrent studies, Bround and Scherfig (1994) and Scherfig et al. (1996) used Argaman and Adams' equations except energy exchange from aeration which was adapted from Talati and Stenstrom (1990). Both dynamic models were able to predict the hourly temperature changes in activated sludge basins within ±0.5 °C.

Wells (1990) modelled the temperature regime of a typical wastewater treatment plant using numerical and analytical techniques but did not account for mechanical or biochemical energy inputs in the activated sludge reactor. The results indicated that, under winter conditions, covering the aeration basins would cause a 2 °C increase in temperature.

La Cour Jansen et al. (1992) developed a steady-state temperature model based on a simple energy balance including the significant energy contributions (Table 2.10) and using the equations from Szeicz et al. (1969) and Wilson (1974).

Although inlet temperature had the greatest impact on the plant temperature, deep tanks and good protection against wind exposure were shown to be attractive means to avoid low process temperatures during winter periods.

Table 2.10 Typical range of contributions to temperature changes in WWTPs (la Cour Jansen *et al.* 1992)

Energy transfer phenomena	Temperature change (°C/d)
Significant energy contributions:	
- Short-wave radiation (increase)	0.5–2.5
- Long-wave radiation (decrease)	0.5–1.0
- Sensible heat (decrease/increase)	0.5–3.5
- Evaporation (decrease)	0.5–2.5
- Process energy (increase)	0.5–2.0
Insignificant energy contributions:	
- Mechanical energy (increase)	<0.1
- Geothermal energy (decrease/increase)	<0.05
- Precipitation (rain/snow) (decrease/increase)	<0.2

Gillot and Vanrolleghem (2003) compared the model of Talati and Stenstrom to a simplified model (van der Graf 1976) and demonstrated that the predictions were comparable for specific conditions.

Makinia *et al.* (2005b) developed a dynamic temperature model for the bioreactor at the Rock Creek WWTP in Hillsboro OR (USA). The effects of different hydrodynamic models (1-D ADE vs. tanks-in-series model) and the contribution of specific heat components on temperature predictions were evaluated. In the case of 1-D ADE, the absolute mean error between the observations and predictions was 0.4 °C (with a standard deviation of the absolute errors of 0.34 °C) for one year period (Figure 2.14a). A simple uncertainty analysis revealed that the model predictions were not extremely sensitive (±0.1-0.2 °C) to changes (±10%) in the net heat flux. Neglecting biochemical heat energy inputs in the activated sludge reactor under-estimated temperatures by up to 0.5 °C. The biochemical heat energy inputs were the most important heat flux contributor accounting for 30-40% of the total heat flux throughout the year (Figure 2.14b). During summer months the gain from solar radiation (30% of the total flux) was approximately equal to the gain from biological processes, but during winter months it became a minor component falling below 8% of the total flux.

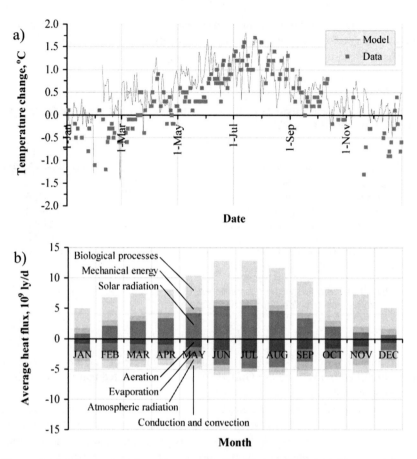

Figure 2.14 Observed and predicted temperature changes from bioreactor inlet to bioreactor outlet (a) and contributions of the individual components to the overall heat balance for the bioreactor (b) at the Rock Creek WWTP, Hillsboro OR (USA) (Makinia *et al.* 2005b).

Lippi *et al.* (2009) presented a refined steady-state temperature model with new functions for selected heat exchange terms (surface convection and surface evaporation) and improved approaches for predicting the effects of covering aeration tanks. The model was validated with the results of 15 published case studies and deviations between the data and model predictions remained in the range ±0.8 °C. It was found that covering the bioreactor could reduce heat losses and increase the process temperature by 3 °C under winter conditions in a warm climate (with the influent temperature 19.3 °C).

2.4.4.3   Temperature model components

A governing equation for the process temperature model in activated sludge bioreactors is composed of a hydrodynamic model combined with an appropriate source term for the overall energy balance (schematically shown in Figure 2.15). The latter term can be given by:

$$\frac{dT}{dt} = \frac{\Phi_n}{\rho_l C_p V} \qquad (2.60)$$

where
Φ_n - Net heat exchange, ET^{-1}
ρ_l - Liquid density, ML^{-3}
C_p - Specific heat of water at constant pressure, $EM^{-1}deg^{-1}$

The net heat exchange (gain or loss), Φ_n, in Equation 2.60 can be expressed as a sum of several components contributing to the overall heat balance:

$$\Phi_n = \Phi_{sr} \pm \Phi_{ar} \pm \Phi_c \pm \Phi_e \pm \Phi_a + \Phi_m + \Phi_{bp} \pm \Phi_w \qquad (2.61)$$

where
Φ_{sr} - Short-wave (solar) radiation, ET^{-1}
Φ_{ar} - Long-wave (atmospheric) radiation, ET^{-1}
Φ_c - Surface convection and conduction, ET^{-1}
Φ_e - Surface evaporation, ET^{-1}
Φ_a - Aeration heat transfer, ET^{-1}
Φ_m - Mechanical power heat exchange, ET^{-1}
Φ_{bp} - Biological processes heat exchange, ET^{-1}
Φ_w - Reactor wall heat exchange, ET^{-1}

The positive terms represent heat gains (increase in temperature) and the negative terms represent heat losses (decrease in temperature). Short-wave (solar) radiation (Φ_{sr}), mechanical energy (Φ_m), and biochemical process energy (Φ_{bp}) are always positive, whereas long-wave (atmospheric) radiation (Φ_{ar}), surface convection (Φ_c), surface evaporation (Φ_{ev}), aeration (Φ_a), heat exchange through reactor walls (Φ_w) can be either positive or negative depending upon local conditions. The approaches to modelling individual contributions from each component are discussed below.

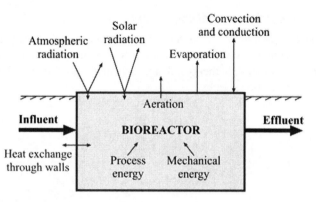

Figure 2.15 Heat exchange components included in process temperature models.

Short-wave (solar) radiation, Φ_{sr}. The net energy input from the sun to a bioreactor is a function of the time of the year, site location (latitude), and meteorological conditions (cloud cover). The input can be taken as a direct measurement from a meteorological station (Makinia *et al.* 2005b) or calculated. Several authors (Argaman and Adams 1977; Talati and Stenstrom 1990; Sedory and Stenstrom 1995; Scherfig *et al.* 1996) used Raphael's (1962) approach, which correlates solar insulation versus solar altitude:

$$\Phi_{sr} = \Phi_{sr,o}(1 - 0.0071C_c^2)A_S \qquad (2.62)$$

where
$\Phi_{sr,o}$ - Clear sky solar radiation, ET^{-1}
C_c - Cloud cover (0-10)
A_S - Surface area of bioreactor, L^2

La Cour Jansen *et al.* (1992) included in their model the equation for solar radiation proposed by Szeicz *et al.* (1969) and Wilson (1974):

$$\Phi_{sr} = \Phi_{sr,o}(1 - r_{sr})\left(K_1 + K_2 \frac{n_s}{N_s} \right)A_S \qquad (2.63)$$

where
r_{sr} - Albedo
K_1, K_2 - Empirical coefficients
n_s - Number of hours with sunshine per day
N_s - Maximum number of hours with sunshine per day

Clear sky solar energy, $\Phi_{sr,o}$, varies with latitude throughout the sunlight hours and during the year, and must be estimated, if meteorological data are not available. Argaman and Adams (1977), Talati and Stenstrom (1990), Scherfig *et al.* (1996) used Thackston and Parker's (1972) correlation for the daily averaged $\Phi_{sr,o}$:

$$\Phi_{sr,o} = a_1 - a_2 \sin\left(\frac{2\pi d}{366} + a_3\right) \tag{2.64}$$

where
a_1, a_2, a_3 - Empirical constants dependent on latitude of the site
d   - Day of the year

Sedory and Stenstrom (1995) developed a new polynomial function to calculate $\Phi_{sr,o}$ (cal/day):

$$\Phi_{sr,o} = \left(\begin{array}{l} -0.06401 + 1.3341alt + 0.2008alt^2 - 0.0043alt^3 \\ +3.79e^{-5}alt^4 - 1.37e^{-7}alt^5 \end{array}\right) 65102.26 \tag{2.65}$$

where
alt   - Solar altitude, deg

Long-wave (atmospheric) radiation, Φ_{ar}. Radiation from the surface of the aeration basin can be classified as long-wave radiation. Raphael (1962) described the effective long-wave back radiation based on the Stefan-Boltzman's fourth power radiation law as the difference between incoming and back radiation, as follows (Argaman and Adams 1977; Talati and Stenstrom 1990; Sedory and Stenstrom 1995; Scherfig *et al.* 1996; Makinia *et al.* 2005b):

$$\Phi_{ar} = \Phi_{ar,w} - \Phi_{ar,a} \tag{2.66}$$

The fluxes $\Phi_{ar,w}$ and $\Phi_{ar,a}$ are given by:

$$\Phi_{ar,,w} = \varepsilon_{ar}\sigma T^4 A_S \tag{2.67}$$

$$\Phi_{ar,a} = (1 - \lambda_{ar})\sigma T_a^4 \beta_{ar} A_S \tag{2.68}$$

where
ε_{ar} - Water surface emissivity
λ_{ar} - Water surface reflectivity

Argaman and Adams (1977) linearized Equation 2.69, assuming that both T_a and T are generally much smaller than 273 °C, as follows:

$$\Phi_{ar} = \left[695 \cdot 10^4(1-\beta_{ar}) + 10.18 \cdot 10^4(T - T_a) + 10.18 \cdot 10^4(1-\beta_{ar})T_a\right]A_S \tag{2.69}$$

The atmospheric radiation factor (β_{ar}) is a linear function of cloud cover, cloud height, and vapour pressure (Sedory and Stenstrom 1995). Under most atmospheric conditions encountered in practice β_{ar} ranges from 0.75 to 0.95 (Argaman and Adams 1977; Talati and Stenstrom 1990; Sedory and Stenstrom 1995) with a typical value $\beta_{ar}=0.87$ (Scherfig et al. 1996).

Other models also contain modifications to Equation 2.69. La Cour Jansen et al. (1992) introduced a parameter which is a function of number of hours with sunshine:

$$\Phi_{ar} = -\left[\sigma T - \sigma T_a(0.53 + 0.077\sqrt{e_a})\right]\left(0.1 + 0.9\frac{n_s}{N_s}\right)A_S \tag{2.70}$$

where
e_a - Vapour pressure of water at air temperature, $ML^{-1}T^{-2}$

Wall and Petersen (1986) assumed that the radiation heat loss is a linear function of relative temperature differences between the liquid and air and the emissivity (ε_{ar}) of the water surface:

$$\Phi_{ar} = 1.71 \cdot 10^{-9}\varepsilon_{ar}A_S(T^4 - T_a^4) \tag{2.71}$$

where temperatures are given in absolute Rankine degrees (°F+459.69°).

Surface convection, Φ_c. Convection heat transfer at the boundary between air and the surface of the bioreactor is associated with air movement. Convection heat transfer is composed of two parts including natural convection and force convection (Wall and Petersen 1986). The natural convection heat loss, Φ_{nc}, occurs as a result of changing air temperatures near the liquid surface (Wall and Petersen 1986):

$$\Phi_{nc} = U_c(T - T_a)A_S \qquad (2.72)$$

where

$$U_c = 0.22(T - T_a)^{1/3} \qquad (2.73)$$

In the same model (Wall and Petersen 1986), the forced convection heat loss, Φ_{fc}, due to the wind was calculated from the following relationship:

$$\Phi_{fc} = hca(T - T_a)A_S \qquad (2.74)$$

where
hca   - Heat coefficient dependent on the wind velocity, $EL^{-2}T^{-1}deg^{-1}$

In other models (Argaman and Adams 1977; Talati and Stenstrom 1990; Sedory and Stenstrom 1995; Scherfig et al. 1996; Makinia et al. 2005b), there is no distinction between the two mentioned parts and the relation originally developed by Novotny and Krenkel (1973) was used:

$$\Phi_c = \rho_a C_{p,a} h_v A_S(T - T_a) \qquad (2.75)$$

where

$$h_v = 392A_S^{-0.05}u_w \qquad (2.76)$$

where
$C_{p,a}$ - Specific heat of air at constant pressure, $EM^{-1}deg^{-1}$
u_w - Wind velocity, LT^{-1}

According to Equation 2.75, heat loss by surface convection is caused by the temperature difference between air and the liquid surface. The Φ_c term is also influenced by the convective (vapour) transfer coefficient (h_v), which is a function of wind velocity (Equation 2.76).

Lippi *et al.* (2009) proposed a method of calculating h_v based on the dimensionless analysis of convection between a horizontal air stream and a plane water surface:

$$h_v = \frac{Nu \cdot k_{th}}{(ab)^{0.5}}, \, (WK^{-1}m^{-2}) \tag{2.77}$$

The Nusselt, Reynolds and Prandtl numbers are defined as follows:

$$Nu = (0.037\,Re^{0.8} - 871)\,Pr^{0.33} \tag{2.78}$$

$$Re = \frac{u_w\,(ab)^{0.5}}{v_L} \tag{2.79}$$

$$Pr = \frac{C_{pv}\mu_L}{k_{th}} \tag{2.80}$$

where
k_{th} - Thermal conductivity coefficient, $EL^{-1}T^{-1}deg^{-1}$
$(ab)^{-0.5}$ - Specific length of reactor, L
μ_L - Dynamic viscosity of liquid, $ML^{-1}T^{-1}$
v_L - Kinematic viscosity of liquid, L^2T^{-1}
C_{pv} - Specific heat of saturated water vapour, $EM^{-1}deg^{-1}$

Surface evaporation, Φ_e. Surface evaporation is changing of water from a liquid to water vapour. While evaporating, water loses heat. The heat transfer by surface evaporation is a function of wind velocity, relative humidity, water temperature, and air temperature (Talati and Stenstrom 1990). Novotny and Krenkel (1973) determined the following relationship for the heat loss by evaporation:

$$\Phi_e = \left[1.145 \cdot 10^6\left(1 - \frac{Rh}{100}\right) + 6.86 \cdot 10^4\,(T - T_a)\right]e^{0.0604 \cdot T_a}u_w A_S^{0.95} \tag{2.81}$$

where
Rh - Relative humidity

The above equation was used in most of the later models (Argaman and Adams 1977; Talati and Stenstrom 1990; Sedory and Stenstrom 1995; Scherfig et al. 1996; Makinia et al. 2005b). La Cour Jansen et al. (1992) and Wall and Petersen (1986) used other equations replacing relative humidity by vapour pressure of the water and air above the water surface. The equation of Wall and Petersen (1986) is relatively simple:

$$\Phi_e = 200hca(P_w - P_a)A_S \tag{2.82}$$

However, values of the heat coefficient (hca) vary differently in five ranges of wind velocity between 0 and 11.18 m/s (0-25 mph).

La Cour Jansen et al. (1992) used the following equation:

$$\Phi_e = -\frac{\rho_a C_{p,a}(e_{sat} - e_a)\kappa^2 u_w}{\gamma\left[\ln\left(z_w \middle/ z_0\right)\right]}A_S \tag{2.83}$$

where
e_{sat} - Saturated vapour pressure of water at air temperature, $ML^{-1}T^{-2}$
κ - Von Karman's constant
γ - Psychrometer constant, $ML^{-1}T^{-2}deg^{-1}$
z_w - Measurement height for wind velocity, L
z_0 - Roughness length, L

Lippi et al. (2009) proposed a new equation for Φ_e which is based on the dimensionless analysis of water evaporation from a plane water surface:

$$\Phi_e = \phi_1 ShD_{AB}(ab)^{-0.5}(\rho_{v,Ta} - \rho_{v,Tw})A_S \tag{2.84}$$

The Sherwood (Sh) and Schmidt (Sc) numbers are defined as follows:

$$Sh = (0.037\,Re^{0.8} - 871)Sc^{0.33} \tag{2.85}$$

$$Sc = vD_{AB} \tag{2.86}$$

where

ϕ_l - Latent heat of vaporization, EM^{-1}

D_{AB} - Binary mass diffusion coefficient, $L^{-2}T$

$\rho_{v,Ta}$ - Water vapour density in air, ML^{-3}

$\rho_{v,Tw}$ - Water vapour outstream density, ML^{-3}

Aeration, Φ_a. Heat exchange resulting from aeration is a combination of convection (sensible heat loss), Φ_{as}, and evaporation (latent loss), Φ_{al} (Talati and Stenstrom 1990; Sedory and Stenstrom 1995; Scherfig *et al.* 1996; Makinia *et al.* 2005b):

$$\Phi_a = \Phi_{as} + \Phi_{al} \qquad (2.87)$$

Therefore, Novotny and Krenkel (1973) suggested that expressions similar to those developed for surface evaporation and conduction should be applied.

Aeration heat loss depends to a large extent on the type of aeration equipment installed. The sensible heat loss for surface and diffused aerators must be calculated differently since for diffused aeration the gas flow rate is known, while the exposure of the spray from a surface aerator to air must be estimated (Talati and Stenstrom,1990). For surface aeration, the sensible heat loss, $\Phi_{as,s}$, can be calculated from the following relationship (Talati and Stenstrom 1990; Sendory and Stenstrom 1995):

$$\Phi_{as,s} = \rho_a h_v C_{p,a} Q_A (T - T_a) \qquad (2.88)$$

The term h_v is defined as:

$$h_v = 392(n_a A_a)^{-0.05} u_w \qquad (2.89)$$

where

A_a - Aerator spray area, L^2

n_a - Number of aerators

The aerator spray area (A_a) can be determined experimentally or obtained from manufacturer's data (Talati and Stenstrom 1990). The overall equation contains the reactor surface area which is an approximation of the true surface area of all water droplets in the spray (Sedory and Stenstrom 1995).

For diffused aeration systems, assuming that the exit air temperature is equal to the liquid temperature in the aeration tank, the sensible heat loss, $\Phi_{as,d}$, is calculated as follows (Talati and Stenstrom 1990; Sedory and Stenstrom 1995; Makinia *et al.* 2005b):

$$\Phi_{as,d} = \rho_a C_{p,a} Q_A (T - T_a) \qquad (2.90)$$

Scherfig *et al.* (1996) presented the above equation using numerical values for ρ_a and $C_{p,a}$:

$$\Phi_{as,d} = 26 \cdot 10^6 Q_A (T - T_a) \qquad (2.91)$$

The equation for the evaporation term was originally proposed by Novotny and Krenkel (1973), but Talati and Stenstrom (1990) developed further that equation obtaining the following form:

$$\Phi_{al} = \frac{M_w Q_A \phi_l}{100R} \left\{ \frac{e_w \left[Rh + h_f (100 - Rh) \right]}{(T + 273)} - \frac{e_a Rh}{(T_a + 273)} \right\} \qquad (2.92)$$

where
e_w - Vapour pressure of water at reactor temperature, $ML^{-1}T^{-2}$
h_f - Exit air humidity factor
M_w - Molecular weight of water, Mmole

The same equation was used by Sedory and Stenstrom (1995), Scherfig *et al.* (1996) and Makinia *et al.* (2005b), but Scherfig *et al.* (1996) used numerical values for M_w, ϕ_l, R, h_f, e_w, and e_a:

$$\Phi_{al} = 148.4 \cdot 10^6 Q_a \left[\frac{750.5}{(T + 273)} - \frac{6.53Rh}{(T_a + 273)} \right] \qquad (2.93)$$

Talati and Stenstrom (1990) estimated the gas flow rate for surface aerators as a function of the spray area and wind velocity:

$$Q_a = n_a A_a u_w \qquad (2.94)$$

The exit humidity factor (h_f) is a measure of the humidity of the air that exits the aerator spray area. For diffused aeration systems, it was assumed to be 1.0 (the air was water saturated), whereas for the surface aerators it was less than 1.0 (Sedory and Stenstrom 1995).

Based on Novotny and Krenkel's (1973) studies, Argaman and Adams (1977) calculated total heat loss by surface aeration from the following formula:

$$\Phi_a = 4.32 \cdot 10^4 \, n_a A_a u_w \begin{bmatrix} 300(T - T_a) + 2920 e^{0.0604 \cdot T_a} \left(1 - \dfrac{Rh}{100}\right) \\ +175 e^{0.0604 \cdot T_a}(T - T_a) \end{bmatrix} \tag{2.95}$$

La Cour Jansen et al. (1992) considered only the sensible heat component:

$$\Phi_{as} = \rho_a C_{p,a} \frac{T_a - T}{\ln\left(\dfrac{z_w}{z_o}\right)} \kappa^2 u_w A_S \tag{2.96}$$

Mechanical energy, Φ_m. In surface aeration systems, the heat gained from mechanical energy can be calculated by assuming that the entire aeration energy is dissipated as heat which is absorbed by the water of the adjacent air (Argaman and Adams 1977). Hence, all the power supplied to the impellers in such aerators is available in the form of heat energy (Talati and Stenstrom 1990). The corresponding equation for surface aeration becomes (Sedory and Stenstrom 1995):

$$\Phi_m = P \frac{\eta_e}{100} \tag{2.97}$$

where
P   - Power of aerator/compressor, E
η_e - Efficiency of aerator/compressor, %

In diffused aeration systems, heat is generated during the process of compression, and the portion added to the reactor is represented by the blower inefficiency (Talati and Stenstrom 1990; Sendory and Stenstrom 1995; Makinia et al. 2005b):

$$\Phi_m = P\left(1 - \frac{\eta_e}{100}\right) \tag{2.98}$$

Scherfig et al. (1996) assumed that the energy loss to the atmosphere directly from compression can be neglected, and the expression for the mechanical energy transfer became:

$$\Phi_m = 24 \cdot 860 \cdot 10^3 P \tag{2.99}$$

Biochemical process energy, Φ_{bp}. Biochemical processes are exothermic and therefore contribute to heat gains in the system heat budget. La Cour Jansen *et al.* (1992) estimated heat released during biochemical processes based on Gibb's free energy terms, ΔG, for aerobic respiration (ΔG_1=-110 kJ/e), nitrification (ΔG_2=-43 kJ/e) and denitrification (ΔG_3=-104 kJ/e):

$$\Phi_{bp} = -10^6 \left(\Delta G_1 \frac{\Delta S}{32} + \Delta G_2 \frac{8 S_{ND,in}}{14} + \Delta G_3 \frac{5 \cdot 0.8 S_{ND,in}}{14} \right) Q \qquad (2.100)$$

where
ΔG_1 - Gibb's free energy for aerobic respiration, Ee^{-1}
ΔG_2 - Gibb's free energy for nitrification, Ee^{-1}
ΔG_3 - Gibb's free energy for denitrification, Ee^{-1}
ΔS - Mass of substrate (as COD) removed per day, MT^{-1}
$S_{ND,in}$ - Soluble biodegradable organic nitrogen concentration in inlet to reactor, $M(N)L^{-3}$

This approach appears to be crucial for advanced wastewater treatment systems with nitrogen removal and was adopted by Makinia *et al.* (2005b) for a pre-denitrification system. So far, the other temperature models (Argaman and Adams 1977; Talati and Stenstrom 1990; Sedory and Stenstrom 1995; Scherfig *et al.* 1996) used the equation based only on the organic substrate removal, which did not account for the impact of nutrient removal processes. Argaman and Adams (1977) assumed a release of approximately 1800 cal of heat per 1 g COD removed at the net (observed) biomass yield, Y_{obs} = 0.25 g VSS/g COD removed. Thus, the contribution of biodegradation of organic matter to the overall heat budget was given as:

$$\Phi_{bp} = 1.8 \cdot 10^6 \Delta S \qquad (2.101)$$

Talati and Stenstrom (1990) and Scherfig *et al.* (1996) used the same equation, whereas Sedory and Stenstrom (1995) modified it by introducing a variable Y_{obs}:

$$\Phi_{bp} = (3.3 - 5.865 Y_{obs}) \Delta S \qquad (2.102)$$

where
Y_{obs} - Net (observed) biomass yield coefficient, $M\,M(COD)^{-1}$

Heat exchange through reactor walls, Φ_w. The rate of exchange from conduction and convection through reactor walls and bottom depends on materials of construction, wall thickness, and temperature difference between the internal and external surfaces (Argaman and Adams 1977). Since accurate calculation of heat exchange through the walls and bottom is only possible when the exact configuration of the basin is specified, a simplified equation involving a single heat transfer coefficient can be used (Argaman and Adams 1977):

$$\Phi_w = U_g A_{w,g} (T - T_g) \qquad (2.103)$$

The same equation was used by Talati and Stenstrom (1990), whereas Sedory and Stenstrom (1995) separated area exposed to air and to ground:

$$\Phi_{w,a} = U_a A_{w,a} (T - T_a) \qquad (2.104)$$

$$\Phi_{w,g} = U_g A_{w,g} (T - T_g) \qquad (2.105)$$

where

$\Phi_{w,a}$ - Heat exchange through reactor wall exposed to air, ET^{-1}
U_a - Overall heat transfer coefficient for conduction and convection to air, $EL^{-2}T^{-1}deg^{-1}$
$A_{w,a}$ - Reactor wall area exposed to air, L^2
$\Phi_{w,g}$ - Heat exchange through reactor wall exposed to ground, ET^{-1}
U_g - Overall heat transfer coefficient for conduction to ground, $EL^{-2}T^{-1}deg^{-1}$
$A_{w,g}$ - Reactor wall area exposed to ground, L^2

Due to a high heat capacity of the ground and the low thermal conductivity, Scherfig *et al.* (1996) assumed that the heat exchange through the walls is negligible compared to the other terms and was set to zero in the model. This component also did not occur in the models of Wall and Petersen (1986), la Cour Jansen *et al.* (1992) and Makinia *et al.* (2005b).

2.5 SEDIMENTATION/CLARIFICATION MODEL

Secondary clarifiers (settling tanks or sedimentation basins) are an integral part of the activated sludge systems providing such functions as clarification, solids thickening and storage. Although the primary function of secondary clarifiers is the solid-liquid separation, some biological processes, primarily denitrification, may also occur. In a properly functioning secondary clarifier, the suspended solids concentration in the RAS lane, $X_{T,RAS}$, may be approximated from the mass balance upon the entire clarifier (Novotny *et al.* 1991). If the wastage flow rate is very small and neglected in the mass balance, then the following expression may be used to determine $X_{T,RAS}$:

$$X_{RAS} = X\left(\frac{1+\alpha_{RAS}}{\alpha_{RAS}}\right) - X_{eff}\left(\frac{1}{\alpha_{RAS}}\right) \qquad (2.106)$$

where

X_{RAS} - Suspended solids concentration in the RAS lane, L^3T^{-1}
X - Suspended solids concentration in bioreactor (influent to clarifier), L^3T^{-1}
α_{RAS} - RAS recirculation ratio
X_{eff} - Effluent suspended solids concentration, L^3T^{-1}

The effluent suspended solids (SS) concentration, X_{eff}, may be modelled using one of the empirical equations listed in Table 2.11. Vitasovic (1989) found that empirical equations do not provide reasonable predictions for the effluent SS concentrations lower than 10 mg/dm^3. At such low values of $X_{T,eff}$, however, the last term in Equation 2.106 becomes insignificant for the solids balance in the clarifier. As an example, Figure 2.16 shows the observed vs. predicted values of $X_{T,eff}$ for one-year data from the Rock Creek WWTP in Hillsboro, OR (USA). With one set of model parameters, the two examined models (Busby and Andrews (1975) and Pflanz (1969)) were not able to predict simultaneously low and high effluent SS concentrations during the dry season (April-October) and wet season (November-March), respectively (Makinia 1998).

Table 2.11 Empirical formulae for the effluent suspended solids concentrations

Reference	Formula
Pflanz (1969)	$K_1 + \dfrac{K_2 Q X}{A_s}$
Busby and Andrews (1975)	$\dfrac{K_1 Q X}{A_s}$
USEPA (1972)	$K_1 + \dfrac{K_2 Q}{A_s} + K_3 X$
Chapman (1983)	$K_1 + \dfrac{K_2 Q}{A_s} + K_3 X + H_s\left(K_4 - \dfrac{K_5 Q}{A_s}\right)$
Dupont and Henze (1992)	$K_1 + K_2\dfrac{S_{NO}}{K_{NO} + S_{NO}} + K_3\dfrac{X\,SVI\left(Q/A_s\right)}{K_4 + X\,SVI\left(Q/A_s\right)}$

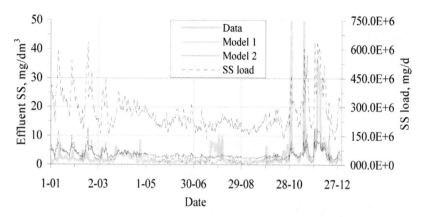

Figure 2.16 Observed vs. predicted effluent SS concentrations the Rock Creek WWTP in Hillsboro, OR (USA) (Model 1 - Busby and Andrews (1975), Model 2 - Pflanz (1969)).

2.5.1 Solids flux theory

Coe and Clevenger (1916) were the first who recognized that the batch settling process of a flocculent suspension results in a well-known distribution of different settling zones. The authors also suggested that a layer in a suspension has only a certain solids-handling capacity. The layer will grow in thickness if solids are not discharged as rapidly as they are received from the overlaying layer (limiting flux). No underlying sedimentation "*theory*", in the modern sense of a partial differential equation, was formulated in the following decades to explain the observed sedimentation behaviour (Burger and Wendland 2001). The theoretical principles of sedimentation were proposed by Kynch (1952) based on the propagation of kinematic waves in an idealized suspension. That paper may be considered "*the origin of the solids-flux theory*" (Diehl 2008). Besides constant horizontal concentration layers, Kynch's principal assumption was that the local settling velocity depends on the solids concentration only and it is decreasing with the increasing concentration. Numerous studies have contributed to the continued development (revisions, modifications, extensions, identifying limitations, etc.) of the theory (e.g. Dick and Ewing 1967; Vesilind 1968; Keinath *et al.* 1977; Tiller 1981; Fitch 1983; Laquidara and Keinath 1983; Keinath 1985, Concha and Bustos 1987; Diplas and Papanicolaou 1997; Diehl 2001; Wett 2002; Ekama and Marais 2004; De Clercq *et al.* 2008; Diehl 2008).

Basically, the flux theory assumes that the downward movement in the thickening zone of the clarifier (the region of hindered settling) is driven by two forces: gravity settling and bulk movement (sludge recirculation). Therefore, the total solids flux (G_{tot}) is a sum of two components related to those processes:

$$G_{tot} = G_s + G_u \qquad (2.107)$$

where
G_s - Gravity settling flux in clarifier, $ML^{-2}T^{-1}$
G_{tot} - Total solids flux in clarifier, $ML^{-2}T^{-1}$
G_u - Solids flux in clarifier due to underflow, $ML^{-2}T^{-1}$

The solids flux due to underflow (G_u) is a linear function of the suspended solids concentration (X) in the clarifier influent:

$$G_u = \upsilon_u X \qquad (2.108)$$

where
υ_u - Underflow velocity in clarifier, LT^{-1}

The underflow velocity, υ_u, is calculated from the following relationship:

$$\upsilon_u = \frac{Q_{RAS}}{A_{clar}} \qquad (2.109)$$

where
A_{clar} - Cross-sectional area of clarifier, L^2
Q_{RAS} - Returned activated sludge (RAS) flow rate, L^3T^{-1}

The solids flux due to gravity settling (G_g) is a linear function of X in the clarifier influent and the settling velocity (υ_s):

$$G_s = \upsilon_s X \qquad (2.110)$$

where
υ_s - Settling velocity of sludge in clarifier, LT^1

According to Kynch's theory, the actual relationship for G_s becomes more complicated because υ_s and X are related to each other. Estimating this relationship plays a critical role and several expressions for v_s as a function of X have been reported in literature (Table 2.12). In the early stages of research, two functions were used most often for practical applications including the power function (also known as Dick and Young's equation) and the exponential function (also known as Vesilind's equation) (Cho *et al.* 1993). Both expressions include two parameters: the maximum settling velocity of sludge in clarifier, $v_{s,max}$, and another empirical constant, n_s.

Table 2.12 Formulae of activated sludge settling velocity (Cho *et al.* 1993)

Reference	Formula for v_s
Thomas (1963), Vesilind (1968)	$v_{s,max} e^{-n_s X}$
Yoshioka *et al.* (1955), Dick and Young (1972)	$v_{s,max} X^{-n_s}$
Richardson and Zaki (1954)	$v_{s,max}(1-n_s X)^{4.65}$
Scott (1966), Steinour (1944)	$v_{s,max} \dfrac{(1-n_s X)^3}{X}$
Cho *et al.* (1993) (model 1)	$v_{s,max} \dfrac{(1-n_s X)^4}{X}$
Cho *et al.* (1993) (model 2)	$v_{s,max} \dfrac{e^{-n_s X}}{X}$
Scott (1968)	$v_{s,max} X(1-X)$
Cho *et al.* (1993) (model 3)	$v_{s,max} \dfrac{(1-n_{s,1} X)^4}{X} e^{-n_{s,2} X}$
Steinour (1944)	$v_{s,max}(1-n_s X)^2 e^{-4.19 X}$
Vand (1948)	$v_{s,max}(1-n_{s,1} X)^2 \exp\left(\dfrac{-n_{s,2} X}{1-n_{s,3} X}\right)$
Shannon *et al.* (1963)	$v_{s,max}(1-n_{s,1} X + n_{s,2} X^2 + n_{s,3} X^3 + n_{s,4} X^4)$
Vaerenbergh (1980)	$v_{s,1}(1-n_{s,1} X)^{n_{s,2}} + v_{s,2}$

Table 2.13 Selected correlations for relating $v_{s,0}$ and n_s to various SVI-type settleability measures (Hartel and Popel 1992; Ekama et al. 1997; Bye and Dold 1999, Giokas et al. 2003)

Test	$v_{s,0}$ [m/h]	n_s [dm³/g]	Reference
DSVI	$13.53\ e^{0.00365\ DSVI}$	$0.249 + 0.002191\,DSVI$	Koopman and Cadee (1983)
SSVI	$(v_{s,0}/n_s) = 67.9\ e^{-0.016\ SSVI}$	$0.88 - 0.393\ \log(v_{s,0}/n_s)$	Pitman (1984), Ekama and Marais (1986)
SVI	7.8	$0.148 + 0.0021\,SVI$	Daigger and Roper (1985)
SVI	$18.2\ e^{-0.00502\ SVI}$	$0.351 + 0.00058\,SVI$	Wahlberg and Keinath (1988)
SSVI	$15.3 - 0.0615\,SSVI$	$0.426 - 3.84 \cdot 10^{-3}\ SSVI + 5.43 \cdot 10^{-5}\ SSVI^2$	Wahlberg and Keinath (1995)
SVI	$17.4\ e^{-0.0113\ SVI} + 3.931$	$0.9834\ e^{-0.00581\ SVI} + 1.043$	Hartel and Popel (1992)
SVI	$28.1\,SVI^{-0.2667}$	$0.177 + 0.0014\,SVI$	Akca et al. (1993)
SVI	6.495	$0.1646 + 0.001586\,SVI$	Daigger (1995)
SSVI	7.973	$0.0583 + 0.00405\,SSVI$	Daigger (1995)
DSVI	7.599	$0.1030 + 0.002555\,DSVI$	Daigger (1995)
SVI	$8.53094\ e^{-0.00165\ SVI}$	$0.20036 + 0.00091\,SVI$	Ozinsky and Ekama (1995)
SSVI	$11.59936\ e^{-0.00165\ SSVI}$	$0.15128 + 0.00287\,SSVI$	Ozinsky and Ekama (1995)
DSVI	$6.35543\ e^{-0.00084\ DSVI}$	$0.19818 + 0.00123\,DSVI$	Ozinsky and Ekama (1995)
SVI	7.27	$0.0281 + 0.00229\,SVI$	Mines et al. (2001)

Both relationships are only applicable to the hindered (zone) settling (Cho et al. 1993). The exponential function is considered to be more accurate, but it requires numerical procedures for further mathematical analysis (Grijspeerdt et al. 1995). A major difficulty is to determine the settling velocity function in terms of the actual settling characteristics of the sludge. Many studies have evaluated the effects of various factors on υ_s including organic loading, filament content, extracellular polymers, temperature, pH and density (Schuler and Jang 2007). The most common approach is to relate $\upsilon_{s,0}$ and n_s to various settleability measures, such as sludge volume index (SVI), stirred SVI (SSVI) and diluted SVI (DSVI). Hartel and Popel (1992), Ekama et al. (1997), Bye and Dold (1999), and Giokas et al. (2003) reviewed the expressions reported in literature (Table 2.13). Bye and Dold (1998, 1999) identified three major problems related to the approach of correlating $\upsilon_{s,0}$ and n_s to SVI-type settleability measures:

- A correlation based on SVI may predict extremely different pairs of $\upsilon_{s,0}$ and n_s values depending on the solids concentration at which SVI is measured;
- The DSVI test "*forces*" SVI test conditions into the region where SVI becomes independent of solids concentration. Consequently, a correlation based on DSVI would predict the same values of $\upsilon_{s,0}$ and n_s for samples with different settling characteristics.
- Two parameters ($\upsilon_{s,0}$ and n_s) in the Vesilind equation cannot be accurately estimated based on a single SVI value from a test carried out at a single X value.

Takacs et al. (1991) modified the Vesilind equation to account for the fact that the settling velocity decreases as the solids concentration approaches zero:

$$\upsilon_s = \upsilon_{s,max} \; e^{-r_h X(1-f_{ns})} - \upsilon_{s,max} \; e^{-r_f X(1-f_{ns})} \qquad (2.111)$$

where
f_{ns} - Non-settleable fraction of influent total suspended solids (TSS)
r_f - Flocculent zone settling parameter, $L^3 M^{-1}$
r_h - Hindered zone settling parameter, $L^3 M^{-1}$

Grijspeerdt et al. (1995) performed a comparative evaluation of several one-dimensional sedimentation models. The authors concluded that the model of Takacs et al. (1991) was the most reliable to fit the data, both at steady state and under dynamic conditions, but the model required a relatively long calculation time of convergence. The comparison did not include the Cho ("*hybrid Vesilind*") model (Table 2.13) and the values of r_f and r_h parameters found by Grijspeerdt *et al.* (1995) were rather close to each other (0.4-0.6 vs. 1.0-2.2) indicating that both exponential terms interfere with each other over a relatively broad range. In another comparative studies, Vanderhasselt and Vanrolleghem (2000), and De Clercq *et al.* (2003) found that model to be more effective compared to the Vesilind and Takacs models in fitting the observed settling curves. It should be realised, however, that special care must be taken when using the Cho model at lower solids concentrations. When the sludge concentration approaches zero, the settling velocity goes to infinity, which is not physically possible. Another evaluation of several settling velocity models was carried out by Koehne *et al.* (1995). They showed that under different dynamic conditions, simulations with models that incorporate the settling velocities of Vesilind (1968), Takacs *et al.* (1991), Hartel and Popel (1992), and Otterpohl and Freund (1992) failed to give consistent results when compared to the pilot-scale experimental data.

Table 2.14 General categories of the settling velocity models (Zeidan *et al.* 2003)

Category of models	Description
Fluid-particle interaction	Based on the assumption that a particle settling down replaces a certain fluid volume and forces it to move in the opposite direction.
Analytical	Based on pure theoretical solutions of the Navier-Stockes equations.
Particle-particle interaction	Use modified or extended equations of fluid-particle interaction to describe settling in a suspension that contains different particle types.
Macroscopic level velocity (empirical)	Predict the average velocity of the suspension in the hindered (zone) settling regions regardless of the size and density distribution (these effects are included in the empirical constants).
Microscopic level velocity	Describe the dynamics of interaction between particles at the microscopic level (where the dynamics of interaction are observed in details).

All the above mentioned settling velocity models are empirical in nature and describe either solely hindered settling, hindered and compression settling, and hindered settling coupled with settling at low solids concentrations. However, such models cannot predict the settling velocity based on the physical properties of activated sludge flocs (e.g. size, density) or solid-liquid interactions. More fundamental approaches to modelling batch settling of flocculated suspensions have been investigated in other application areas (e.g. flocculated slurries) (De Clercq et al. 2008). Zeidan et al. (2003) reviewed the available settling velocity models and grouped them under five general categories (Table 2.14). Based on the conservation of mass and momentum of both liquid and solids, De Clercq et al. (2008) proposed the first mechanistic model to describe the settling behaviour of activated sludge. The forces acting on the activated sludge flocs included gravity, buoyancy, liquid pressure, friction and effective solids stress.

2.5.2 Approaches to dynamic modelling clarifier operation

The flux theory consisted a framework for the construction of numerous one-dimensional (1-D), dynamic models of clarifier operation. Such models were reported in the literature, for example, by Stenstrom (1975), Attir and Denn (1978), Vitasovic (1989), Takacs et al. (1991), Dupont and Henze (1992), Hartel and Popel (1992), Hamilton et al. (1992), Otterpohl and Freund (1992), Lessard and Beck (1993), Jeppsson and Diehl (1996), Wett (2002), De Clercq et al. 2008). A general governing equation for the solids flow in the secondary clarifier is 1-D advection-dispersion with a source term, which can be presented as (Vitasovic 1989; Hamilton et al. 1992, Watts et al. 1996):

$$\frac{\partial X}{\partial t} = -\frac{\partial G_s}{\partial z} + \upsilon_u \frac{\partial X}{\partial z} + E_z \frac{\partial^2 X}{\partial z^2} - r_{MLSS} \qquad (2.112)$$

where
E_z - Dispersion coefficient along the vertical axis of clarifier, L^2T^{-1}
r_{MLSS} - Reaction (formation/loss) rate of solids (MLSS), $ML^{-3}T^{-1}$
z - Distance along the vertical axis of clarifier, L

Most often the last two terms (i.e. dispersion mixing and reaction rate of solids) of Equation 2.112 are neglected and a hyperbolic conservation law (the continuity equation) is applied. This only takes into account transport of solids by advection (Verdict et al. 2006; Diehl 2008).

A common approach is to discretise the continuity equation by dividing the clarifier into a number of horizontal layers of equal height, within which the solids concentration is assumed to be constant. The updates of the concentration are made in accordance to a mass balance at the layer interfaces imposing the minimum flux restriction that restrains the gravitational flux from a certain layer to the layer below to the minimum gravitational flux of both layers (Stenstrom

1975; Vitasovic 1989; Ekama *et al.* 1997; Vanderhasselt and Vanrolleghem 2000). As a result, a discrete solids profile is obtained rather than a settling curve.

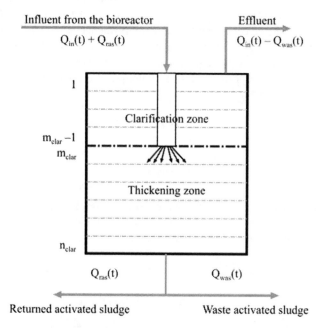

Figure 2.17 A schematic diagram representing 1-D models of the secondary clarifier.

The simplest layer model considers two well-mixed compartments, one (clarification) above and one (sedimentation) below the sludge blanket level. Wett (2002) proposed a conceptual three-layer model which differentiates clarification, hindered settling and compression zones. The most common models consider the clarifier vessel as a number (n_{clar}) of horizontal layers (typically 10 to 100) with the feed into layer m_{clar} as presented in Figure 2.17 (Olsson and Newell 1999). A limited number of studies were conducted to estimate n_{clar} which is required to obtain mesh-independent results (Verdickt *et al.* 2006). Jeppsson and Diehl (1996) concluded that a 10-layer model is too crude approximation to capture the detailed dynamic behaviour of the settler. Hamilton *et al.* (1992) and Lee *et al.* (1999a) found that their models yielded accurate results when at least 20-24 layers were used. In theory, as the number of layers increases, a model should resolve the detailed behaviour of the settling process and improve the accuracy of the predicted settling curve (Jeppsson and Diehl 1996; Ekama *et al.* 1997;

Vanderhasselt and Vanrolleghem 2000). In practice, however, this statement is not always correct. For example, Takacs *et al.* (1991) originally modelled the clarifier as consisting of 10 layers. Watts *et al.* (1996) found, however, that when n_{clar} in that model was increased to 20 without changing model parameters, the model predictions deteriorated considerably. Furthermore, model performance was worse at finer discretisations (e.g. 50 layers) compared a discretisation of 10 layers, even with parameters optimally fitted for that level of discretisation (Figure 2.18).

Verdickt *et al.* (2006) proposed the following exponential function to describe the observed relationship between $n_{clar,min}$ and E_z:

$$n_{clar,min} = 75 + e^{(-1.409\,E_z + 6.12)} \qquad (2.113)$$

where E_z is expressed in m^2/d.

In addition to the E_z value, the results presented by Verdickt *et al.* (2006) indicated that the operational status of the clarifier strongly influenced the required number of layers. Therefore, the authors emphasized that caution should be taken when using values for $n_{clar,min}$ under different conditions than those for which they were derived.

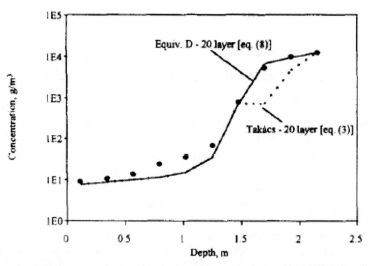

Figure 2.18 Comparison of concentration profiles obtained by 20-layer versions of the Takacs *et al.* (1991) model (equation (3)) and the model with dispersion (equation (8)) to experimental data (case 1 of the Pflanz data). Parameters reported for the 10-layer model of Takacs *et al.* (1991) were employed to generate model fits. (Watts *et al.* 1996).

Hamilton *et al.* (1992) included the dispersion term in their model and suggested the spatially constant value 0.54 m²/h for E_z. Grijspeerdt *et al.* (1995) demonstrated that the model of Hamilton was very sensitive to the changes in values of that parameter. This is an advantage for the identification process, but causes certain dangers in terms of physical interpretation. Verdickt *et al.* (2006) investigated in depth the role of the E_z coefficient in modelling sedimentation/thickening processes in secondary clarifiers. It was demonstrated that E_z is a crucial parameter in Equation 2.112 and has a great effect on both steady state predictions and numerical behaviour of the model (to minimise the latter effect Equation 2.113 was derived). Ekama *et al.* (1997) assigned a physical meaning to E_z which implicitly accounts for several factors, such as *"turbulent diffusivity, 2-D and 3-D dispersion, anomalies in the particles transport, errors introduced by the numerical method and the sludge removal procedure"*. Several attempts were made to use a space-dependent E_z coefficient:

- maximum E_z at the feed inlet and decreasing to the effluent and underflow boundaries (Ozinsky *et al.* 1994);
- E_z dependent on a concentration of solids and feed velocity (Watts *et al.* 1996; Plosz *et al.* 2007);
- E_z as a function of the local flow rate and the Peclet number (Lee *et al.* 1999);
- E_z dependent on the eddy viscosity of the fluid phase (Verdickt *et al.* 2006).

A more complex approach to clarifier modelling, applied for the first time by Larsen (1977), includes two-dimensional (2-D) hydrodynamic models which allow to analyse both the internal flow field (e.g. flow velocities, turbulent eddy viscosity distribution) and solids concentration field. These models account for variations in both axial and radial directions, often encountered in full-scale and resulted from such problems as the construction and dimensions of the clarifier, the effects of the different sludge collection systems, density currents, and compression limitation problems of the activated sludge flocs (Ekama *et al.* 1997). In particular, the 2-D models can be applied to clarifiers with increasing geometrical and process complexity (e.g. the buoyancy forces) including special arrangements, such as inlet deflectors, inlet vanes, sludge hoppers and density effects (Stamou 1995).

Krebs (1995) noted that 2-D hydrodynamic models have great development potentials as a research tool for improvement of the understanding of internal processes. On the other hand, they are very complex and require a lot of knowledge to be applied including a better characterization of sludge properties, which limits their practical applications. Furthermore, they are difficult to implement in combination with the models of activated sludge bioreactors. It is not surprising that *"due to the gaps in knowledge in settling mechanisms and the complexity involved in 2-D modelling and simulations, 1-D settler models can still be used with adjustments for simplicity*" (Lee *et al*. 1999) or "*the complexity of two-and three-dimensional models makes, however, a one-dimensional model still interesting for the design of full-scale secondary settling tanks*" (Diehl 2008).

Ekama and Marais (2004) used a hydrodynamic 2-D model to evaluate the applicability of the 1-D flux theory for design of secondary clarifiers by comparing predicted maximum surface overflows and solids loading rates (SLRs) from 35 full-scale hydraulic stress tests conducted on different clarifiers. From the simulations, a relatively consistent pattern appeared, i.e. that the flux theory can be used for design but its predicted maximum SLR needs to be reduced by an appropriate flux rating (0.75-0.85), which primarily depends on the depth and hydraulic loading rate of secondary clarifiers.

A comparison of the capabilities of various clarifier models is presented in Table 2.15. Very simple models are considering ideal solids/liquid separation without sludge storage are applicable at "*pseudo stationary*" situations, but if a rain event or other significant perturbations causing shifts between bioreactor and clarifier occur then dynamic models should be applied (Langergraber *et al*. 2004). From a control perspective, the important parameters that should be predicted by the models include the sludge blanket level, sludge inventory, and overflow and underflow concentrations (Olsson and Newell 1999).

Table 2.15 Capabilities of the secondary clarifier models (Krebs 1995)

Purpose	Conceptual	Flux model	Sophisticated layer model	Analytical 1-D modelling	Hydrodynamic 2-D modelling
Dimensioning	+	o	p	p	o
Operation[1]	–	o/–	+/p	p	p
Effluent[1]	–	–/–	o/–	o/–	p
Recycle[1]	o	+/o	+/o	+/+	p
Sludge blanket[1]	o	o/–	+/o	+/p	p
Flocculation	–	–	–	–	p
Combination with ASM	–	–	+	o	–
Control	–	–	+	p	o
Design (inlet, outlet)	–	–	–	–	+
Process understanding	–	o	o	+	+
Ready for applicability	+	+	+	o	–
Computer capacity	+	+	+	o	–

+ positive, o neutral, – negative, p development potential
[1] Two grades are distinguished for steady state/dynamic case

2.5.3 Biological processes in the secondary clarifier

2.5.3.1 Occurrence of denitrification and secondary phosphate release

Although the primary function of secondary clarifiers is the solid-liquid separation, some biological processes may also occur there. Denitrification in secondary clarifiers has received much attention due to possible sludge rising as a result of this process (Henze et al. 1993). For example, 1 cm^3 of nitrogen gas would be sufficient to allow 53–25 mg/dm^3 of activated sludge solids to rise, and such a volume of gas is produced by the denitrification of only 1.25 mg NO_3-N (Sarioglu and Horan 1996). Henze et al. (1993) identified the most important factors affecting that phenomenon including the denitrification rate in combination with the HRT in the thickening zone. The critical NO_3-N concentration for the development of sludge rising was 6-8 mg N/dm^3 at T = 20 °C. In contrast, dissolved oxygen concentration at the inlet to the clarifier and clarifier depth played only a marginal role.

Denitrification in secondary clarifiers may also contribute significantly to the overall nitrogen balance in activated sludge systems, especially if the clarifier volume/anoxic zone volume ratio is large or if a substantial amount of sludge is maintained in the thickening zone of the clarifier (Siegrist et al. 1995).

One of the first studies related to this issue was performed by Crabtree (1983). The observed denitrification rates (0.2-2.1 mg N/(g SS·h)) in the clarifier were in the range associated with endogenous denitrification. Siegrist et al. (1995), Koch et al. (1999) and Mikola et al. (2009) estimated that denitrification in the clarifier could account for up to 40% of the total denitrification. Stephenson and Luker (1994) found that the average NO_3-N concentration in the RAS line was 2.9 mg N/dm^3, whereas the average effluent NO_3-N concentration was 10.8 mg N/dm^3. The calculated specific NUR was 0.93 mg N/(g VSS·h), which is close to the observations made by Henze (1986) for denitrification with endogenous respiration as a carbon source.

The secondary release of phosphate in secondary clarifiers has received less attention than denitrification, but the phenomenon was reported in several studies (e.g. Schonberger 1990; Wouters-Wasiak et al. 1996; Mulkerrins et al. 2000; Hughes et al. 2004; Batista et al. 2005; Barnard and Abraham 2006; Mikola et al. 2009) Substantial amounts (up to 60%) of the phosphate taken up in the aerobic stage of the bioreactor were subsequently released in the clarifier.

2.5.3.2 Approaches to modelling biochemical processes
         in secondary clarifiers

In the review paper, Stamou (1995) concluded that the clarifier models should be extended with biological processes in order to be capable of predicting of important problems encountered in practice, such as sludge rising, sludge bulking and biological phosphorus removal. Also Watts *et al.* (1996) noted that the clarifier model should be integrated with a biokinetic model to be most valuable and simulate the effects of varying flow rates and feed compositions on both biochemical and sludge thickening performance. Models of denitrification in secondary clarifiers, which included the source term based on ASM1, were presented by Hamilton *et al.* (1992) and Siegrist *et al.* (1995). In some of the simulation platforms (see: Section 5.7), so-called "*reactive clarifier*" models can be easily implemented. Alternatively, Hulsbeek *et al.* (2002) suggested that denitrification in the RAS lane can be calibrated by applying a virtual tank in that lane. This approach has been reported in several simulation studies (e.g. Brdjanovic *et al.* 2000; Meijer *et al.* 2001; Salem *et al.* 2002; Wichern *et al.* 2003).

Gernaey *et al.* (2006) performed a theoretical simulation study with ASM1 using the IWA/COST simulation benchmark plant (Copp 2002). Biological activity in the clarifier was described by two methods:

- Including an extra model block in the RAS lane, allowing heterotrophic biomass growth to consume oxygen and a user-defined fraction of NO_3-N;
- Using a reactive one-dimensional settler model by coupling the settler model of Takacs *et al.* (1991) with ASM1.

The non-reactive model of Takacs *et al.* (1991) was used as a reference. When simulating nitrogen removal, the models considering biological activity in the settler resulted in a similar (10-15%) improvement of predictions. The authors concluded that, for ASM1, the use of an extra model block in the RAS lane would probably be a better approach compared to a reactive settler model. It was computationally more efficient and did not overrate the importance of decay processes in the settler. Also Makinia *et al.* (2005a) found that the reactive clarifier model did not improve predictions of the non-reactive model while considering NO_3-N concentrations in the sludge denitrification (SDN) zone of a pilot-scale Johannesburg process (shown in Figure 1.3). It should be emphasized, however, that the observed NO_3-N concentrations were very low (<0.4 mg N/dm^3) in that case.

Chapter 3

Modelling specific biochemical processes occurring in activated sludge systems

Complex biokinetic models describe a range of biochemical (and chemical) processes occurring simultaneously in activated sludge systems. This chapter provides insight into the approaches to modeling principal unit processes (including the recent developments, e.g. growth of *Microthrix parvicella* and Anammox bacteria), which can be used alone or incorporated in complex models as an extension or modification. Moreover, the ranges of kinetic and stoichiometric coefficients reported in literature are provided.

3.1 GROWTH OF MICROORGANISMS

Bacteria can reproduce in binary fission, i.e. by dividing the original cell into two new microorganisms. Therefore, the exponential growth of bacteria may be expressed in terms of their doubling (or generation) time with respect to either their number or concentration (Herbert *et al*. 1956):

$$\frac{dX}{dt} = \mu X \qquad (3.1)$$

where

X - Concentration (or number) of microorganisms, ML^{-3}

μ - Specific growth rate of microorganisms, T^{-1}

The equation that provides a reasonable description of the kinetics of microbial growth with greatest acceptance in modelling activated sludge systems is the one proposed by Monod (1949). This equation relates the growth rate of microorganisms in terms of the concentration of a growth-limiting factor (Equation 1.5), which can be the carbon source, the electron donor, the electron acceptor, nutrient (phosphorus or nitrogen) or any other factor needed by the microorganisms for growth. The Monod equation has been used to describe growth of many bacterial species forming a "*heterogeneous assemblage*", but simply expressed as a microbial "*biomass*", on a complex substrate consisting of a mixture of hundreds of organic compounds measured by a single, non-specific test like COD (Grady *et al.* 1999). The authors divided the growth-limiting factors into two categories: complementary and substitutable. Complementary factors are those that meet entirely different needs by growing microorganisms (e.g. ammonia as the nitrogen source for cell synthesis and organic substrates as the carbon and energy sources). On the other hand, if factors meet the same need (e.g. ammonia and nitrate as the nitrogen source for cell synthesis), they are called substitutable. Among the complex activated sludge models, which are outlined in Section 2.4.1, the models of Barker and Dold (1997a) and Hu *et al.* (2007) incorporate substitutable factors for growth of "*ordinary*" heterotrophs and phosphate accumulating organisms (PAOs) depending on whether ammonia or nitrate is the nitrogen source.

If complementary factors are considered at a time the relationships representing them have been classified as interactive and non-interactive (Grady *et al.* 1999). A non-interactive relationship assumes that the growth rate of a microbial culture can only be limited by one nutrient at the same time:

$$\mu = \mu_{max} \ min\left(\frac{S_1}{K_{S,1} + S_1}, \frac{S_2}{K_{S,2} + S_2}\right) \qquad (3.2)$$

In contrast, an interactive relationship assumes that two (or more) complementary relationship can influence the growth rate at the same time:

$$\mu = \mu_{max} \frac{S_1}{K_{S,1} + S_1} \frac{S_2}{K_{S,2} + S_2} \qquad (3.3)$$

All process rates in the complex activated sludge models are described by interactive relationships. This type of relationship is continuous and thus mathematically preferable for dynamic modelling. However, it may produce erroneous (too low) predictions of growth rates when both terms, i.e. $S_1/(K_{S,1}+S_1)$ and $S_2/(K_{S,2}+S_2)$, are small (Grady et al. 1999).

Clara et al. (2005) described a mixed substrate growth by the multisubstrate Monod growth relationship, where μ_{tot} is the total specific growth rate and μ_i is the specific growth rate on substrate i, and the summation is taken over the n substrates:

$$\mu_{tot} = \sum_{i=1}^{n} \mu_{max,i} \frac{S_i}{K_{i,S} + \sum_{j=1}^{n}\left(K_{i,S}/K_{j,S}\right)S_j} \qquad (3.4)$$

Equation 3.4 describes biodegradation of micropollutants present only in trace levels. In such a case, the removal of one compound does not result in any significant biomass growth but a group of compounds is transformed by a cometabolism (mixed substrate growth).

It should also be noted that apart from the Monod equation, several other formulae for microbial growth have been proposed in literature and reviewed by Vavilin (1982) (see: Section 1.2.2). One well-known equation, not included in that review, but worth to mention is the Haldane equation (also known as the Andrews model (Andrews 1968)):

$$\mu = \mu_{max} \frac{S}{K_S + S + S^2/K_I} \qquad (3.5)$$

where
K_I - Inhibition coefficient, ML^{-3}

Equation 3.5 describes the inhibitory effects of substrate on the microbial growth rate. Therefore it has been used to model the nitrification process at high initial ammonia concentrations (see: Section 3.7). Other inhibition models were proposed by Edwards (1970) and Luong (1987). In the latter model, the Monod term was extended with an additional empirical relationship to correlate substrate inhibition:

$$\mu = \mu_{max} \frac{S}{K_S + S}\left(1 - \frac{S}{S_m}\right)$$ (3.6)

where

S_m - maximum substrate concentration above which growth is completely inhibited, ML^{-3}

Below are discussed the stoichiometric (Y_H) and kinetic ($\mu_{H,max}$, $K_{S,H}$) parameters for "*ordinary*" heterotrophs including corrections for their growth under anoxic conditions. Other aspects of modeling microbial growth are covered in Section 3.2 (observed vs. "*true*" Y_H), Section 3.3 (the effect of storage on Y_H), and Sections 3.7–3.11 (growth of other groups of microorganisms).

3.1.1 Maximum specific growth rate for heterotrophic biomass, $\mu_{H,max}$

The typical values of $\mu_{H,max}$ reported by Henze *et al.* (1987) were within the range $3.0 - 13.2 \ d^{-1}$ with the default value of $6.0 \ d^{-1}$ used in ASM1. In many later studies (Sollfrank and Gujer 1991; Kappeler and Gujer 1992; Siegrist and Tschui 1992; Brands *et al.* 1994, Orhon *et al.* 1994b), the reported values of $\mu_{H,max}$ were below the typical range given above, even as low as $1.0 \ d^{-1}$ for systems with short solids retention times (SRTs) (Kappeler and Gujer 1992). Blok and Struys (1996) conducted the biodegradation tests on 52 different compounds encountered in wastewater using respirometric measurements. The estimated values of $\mu_{H,max}$ were within the range $1.0 - 6.0 \ d^{-1}$. Insel *et al.* (2002) claimed that a typical range of this coefficient for municipal wastewater is $5.0 - 6.0 \ d^{-1}$. The default value of $\mu_{H,max}$ in ASM2 and ASM2d is $6.0 \ d^{-1}$ (Henze *et al.* 2000). In ASM3, the default value of $\mu_{H,max}$ on cell internal storage compounds, X_{STO}, is only $2.0 \ d^{-1}$ (Gujer *et al.* 1999). This value was increased by Rieger *et al.* (2001) to $3.0 \ d^{-1}$ in ASM3P, which is similar to $\mu_{H,max} = 3.2 \ d^{-1}$ used in the model of Barker and Dold (1997a).

Certain factors should be taken into consideration while seeking an appropriate estimation method for $\mu_{H,max}$. Since the actual specific growth rate of heterotrophic biomass, μ_H, is limited by low concentrations of one of the growth nutrients such as carbon source, nitrogen, dissolved oxygen, etc., the $\mu_{H,max}$ can be estimated only under unlimited growth conditions where μ_H approaches $\mu_{H,max}$ (Grady and Lim 1980). Williamson and McCarthy (1975) developed a method called the infinite dilution procedure to estimate the substrate saturation coefficient, K_S, along with the maximum substrate removal rate, $r_{S,max}$. However, the main function of $\mu_{H,max}$ in biokinetic models is to predict the maximum OUR. This suggests that estimation of $\mu_{H,max}$ should be based on oxygen uptake measurements rather than cell growth or substrate removal (Henze et al. 1987). Ekama et al. (1986) and Orhon et al. (1994b) proposed to use batch tests for determining $\mu_{H,max}$. Both methods included the active biomass concentration of heterotrophs, X_H, which is unknown in the experiment.

Kappeler and Gujer (1992) developed another batch procedure independent of X_H, where a linear relationship was determined between the relative OUR ($OUR_H(t)/OUR_{H,i}$) and time, with the slope equal to the difference ($\mu_{H,max}-b_H$). In order to perform the test, centrifuged wastewater and a very small amount of activated sludge biomass should be mixed together to obtain the S/X_V ratio between 1 and 20 (approximately 4). During the first period of the test, the OURs increase exponentially due to unlimited heterotrophic growth on readily biodegradable substrate, S_S. At the point where the growth is starting to be limited by low concentrations of S_S, the OURs decrease to a level where the growth is dominated by substrate released in the hydrolysis process. For the estimation of $\mu_{H,max}$ only the OURs measured during exponential growth phase should be taken into account.

Assuming no substrate and oxygen limitation in the batch reactor, the oxygen respiration (Equation 3.7) and the mass balance for heterotrophic biomass (Equation 3.8) can be written:

$$OUR_H(t) = -\left(\frac{1-Y_H}{Y_H}\right)\mu_{H,max}\,X_H(t) - (1-f_p)\,b_H\,X_H(t) \qquad (3.7)$$

where
OUR_H - Heterotrophic oxygen uptake rate, $M(O_2)L^{-3}T^{-1}$
Y_{HN} - "*True*" growth yield coefficient for heterotrophic organisms, $M(COD)$ $M(COD)^{-1}$
f_P - Fraction of inert COD generated in biomass lysis (decay)
b_H - Specific lysis (decay) rate constant for heterotrophic organisms, T^{-1}

$$\frac{dX_H}{dt} = (\mu_{H,max} - b_H)X_H(t) \qquad (3.8)$$

Integrating Equation 3.7 and introducing it into Equation 3.8 gives:

$$OUR_H(t) = \left[-\left(\frac{1-Y_H}{Y_H}\right)\mu_{H,max} - (1-f_P)b_H\right]X_{H,0}\,exp\left[(\mu_{H,max}-b_H)t\right]$$

$$(3.9)$$

$$OUR_H(t) = OUR_H(t = 0)\exp\left[(\mu_{H,max} - b_H)t\right] \qquad (3.10)$$

where

$X_{H,0}$ - Initial concentration of heterotrophic organisms, $M(COD)L^{-3}$

Equation 3.10 can be presented in a logarithmic form:

$$\ln\left(\frac{OUR_H(t)}{OUR_H(t = 0)}\right) = (\mu_{H,max} - b_H)t \qquad (3.11)$$

If the value of b_H is unknown it can be assumed as approximately 5% of $\mu_{H,max}$ (Kappeler and Gujer 1992). The initial concentration of heterotrophic biomass, $X_{H,0}$, can be determined from Equation 3.9 provided that Y_H, and f_p are known:

$$X_{H,0} = \frac{OUR_H(t = 0)}{\left(\dfrac{1 - Y_H}{Y_H}\right)\mu_{max} - (1 - f_p)b_H} \qquad (3.12)$$

Novak *et al.* (1994) determined the value of $\mu_{H,max}$ by means of a combined technique of mathematical modelling and batch cultivations. The active fraction of heterotrophic organisms in the activated sludge ($f_{X,H}$) was estimated accurately enough using a numerical simulation of a continuously operating system. Provided that the simulation results give coherent values of the MLVSS concentration and the actual OURs, the concentration of heterotrophic biomass could be considered to be correct. Once this concentration is estimated, the value of $\mu_{H,max}$ can be calculated as a function of the maximum heterotrophic OUR, $OUR_{H,max}$, obtained from the from batch tests:

$$\mu_{H,max} = \frac{OUR_{H,max}}{f_{X,H}X_V}\frac{Y_H}{1 - Y_H} \qquad (3.13)$$

where

$OUR_{H,max}$ - Maximum heterotrophic oxygen uptake rate, $M(O_2)L^{-3}T^{-1}$

The $OUR_{H,max}$ is calculated as a difference between the maximum total OUR ($OUR_{T,max}$) and maximum autotrophic OUR ($OUR_{A,max}$). The value of $OUR_{T,max}$ may be estimated from a batch test when a sample of the activated sludge withdrawn from the nitrification tank is mixed with filtered wastewater (Whatman GF/C). The S/X_V ratio used in the test should be in the range 0.1-0.2 g COD/g

VSS. The maximum autotrophic OUR ($OUR_{A,max}$) can be calculated from the maximum ammonia utilization rate (AUR_{max}) (see: Section 3.7).

Comparing the combined technique with the method of Kappeler and Gujer, Novak *et al.* (1994) obtained different results for $\mu_{H,max}$, i.e. 4 d^{-1} and 10 d^{-1}. A possible discrepancy between both methods could be explained by the fact that the batch cultivation conditions are totally different from conditions in which the activated sludge microorganisms grow in a continuously operated system. Based on the literature data, it was hypothesized that during a batch test under high S/X_V ratio the fast growing group of microorganisms in the activated sludge can be favoured. As a consequence, the results obtained in the batch test specify only the group of microorganisms that posses higher growth rates.

3.1.2 Substrate saturation coefficient for heterotrophic biomass, $K_{S,H}$

The substrate half-saturation coefficient for heterotrophs, $K_{S,H}$, determines how rapidly μ_H approaches $\mu_{H,max}$ in the Monod equation in terms of the substrate concentration, and is defined as the substrate concentration at which μ_H is equal to half of $\mu_{H,max}$ (Grady and Lim 1980). The $K_{S,H}$ coefficient can be related to diffusion limitation in the microbial flocs. Due to higher turbulence and smaller flocs, the $K_{S,H}$ values in pilot-scale experiments tend to be low in comparison to full-scale facilities (Henze *et al.* 1995a). The typical values of $K_{S,H}$ reported by Henze *et al.* (1987) were within the range 10-180 mg COD/dm^3. However, the actual values of $K_{S,H}$ (2.5-5.0 mg COD/dm^3) used in a number of later studies (e.g. Gujer and Henze 1991; Oles and Wilderer 1991; Sollfrank and Gujer 1991; Dipankar an Randall 1992; Kappeler an Gujer 1992; Pedersen and Sinkjaer 1992; Siegrist and Tschui 1992, Brands *et al.* 1994; Barker and Dold 1997a) were well below the low range recommended for use in ASM1. Blok and Struys (1996) tested 52 different compounds encountered in wastewater and found that the values of $K_{S,H}$ even less than 1.0 mg COD/dm^3 were common. In ASM3, the $K_{S,H}$ default is 2.0 mg COD/dm^3 (Gujer *et al.* 1999) but this value was increased by Rieger *et al.* (2001) to 10 mg COD/dm^3 in ASM3P. Barker and Dold (1997a) used $K_{S,H}$ = 5.0 mg COD/dm^3 in their model. In ASM2 and ASM2d, two saturation coefficients, $K_{F,H}$ and $K_{A,H}$, occur depending on whether fermentable organic compounds, S_F, or fermentation products, S_A, are the substrate for heterotrophic growth. In both cases the default value is 4.0 mg COD/dm^3 (Henze *et al.* 2000).

In the case of multiple substrate, such as municipal wastewater, the value of $K_{S,H}$ is difficult to estimate accurately (Henze *et al.* 1987). Two early methods for the combined determination of K_S and $r_{S,max}$ were presented by Williamson and McCarthy (1975) and Cech at al. (1985).

3.1.3 Yield coefficient for heterotrophic biomass, Y_H

The true growth yield, Y_H, used in the biokinetic models, is defined as the ratio of the rate of cell growth in the absence of maintenance energy requirements (Henze et al. 1987). The typical values of Y_H reported by these authors were within the range 0.46–0.69 g cell COD synthesized/g COD removed with the default value equal to 0.67 g cell COD/g COD used in ASM1 and the model of Barker and Dold (1997a). Wanner et al. (1992) concluded that although Y_H was the most sensitive, it was also the most stable parameter among all kinetic and stoichiometric coefficients in ASM1. The Y_H coefficient influences not only the estimation of sludge production and oxygen demand but also has an effect on the values of other parameters (e.g. $\mu_{H,max}$) to be estimated (Vanrolleghem et al. 1999). The default value of Y_H used in ASM2 and ASM2d was 0.63 g cell COD/g COD. The same value was assumed as the ASM3 default for heterotrophic growth on cell internal storage products, X_{STO}, but Rieger et al. (2001) increased the yield to 0.8 g cell COD/g COD in ASM3P. In contrast, Brands et al. (1994) found Y_H within the range 0.2–0.43 g cell COD/g COD for high loaded systems vs. 0.6–0.67 g cell COD/g COD for low loaded systems. This difference could have resulted from different storage capabilities of the examined sludges (see: Section 3.3).

The true yield coefficient cannot be measured directly. During periods of rapid, unrestricted growth, the observed yield approaches the true yield and can be used as its approximation (Grady and Lim 1980). Two methods (Henze et al. 1987; Sollfrank and Gujer 1991) can be applied to estimate the value of Y_H. It can be estimated based on batch test results with filtrate of an aliquot of wastewater and acclimated biomass. When the soluble COD and the total COD are measured periodically, the yield Y_H can be determined (Henze et al. 1987):

$$Y_H = \frac{\Delta \text{ cell COD}}{\Delta \text{ soluble COD}} = \frac{\Delta \left(\text{total COD} - \text{soluble COD} \right)}{\Delta \text{ soluble COD}} \qquad (3.14)$$

The authors suggested that the value of Y_H is 0.67 g COD/g COD in the temperature range from 10 to 20 °C, which is equivalent to 0.47 using g VSS/g COD units (under the assumption that 1 g VSS = 1.42 g COD).

Sollfrank and Gujer (1991) determined Y_H from the net respiration rate for the degradation of the added organic material:

$$Y_H = \frac{COD_{deg r.} - \int\limits_{t_0}^{t_E} OUR_{H,net} V \, dt}{COD_{deg r.}} \qquad (3.15)$$

where
$COD_{degr.}$ - Biodegradable COD in the added material, $M(COD)L^{-3}$

The amount of degradable COD can be calculated as a difference between the concentration of COD in the filtered wastewater (COD_{spike}) and the inert portion (S_I), as determined in secondary effluent:

$$COD_{degr.}(t = t_0) = (COD_{spike} - S_I)V_{ww} \qquad (3.16)$$

where
COD_{spike} - COD in the material added to the reactor, $M(COD)L^{-3}$
S_I - Concentration of soluble inert organic material, $M(COD)L^{-3}$

3.1.4 Correction factors for anoxic kinetics and stoichiometry

Under anoxic conditions, nitrate and nitrite serve as electron acceptors in the respiratory electron transport chain in the same manner as oxygen, with only small modification to the metabolic system (i.e. the enzymes) of the bacteria (Zhao et al. 1999). It is commonly observed, however, that the rate of substrate utilization is always lower under anoxic conditions compared to aerobic conditions. The actual factors affecting the phenomenon of a lower substrate removal rate under anoxic conditions cannot be identified individually. The current approach is based on the simplifying assumption to adopt the same aerobic heterotrophic rate expression with the addition of a single empirical coefficient which acts as an overall reduction factor for anoxic conditions (Orhon et al. 1994b). The reduction coefficient, $\eta_{NO3,H}$, accounts for either the change in $\mu_{H,max}$ associated with anoxic conditions or for the fact that only a portion of the biomass is capable to denitrify (Batchelor 1982). The value of $\eta_{NO3,H}$ should always be lower than 1.0 as aerobic respiration is thermodynamically a more efficient process than denitrification. Indeed, the values of $\eta_{NO3,H}$ for domestic wastewater reported in the literature were in the range 0.38–0.94 (Orhon et al. 1996). The ASM1 default of $\eta_{NO3,H}$ was 0.8 and its value remained unchanged in ASM2 and ASM2d (Henze et al. 2000). The ASM3 default of $\eta_{NO3,H}$ was 0.6 but its value again increased to 0.8 in ASM3P (Rieger et al. 2001). The lowest anoxic reduction factor of 0.37 was used in the model of Barker and Dold (1997a).

The laboratory experiments to measure $\eta_{NO3,H}$ should be performed at the same time by evaluating oxygen and nitrate consumption rates in two batch reactors which are equivalent in every aspect except for the terminal electron acceptor (oxygen in the one and nitrate in the other) (Henze et al. 1987). Immediately after bringing biomass into contact with wastewater in a batch reactor, the activity in the reactor will be dominated by growth of heterotrophs on the RBCOD whereas the later activity will be predominantly due to use of substrate released from hydrolysis of the slowly biodegradable COD (SBCOD) (Ekama *et al.* 1986). If OUR_g represents the OUR during the first period in the aerobic reactor and NUR_g represents the nitrate utilization rate in the anoxic reactor, then:

$$\eta_{NO3,H} = \frac{2.86 NUR_g}{OUR_g} \tag{3.17}$$

where

$\eta_{NO3,H}$ - Anoxic reduction factor for growth of heterotrophic organisms,

NUR_g - Nitrate utilization rate during the first period in the anoxic batch reactor, $M(N)L^{-3}T^{-1}$

OUR_g - Oxygen uptake rate during the first period in the aerobic batch reactor, $M(O_2)L^{-3}T^{-1}$

Likewise, if OUR_h represents the OUR during the second period in the aerobic reactor and NUR_h - the corresponding NUR in the anoxic reactor, then:

$$\eta_{NO3,hyd} = \frac{2.86 NUR_h}{OUR_h} \tag{3.18}$$

where

$\eta_{NO3,hyd}$ - Anoxic reduction factor for hydrolysis

NUR_h - Nitrate utilization rate during the second period in the anoxic batch reactor, $M(N)L^{-3}T^{-1}$

OUR_h - Oxygen uptake rate during the second period in the aerobic batch reactor, $M(O_2)L^{-3}T^{-1}$

Alternatively, the $\eta_{NO3,H}$ value can be derived from the following equation (Orhon *et al.* 1994b):

$$\mu_{H,max} = \frac{2.86 Y_H}{1 - Y_H} \frac{\Delta(NO_3 - N)}{\Delta t} \frac{1}{X_H \eta_{NO3,H}} \tag{3.19}$$

The value of $\mu_{H,max}$ can be substituted from the aerobic batch test and the initial NUR associated with RBCOD can be used from the respective anoxic test.

It has long been surmised that the yield of biomass under anoxic conditions with nitrate as electron acceptor is lower than the corresponding aerobic yield for heterotrophic biomass (Barker and Dold 1997a). Orhon *et al.* (1996) were the first who provided the conceptual proof, based on the bioenergetic principles of the related metabolic processes, that the anoxic yield is considerably lower compared to the aerobic one. The theoretically derived ratios of the anoxic and aerobic yields were 0.79, 0.80, 0.80 and 0.85 for municipal wastewater, protein, lactate and carbohydrate, respectively. However, experimental data quantifying the reduced yield have still been limited. Muller *et al.* (2003) summarized the results of earlier studies (with both artificial substrates and municipal wastewater) in which values of the anoxic yield for heterotrophic biomass ($Y_{H,NO}$) were reduced to approximately 0.78-0.85 of the corresponding aerobic yield values.

Lower values of the anoxic growth yield affect predictions of denitrification and sludge production (Barker and Dold 1997a; Muller *et al.* 2003). The effect on sludge production appears to be insignificant in activated sludge systems treating municipal wastewater. Due to the relatively low influent TKN/COD ratios (usually <0.12 mg N/mg COD), the amount of nitrate generated is limited and thus the mass of sludge produced under anoxic conditions is small compared to that produced under aerobic conditions (Barker and Dold 1997a). The effect on denitrification can be more remarkable. For example, by decreasing the value of $Y_{H,NO}$ from 0.67 to 0.54 mg COD/mg COD, approximately 40% more nitrate (0.161 mg N vs. 0.115 mg N) will be denitrified for one unit of COD consumed (Muller *et al.* 2003).

In ASM1, ASM2, ASM2d (Henze *et al.* 2000) and similar models, e.g. Wentzel *et al.* (1992) and van Veldhuizen *et al.* (1999a), a single yield coefficient for heterotrophic organisms was assumed irrespective of the electron acceptor (oxygen or nitrate) conditions. A lower value of the anoxic yield (0.54 mg COD/ mg COD) has been incorporated in some of the activated sludge models (Gujer *et al.* 1999; Rieger *et al.* 2001; Hu *et al.* 2007). In the model of Barker and Dold (1997a), separate anoxic and aerobic yields were included to account for the possibility of a lower anoxic yield. Initially, the same value of 0.666 mg COD/mg COD was assumed for the two coefficients, but while subsequently implementing the model in the BIOWIN computer program the anoxic yield value was decreased first to 0.403 and later to 0.54 mg COD/mg COD (Muller *et al.* 2003). The necessity of a reduced anoxic yield for heterotrophic organisms was also recognized and included in ASM3 (Gujer *et al.* 1999). In this model, it is assumed that all readily biodegradable substrate is utilized by heterotrophic organisms through an intracellular substrate storage which is followed by growth. Both processes have different yields under anoxic and aerobic conditions ($Y_{STO,NO}$ and $Y_{H,NO}$ vs. $Y_{STO,O2}$ and $Y_{H,O2}$). For the storage process, the default $Y_{STO,O2}$ is 0.85 mg COD/mg COD and the default $Y_{STO,NO}$ is 0.80 mg COD/mg COD. For the following growth process, the corresponding yields ($Y_{H,O2}$ and $Y_{H,NO}$) are 0.63 and 0.54 mg COD/mg COD. The respective net aerobic and anoxic yields, calculated as $Y_{STO,O2}Y_{H,O2}$ and $Y_{STO,NO}Y_{H,NO}$, are then 0.54 and 0.43 mg COD/mg COD. This results in the ratio of net anoxic to aerobic yields of 0.81. Muller *et al.* (2003) noted that no guidance was given on the source for these values, except for the statement that it was accepted that the ratio of anoxic to aerobic energy yields was 0.7. Accepting the ASM3 default values for $Y_{STO,NO}$, $Y_{STO,O2}$ and $Y_{H,O2}$, Muller *et al.* (2003) determined experimentally, based on the measurements of nitrate and oxygen utilization rates (NURs and OURs), that the value of $Y_{H,NO}$ should be approximately 0.42 mg COD/mg COD which is significantly lower than the ASM3 default i.e. 0.53 mg COD/mg COD. For comparison, using the same experimental data and accepting in the ASM1 concept that $Y_{H,O2} = 0.67$ mg COD/mg COD, then $Y_{H,NO} = 0.54$ mg COD/mg COD. The obtained anoxic to aerobic growth yield ratio of 0.79 is very close to the literature data reviewed by the authors.

Based on the literature review, Hu *et al.* (2002b) concluded that the growth yield of heterotrophic microorganisms (*"ordinary"* heterotrophs and PAOs) under anoxic conditions should be reduced to 0.79 to 0.84 of that under aerobic conditions. For the typical aerobic yield value of 0.67 mg COD/mg COD, this gives an anoxic growth yield in the range 0.53–0.56 mg COD/mg COD.

3.2 DISAPPEARANCE (LOSS) OF BIOMASS AND CELL INTERNAL COMPONENTS

The amount of biomass actually formed per unit mass of substrate used in biochemical processes, referred to as the observed growth yield, Y_{obs}, is always less than the maximum (true) growth yield, Y. The loss of suspended organic matter in the result of a reduced growth yield can be attributed to several factors such as maintenance energy requirements, decay of cells, endogenous respiration, predation/grazing by higher animals or lysis due to adverse environmental conditions (e.g., pH, toxic substances or temperature) (Grady *et al.* 1999; van Loosdrecht and Henze 1999). Since these mechanisms cannot be distinguished macroscopically, they are thus lumped together under the term *"microbial decay"* (Grady *et al.* 1999). Under the term *"decay"*, van Loosdrecht and Henze (1999) classified all processes that reduce the number of microorganisms and/or the weight and specific activity of biomass. The authors assumed that the decay can be caused by cell external factors, e.g. predation (external decay) or mechanisms inside the microbial cell, such as death and self-oxidation of cell components (internal decay), whereas the term *"lysis"* can refer to solubilisation of biomass, causing the release of secondary substrates into the liquid phase.

The *"microbial decay"* has been modelled using three major concepts including maintenance, endogenous respiration and death-regeneration (Figure 3.1). Comprehensive overviews of these approaches for modelling the loss (disappearance) of biomass and other cell internal components were presented by

Grady *et al.* (1999) and van Loosdrecht and Henze (1999) and these overviews are summarized below.

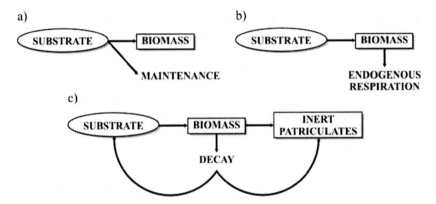

Figure 3.1 Schematic representation of three concepts for modelling the formation and disappearance of biomass (van Loosdrecht and Henze 1999).

Table 3.1 Expressions for the observed yield, "*loss*" rate coefficient and inert fraction from biomass "*loss*" according to different model concepts (adopted from van Loosdrecht and Henze (1999) and Grady *et al.* (1999))

Concept	Observed yield, Y_{obs}	"*Loss*" rate coefficient	Inert fraction from biomass "*loss*"
Maintenance	$Y \dfrac{1}{1+ SRTm_S Y}$		
Endogenous respiration	$Y \dfrac{1}{1+ SRTk_{end}}$	k_{end}	f_P
Death-regeneration	$Y \dfrac{1}{1+ SRTb\left[1-\left(1-f'_P\right)Y\right]}$	$b = \dfrac{k_{end}}{1-\left(1-f'_P\right)Y}$	$f'_P = \left(\dfrac{1-Y}{1-Yf_P}\right)f_P$

The three concepts presented in Figure 3.1 link together the maximum growth yield (Y), the observed growth yield (Y_{obs}), SRT, inert residue formed (f_P) and coefficients for maintenance (m_s), endogenous respiration (k_{end}) or decay (b). The resulting mathematical relationships for the observed growth yield (Y_{obs}), "*loss*" rate coefficient and inert fraction from biomass "*loss*" are presented in Table 3.1.

Maintenance can be understood as energy consumption to satisfy basic metabolic requirements, which is accompanied by the utilization of external primary substrate or storage polymers (PHA and glycogen). The process is modelled as a direct consumption of both types of substrates for maintenance of the cell integrity. In the modelling without structured biomass, the consumption of stored compounds is observed as a reduction in biomass weight, even though the biomass itself does not disappear. A review of cellular processes, both mechanical and chemical, that require energy was presented by Grady *et al.* (1999).

Endogenous respiration, also known as the traditional "*decay*" approach (Grady *et al.* 1999), is modelled as a one-step process of biomass loss coupled with a direct utilization of the electron acceptor (oxygen or nitrate) (van Loosdrecht and Henze 1999). Moreover, it is assumed that during the respiration of biomass inert residues (debris) are formed from a portion (f_P) of the disappearing biomass (Grady *et al.* 1999):

$$\text{Biomass(COD)} + \left[-(1-f_P)\right] O_2 \text{ equivalents of electron acceptor} \\ \rightarrow f_P \text{ inert residue (COD)} \tag{3.20}$$

The endogenous respiration concept was originally developed upon the observation that cell internal storage polymers are used for maintenance purposes when the external substrate is depleted (van Loosdrecht and Henze 1999). In the case of modelling with structured biomass including the storage polymers, the concepts of endogenous respiration of these compounds and maintenance are essentially described in the same way (see: Table 3.2). Grady *et al.* (1999) concluded that the main attribute of the concept is its simplicity, but it can not easily handle cases in which the conditions are changing in terms of the nature of the electron acceptor.

The death-regeneration concept, also known as lysis-regrowth approach (Grady *et al.* 1999), addresses this situation. The concept is more complicated than endogenous respiration and was originally proposed in the model of Dold *et al.* (1980) and adopted later in the series of ASMs (Henze *et al.* 1987; Henze *et al.* 1995a; Henze *et al.* 1999). In the first stage, it assumes a transition of viable microbial cells, which die or are inactivated, into particulate (slowly biodegradable) substrate without the utilization of the electron acceptor (Grady *et al.* 1999):

$$\text{Biomass(COD)} \rightarrow \left(1-f_P'\right) \text{particulate substrate (COD)} \\ + f_P' \text{ inert residue (COD)} \tag{3.21}$$

This transformation was called "*decay*" in ASM1 (Henze *et al*. 1987) or "*lysis*" in ASM2 (Henze *et al*. 1995a) and ASM2d (Henze *et al*. 1999), although both processes are equivalent in terms of the mathematical description. The concept of lysis was introduced to reflect more complex transformations of the PAO cell components (biomass, polyphosphate and PHA) in contrast to the single decay process of heterotrophic and autotrophic biomass included in ASM1. The process of decay (or lysis) is followed by growth (so called cryptic growth) on the secondary substrate generated during the decay/lysis.

Table 3.2 Stoichiometry and simplified kinetics of three concepts for modelling the loss of biomass and cell internal components in the absence of external substrate (van Loosdrecht and Henze 1999)

Process	S_S	S_O	X_{STO}	X_S	X_I	X_H	Simplified rate expression[1]
Maintenance concept							
Maintenance		−1	−1				$m_s\,X_H$
Endogenous respiration concept							
Endogenous respiration of cells		−(1−f_P)			f_P	−1	$k_{end}\,X_H$
Endogenous respiration of storage compounds		−1	−1				$k_{end}\,X_{STO}$
Decay (death-regeneration) concept							
Decay of cells				1−f_P	f_P	−1	$b\,X_H$
Decay of storage compounds			−1	1			$b\,X_H$
Hydrolysis	1			−1			$k_{hyd}\,X_S$

[1] Saturation (Monod) terms are neglected

The stoichiometry and simplified kinetics of the concepts for modelling the loss of biomass and cell internal components in the absence of external substrate are presented in Table 3.2. Both the "*traditional*" decay and death-regeneration concepts are modelled as being first order with respect to the active biomass concentration. However, the decay coefficient in the traditional approach is smaller than the coefficient in the death-regeneration concept, although the fraction of the biomass leading to the formation of inert solids is larger (Grady *et al*. 1999). A typical contribution of the decaying biomass to the inert residues is about 20% and thus the value of f_P is usually 0.20 in the traditional approach (Henze *et al*. 1987).

In the death-regeneration concept, however, the fraction of inert products formed has to be less than 0.2 during each passage through the synthesis-resolubilization cycle. If the observed inert fraction is assumed to be 20%, then the value of f'_p for this model should be around 0.08 (Table 3.1).

The heterotrophic decay rate coefficient (b_H) used for modelling range from 0.05 d^{-1} for domestic wastewater in the USA to highs of 1.6 d^{-1} for some food-processing wastes (Henze et al. 1987). These values are substantially higher than the decay rate coefficients for autotrophic nitrifying organisms (b_A) and PAOs, (b_{PAO}). For comparison, the reported values of b_A (at 20 °C) vary within the range 0.04–0.17 d^{-1} (Henze et al. 1987; Copp and Murphy 1995; Weijers and Vanrolleghem 1997). Based on the findings described by Melcer et al. (2003), it appears that the b_A is significantly higher than 0.04 d^{-1} and an appropriate value of b_A should remain in the range 0.15–0.17 d^{-1}. Applying too low values of b_A may result in a significant underestimate of $\mu_{A,max}$, even by a factor of two.

In general, the decay rate coefficient for heterotrophic bacteria ranges from 0.3 to 0.7 d^{-1}, whereas the values of this parameter for nitrifying bacteria and PAOs are in order of 0.15–0.2 d^{-1} (van Loosdrecht and Henze 1999). The authors provided two theories justifying this difference:

- The maintenance requirements are coupled to the Gibbs free energy of the catabolic reaction (the energy producing reaction) and depend upon the type of growth systems and electron acceptor conditions. According to this theory, nitrifying bacteria decay slower three times than "*ordinary*" heterotrophs, which is in accordance with the observations. This theory does not explain, however, why the rates for pure cultures are lower by a factor of 10 than those reported for activated sludge systems.
- The activity of protozoa contributes significantly to the external decay rate. Different growth patterns of autotrophs and PAOs, which grow in dense, clearly visible micro-colonies, compared to "*ordinary*" heterotrophs, which grow in a loose way or even in suspension, can easily lead to differences in observed overall decay rate due to differences in the predation rate of protozoa. The decay rate of heterotrophs in the absence of protozoa drops from 0.4 to 0.05 d^{-1}, which is similar to that of PAOs in enriched cultures. Since protozoa are not active under anaerobic/anoxic conditions, the difference between aerobic and anaerobic/anoxic decay rates can also be justified.

In the traditional approach, the heterotrophic decay rate, b_H, can be determined by measuring the respiration rate of endogenous sludge in a batch reactor with the nitrification inhibitor added (Henze et al. 1987). For this purpose, OURs are measured many times over a period of several days (Ekama et al. 1986). The "*traditional*" decay coefficient, b_H, is then calculated as the slope of a plot of the natural logarithm $\ln(OUR_t/OUR_{t=0})$ versus time (t). The coefficient obtained by this method differs from that one used in the death-regeneration concept, but the new decay coefficient b'_H can be calculated from b_H according to the relationship presented in Table 3.1.

The value of b_A can be determined along with the maximum specific growth rate constant for autotrophs, $\mu_{A,max}$ (see: Section 3.7). It should be emphasized that in contrast to the decay rate constants for "*ordinary*" heterotrophs, b_H, and PAOs, b_{PAO}, this parameter for autotrophic bacteria in the death-regeneration concept is numerically equivalent to the "*traditional*" decay rate constant. This follows from the fact that the recycling of organic matter through the synthesis-resolubilization cycle only occurs through the activity of the heterotrophic microorganisms and not the autotrophic ones (Grady *et al.* 1999).

Spanjers and Vanrollenghem (1995) proposed a method for the simultaneous determination of the "*traditional*" decay coefficients of heterotrophs and autotrophs. The method is based on the respirometric measurements after addition an adequate amount of readily biodegradable substrate (acetate) and ammonium to a batch reactor. The parameter combinations including $\mu_{H,max}$, $\mu_{A,max}$, X_H and X_A can be estimated by applying a model for heterotrophic and autotrophic degradation in the form of a simplified ASM1 (Henze *et al.* 1987) repeatedly over a period of several days. Assuming the yield coefficients (Y_H, Y_A) and maximum growth rates ($\mu_{H,max}$, $\mu_{A,max}$) are constant, the values of the parameter combinations only become a function of the active biomass concentrations (X_H, X_A). The decrease of these values is governed by the decay coefficients and the logarithm of the respective parameter combinations versus time is a straight line with slopes equal to b_H and b_A, respectively.

3.3 STORAGE OF SUBSTRATES

In activated sludge systems, the biomass grows under dynamic (unbalanced) conditions with successive periods of external substrate availability (feast periods) and absence of external substrate (starvation, famine periods). Under such conditions a storage response is usually established without any additional need for other external limitations, such as lack of nutrients or electron acceptors (Majone *et al.* 1999). Microorganisms which are capable of storing substrate internally consume it in a more balanced way that is independent of the external substrate availability. Such microorganisms have a strong competitive advantage over microorganisms without this capability, which would continuously undergo rapid growth and starvation periods (van Loosdrecht *et al.* 1997). Hanada *et al.* (2002) suggested that both groups of microorganisms, with and without the capability of substrate storage, can exist in reality and the storage rate constant (k_{STO}) may not be a universal parameter and can change from case to case. The authors observed a high variability of the content of microorganisms with the storage capability of sludges originating from several full-scale and laboratory scale activated sludge systems operated under various conditions. The content varied within the range 15-50% of mixed liquor suspended solids (MLSS) and the measured production rate of a storage polymer (poly-β-hydroxyalkanoate (PHA)) was 3.9-75.2 mg C/(g MLSS•h). However, there is no evidence that storage is attributed to just a few particular groups of microorganisms and both filamentous and floc-forming bacteria have been reported to be able to store carbon sources (Majone *et al.* 1999). Although the storage mechanism is difficult to study and many different approaches are possible, the recognition of the importance of storage polymers has stimulated research on the COD conversion to biomass growth accounting also for storage.

Several types of organic storage polymers have been reported in literature. The most common ones are PHA and polysaccharides (Majone *et al.* 1999). In particular, poly-β-hydroxybutyrate (PHB) and glycogen are usually stored when the system is fed with acetate and glucose, respectively (Karahan-Gul *et al.* 2002a). The conceptual models of PHB and glycogen storage is presented in Figure 3.2. PHB metabolism of a pure culture (*Paracoccus pantotrophus*) was described by van Aalst-van Leeuwen *et al.* (1997) and the stoichiometry and kinetics of activated sludge cultures were studied by Beun *et al.* (2000a, 2000b) and Dircks *et al.* (2001a). PHB is probably the most dominant polymer as it is directly formed out of the central metabolite Acetyl-CoA (van Loosdrecht *et al.* 1997). Indeed, Beun *et al.* (2000a, 2000b), Carucci *et al.* (2001) and Dionisi *et al.* (2001) estimated that a prevailing part of acetate consumed was used for storage of PHB. PHA can also be generated from several different substrates including glucose (Majone *et al.* 1999) and ethanol (Beccari *et al.* 2002). Glycogen is implicitly formed only when sugars are present in the influent, but it also plays an essential role in the metabolism of PAOs and glycogen accumulating organisms (GAOs) (van Loosdrecht *et al.* 1997). Dircks *et al.* (2001b) studied the stoichiometry and kinetics of glycogen metabolism in mixed cultures. The possibility of other storage polymers (e.g. lipids) should also be considered, especially when using an undefined medium (Majone *et al.* 1999). The role of lipids is unclear, however, since only a few reports are available without explanation how a distinction was made between cell/membrane lipids and stored lipids (van Loosdrecht *et al.* 1997). The identification and role of storage compounds other than glycogen and PHA in the activated sludge processes still remains an important topic for future research (Dircks *et al.* 2001b).

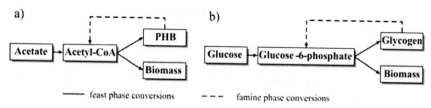

Figure 3.2 Metabolic pathways for storage of PHB (a) and glycogen (b) (Karahan-Gul *et al.* 2003).

There is still little knowledge about storage mechanisms in activated sludge cultures fed with real wastewater, in which the soluble fraction can contain several groups of compounds (see: Section 2.3) and thus the true nature of storage may be difficult to determine and quantify by only direct measurements of intracellular (internal) storage compounds (Goel *et al.* 1999). Moreover, the different COD removal phenomena should also be considered including adsorption, accumulation, oxygen consumption for maintenance or storage unidentified compounds (Majone *et al.* 1999; Carucci *et al.* 2001, Beccari *et al.* 2002). Carucci *et al.* (2001) demonstrated that PHB formation occurred during the high-rate RBCOD removal from wastewater, however, it accounted for only a fraction (18-22%) of the overall COD removal. Beccari *et al.* (2002) observed a low amount of PHB formed from raw and filtered wastewater, which was probably due to a low concentration of acetate, and the sum of storage and growth (estimated directly through ammonia consumption) did not match the overall solids formation. The same authors also studied the effect of other carbon sources on aerobic storage by activated sludge. PHB was stored when the substrate was acetate or ethanol, whereas negligible amounts of the storage compound were detected during the test with glutamic acid. In the study of Dionisi *et al.* (2004), the dynamic storage response of biomass in a SBR was described as a function of single or simultaneous feed of several substrates (acetate, glucose, glutamic acid and ethanol). The removal of every single substrate was affected (positively or negatively) by the presence of the others, demonstrating that the substrates can be also used to some extent by the same metabolism. The only exception was acetate, which removal rate was not affected by the presence of other substrates.

An overview of specific substrate uptake rates (q_{UPT}), specific PHB production rates (q_{PHB}) and the q_{PHB}/q_{UPT} ratios under aerobic conditions was presented by Makinia (2006). These data reveal a wide range of specific PHB production rates, i.e. from negligible up to 0.632 mg COD/(mg COD•h), even if acetate was used as a substrate. The q_{PHB}/q_{UPT} ratios are also highly variable, but two relatively narrow ranges (0.68-0.73 and 0.45-0.56 mg COD/mg COD) are associated with the acetate metabolism. The theoretical thermodynamic yield was reported to be 0.73 mg COD/mg COD for PHB storage from acetate (van Aalst-van Leeuwen *et al.* 1997), which is similar to the first range. The difference for the second range can result from different storage capabilities of the examined sludges or different approaches to the measurement of OUR. In general, the OUR response from activated sludge to the addition of a single organic substrate can be divided into two phases. The first phase reflects the primary metabolism of the added (exogenous) substrate, whereas the second phase implicitly originates from the utilization of stored polymers. Dircks *et al.* (1999) determined the yield coefficient for the activated sludge from two Danish WWTPs using the substrate and oxygen consumption. The yields with acetate as a substrate in the first phase were 0.71–0.72 mg COD/mg COD. Considering both phases together, the yields were substantially lower (0.48–0.52 mg COD/mg COD). Higher rates were obtained with glucose as a substrate, i.e. 0.79–0.91 mg COD/mg COD for the first phase and 0.71–0.82 mg COD/mg COD while considering both phases. For this substrate, the yields reported in the literature varied within the range 0.76–0.90 mg COD/mg COD (Majone *et al.* 1999). For comparison, the theoretical yield for the formation of glycogen from glucose is 0.96 mg COD/mg COD (van Loosdrecht *et al.* 1997). The third substrate examined by Dircks *et al.* (1999) was ethanol for which the yield values of 0.66-0.67 and 0.55 mg COD/mg COD were determined for the first phase and both phases, respectively.

In comparison with the storage under aerobic conditions, the storage of internal polymers under anoxic conditions has received less attention so far. Majone *et al.* (1998) compared the acetate uptake rate and PHB storage rate under different electron acceptor situations. Both anoxic process rates ($q_{UPT,ANOX}$ = 0.092 mg COD/(mg COD$\cdot$h), $q_{PHB,ANOX}$ = 0.037 mg COD/(mg COD$\cdot$h)) were considerably lower in comparison with the aerobic ones reported in the literature (Makinia 2006). Moreover, the storage yield was also reduced by approximately 40%. Reduced acetate uptake rates, 3-4 times and 2 times, respectively, were also observed by Beun *et al.* (2000b) and Dionisi *et al.* (2004). In the latter case, the authors observed that substrate removal rates were significantly affected by the electron acceptor (nitrate or oxygen) only for acetate, of which removal rate almost doubled under aerobic conditions. The other substrates (glucose, glutamic acid and ethanol) were simultaneously removed at a significantly greater nitrate removal rate than when single substrates were present, which revealed that the simultaneous removal was partially due to independent metabolic activities. On the other hand, the removal of every substrate was affected (positively or negatively) by the presence of the others, demonstrating that the substrates can be also used by the same metabolism.

Table 3.3 Summary of kinetic models of storage and growth on stored products (Majone et al. 1999)

Substrate	"Storage" related phenomena	STORAGE PHASE – Specific rate of storage as function of the concentration of: Substrate	Stored product	Simultaneous growth on the substrate	Energy need (COD/COD)	CONSUMPTION PHASE – Specific rate of consumption of the stored product as function of:	Maintenance or decay on X_{STO}
soluble and particulate	adsorption storage		$f_{max,STO}\dfrac{S_S}{K_S+S_S}-\dfrac{X_{STO}}{X}\times X$	no	no	$\dfrac{X_{STO}}{K_{STO}+X_{STO}}$	no
particulate biodegradable	adsorption storage	S_S	$f_{max,STO}-\dfrac{X_{STO}}{X}$	no	0.08	$\dfrac{X_{STO}/X_H}{K_{STO}+X_{STO}/X_H}$	yes
unspecified	sorption storage	S_S	$1-\dfrac{X_{STO}/X_H}{f_{max,STO}}$	no	no	(X_{STO}/X_H)	yes
soluble	biosorption	S_S	$f_{max,STO}-\dfrac{X_{STO}}{X_H}$	no	no	$\dfrac{X_{STO}}{K_{STO}+X_{STO}}$	no
soluble	accumulation storage		$\dfrac{K_{I,STO}}{K_{I,STO}+X_{STO}/X_H}$	no	0.10	$\dfrac{X_{STO}/X_H}{K_{STO}+X_{STO}/X_H}$	no
acetate	storage		$1-\dfrac{X_{STO}/X_H}{f_{max,STO}}$	yes	0.27	$(X_{STO}/X_H)-f_{max,STO}$	yes

Possible conceptual approaches to storage modelling were discussed by Majone *et al.* (1999) and this discussion is summarized below. In general, the storage of organic polymers in the activated sludge systems can be modelled using unstructured and structured models. In unstructured models, the biomass is considered as a black box and the processes occurring inside the black box are ignored. This means that the active biomass and storage products are not distinguished and changes in the internal composition of the biomass are not considered. An overall yield coefficient will lump two phenomena, i.e. storage and growth, even though the actual yields for these phenomena are different. Consumption of storage products will result in a higher rate of the endogenous metabolism. Therefore, the impact of storage can be implicitly modelled by considering a higher observed yield coefficient and a higher decay rate constant. An unstructured model allows for a good description of activated sludge systems, in which the storage does not play a significant role, provided that suitable values for these two parameters are assumed. An example of such a system is a completely mixed configuration treating wastewater with a low content of soluble COD. A well-known example of the unstructured model is ASM1 (Henze *et al.* 1987).

In structured models, changes in the internal composition of microorganism biomass are considered and the activity of microorganisms is related to that composition. Structured models are especially needed in two cases: (i) when the biomass composition changes significantly or (ii) when a specific component of the biomass is to be modelled. Even though a large number of components can be attributed to the biomass, only most relevant variables for the physiological state of the biomass should be considered as the structured approach to the biomass composition increases the number of expressions and parameters to be estimated. For the modelling purposes, usually simplified compartmental models incorporating two or three variables are defined. The simplest approach to modelling storage of internal substrate is to divide the biomass into two fractions: active cells and stored substrate. Compartmental structured models are commonly used in relation to EBPR (see: Section 3.9). Several models have also been available for aerobic storage, revealing a wide range of stoichiometric and kinetic expressions (Table 3.3). In all these models, inhibition (or saturation) terms are introduced in order to decrease the storage rate at the increasing content of the stored substrate. Most of the models introduce no or little energy requirement for the "*storage*" phase, which is in contradiction to the results of experimental studies revealing a significant energy requirement for this process. Some of these models assume that storage is the first step followed by growth on the stored substrate, whereas the other models assume that parallel storage and growth are allowed to occur simultaneously.

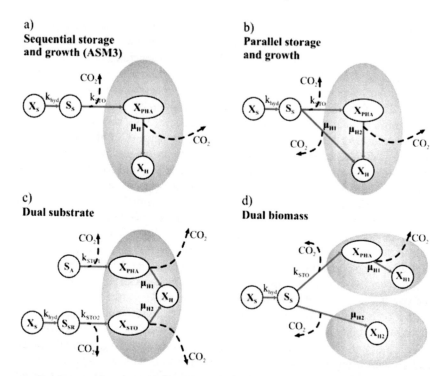

Figure 3.3 Schematic representation of the substrate flux in various models considering the storage mechanisms: (a) sequential storage and growth, (b) parallel storage and growth, (c) dual substrate, (d) dual biomass (Makinia *et al.* 2006a).

A well-known example of the model assuming simultaneous storage and growth on the storage polymers is ASM3 (Gujer *et al.* 1999). This concept, shown schematically in Figure 3.3a, is not valid mechanistically due to experimental evidence that microorganisms utilize the stored polymers as a carbon and energy source only after depletion of primary substrate (Karahan-Gul *et al.* 2003). A more consistent description of substrate conversion could be provided if growth on primary substrate (S_S) in the feast period is additionally incorporated in the model (Figure 3.3b). This approach was originally proposed by Krishna and van Loosdrecht (1999) and used also later by Winkler *et al.* (2001), Beccari *et al.* (2002), Karahan-Gul *et al.* (2003), Sin *et al.* (2005b), Makinia *et al.* (2006a), and Ni and Yu (2007). Karahan-Gul *et al.* (2003) introduced a switching function $K_S/(K_S+S_S)$ for the growth rate on the stored polymers in order to prevent this growth before the growth on the primary substrate is depleted. Carucci *et al.* (2001) developed a dual substrate model (Figure 3.3c) considering two fractions of readily biodegradable COD: acetate (S_A) and all other readily biodegradable substrates (S_{SR}). Accordingly, two storage compounds were considered: PHA (X_{PHA}) generated only from acetate and a generic storage compound (X_{STO}), which can also include PHA not coming from acetate, generated from S_{SR}. The slowly biodegradable COD (X_S) was assumed to be hydrolyzed only to S_{SR}. A single group of heterotrophic biomass (X_H) grows simultaneously on both storage products (X_{PHA} and X_{STO}). In a dual biomass model of Hanada et al. (2002), shown in Figure 3.3d, the heterotrophic biomass is divided into two fractions: the one with a storage capability (X_{H1}) and another one without a storage capability (X_{H2}). The X_{H1} and X_{H2} fractions grow according to the ASM3 concept and ASM1 concept, respectively.

Karahan-Gul et al. (2003) observed that the sequential storage and growth concept used in ASM3 predicted higher storage of the internal compounds, lower ammonia consumption for growth and thus lower biomass production. Due to the single growth process definition, this approach also failed to simulate high rates of oxygen uptake in the first (feast) phase and the lower rates in the second (famine) phase (Avcioglu et al. 2003; Karahan-Gul et al. 2003). When the parallel storage and growth concept was considered, the proposed model gave a better description of the experimental data in terms of oxygen uptake, glycogen generation and biomass production (Karahan-Gul et al. 2003). Beccari et al. (2002) also compared these two concepts (sequential vs. parallel storage and growth) using experimental data from batch tests with synthetic substrates. The authors observed that both concepts could not cope with predicting all of the measured parameters. The ASM3 could well describe the experimental data (substrate, ammonia and OUR profiles) only assuming a stored product formation much higher than the one detected analytically. On the other hand, that discrepancy could not be recovered by simply adopting parallel storage and growth. The model incorporating this concept could well describe the observed stored product profile only assuming a direct contribution of growth much higher than estimated from ammonia consumption.

Experimental procedures for the determination of the stoichiometric and kinetic coefficients associated with the storage process are consistent and limited only with the mechanistic model and experimental conditions adopted for this purpose (Karahan-Gul et al. 2002b). Most of the studies have been carried out for the sequential storage and growth conceptual model used in ASM3 (Table 3.4). The same table also contains the values of k_{STO} and Y_{STO} for the parallel storage and growth model examined along with ASM3. Carucci et al. (2001) calibrated their dual substrate model by adjusting the rate constant of PHB storage, k_{PHB}, to approximately 0.5 d^{-1} for the tests with both filtered and raw wastewater. In comparison with k_{PHB}, the storage rate constant of a generic storage compound,

k_{STO}, increased 4 times for the filtered wastewater and 5 times for the raw wastewater. The storage yields, Y_{PHB} and Y_{STO}, were set to 0.79 g COD/g COD for both types of the substrate. Hanada *et al.* (2002) did not provide numerical values of the kinetic and stoichiometric coefficients in their dual biomass model. The authors only indicated that the k_{STO} and K_{STO} (saturation coefficient for storage of S_S by X_{H1}) should be high, whereas μ_{H2} and K_{S2} (saturation coefficient for growth of X_{H2} on S_S) should be low.

Table 3.4 Examples of Y_{STO} and k_{STO} values in the models incorporating sequential and parallel storage and growth

Reference	Substrate	Y_{STO}, g COD/g COD		k_{STO}, d^{-1}	
		Sequential	Parallel	Sequential	Parallel
Krishna and van Loosdrecht (1999)	acetate	0.73	0.73	10.0	10.0
Koch *et al.* (2000b)	acetate	0.72		5.0	
	soluble COD from primary sludge acidification	0.80		11.0	
	raw wastewater	0.80		13.0	
Beccari *et al.* (2002)	acetate	0.83	0.85	0.72	0.31
	ethanol	0.83	0.75	0.34	0.065
	glutamic acid	0.47	0.55	0.26	0.007
	raw wastewater	0.88	0.70	1.06	0.003
	filtered wastewater	0.54	0.55	0.79	0.003
Avcioglu *et al.* (2003)	acetate	0.75–0.80		10.0–16.0	
Karahan-Gul *et al.* (2003)	acetate	0.80	0.80	16.0	14.0
	glucose	0.88	0.90	12.0	8.0
Sin *et al.* (2005a)	acetate		0.83		0.4
	acetate		0.91		3.3

3.4 ADSORPTION OF SUBSTRATES

Adsorption is a simple physical-chemical interaction that can contribute to fast removal of the soluble substrate from the liquid phase in a similar way to enmeshment (i.e. instantaneous physical entrapment) of the particulate substrate (Majone *et al.* 1999). In the first phase, microorganism cells get into contact with external (exogenous) substrate, whereas in the second phase, the substrate is adsorbed on the surface of cells (Phan and Rosenwinkel 2004). The adsorbed substrate can further be used for growth or/and storage. Majone *et al.* (1999) suggested that the substrate can also be removed quickly by accumulation, i.e. the substrate is transported into the cell and maintained there in almost unchanged form or transformed into low-MW metabolic intermediates. However, due to thermodynamics reasons (unfavourable gradients, osmotic pressure) this mechanism occurs to a much more limited extent than the storage where a molar concentration of substrate or intermediates is reduced by polymerization.

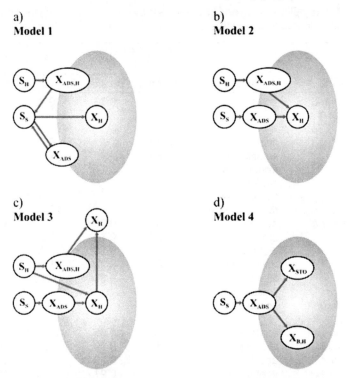

Figure 3.4 Mass flow path diagram of carbonaceous substrate conversion in the biokinetic models incorporating adsorption (Models 1-3 adopted from Novak *et al.* (1995) and Model 4 adopted from Beccari *et al.* (2002)).

Conceptual models considering adsorption as an intermediate step in the carbonaceous substrate conversion are presented in Figure 3.4. Models 1–3 and Model 4 were proposed by Novak *et al.* (1995) and Beccari *et al.* (2002), respectively. Novak *et al.* (1995) developed their models to investigate interactions between the biokinetic model structure and population dynamics of microorganisms in the activated sludge. Two different concepts (Models 1–2) favouring and disfavouring the growth of filamentous microorganisms on readily biodegradable substrates were considered and the final model (Model 3) constituted a compromise between the two initial approaches. Beccari *et al.* (2002) introduced the adsorption (or/and internal accumulation) step to account for "*lacking COD*" in the COD balance, i.e. not all substrate depletion was recovered into analytically detected components. The authors emphasized that the presence of such additional phenomena still require experimental confirmation. The brief characteristics of these models are presented below:

- **(Model 1)** Two soluble substrates, i.e. readily biodegradable, S_S, and rapidly hydrolysable, S_H, are adsorbed and converted to the adsorbed substrates, X_{ADS} and $X_{ADS,H}$, respectively. In the result of reverse mechanisms, the substrates X_{ADS} and $X_{ADS,H}$ are converted to S_S which diffuses back to the liquid phase before it is utilized. These mechanisms include desorption of X_{ADS} and hydrolysis (or solubilisation) of $X_{ADS,H}$. The molecules of X_{ADS} are supposed to be very simple and do not need to be hydrolyzed prior to their utilization by heterotrophic biomass. On the other hand, the $X_{ADS,H}$ molecules contain larger carbon chains that require hydrolysis for their conversion to S_S. The S_S substrate is directly utilized for the growth of heterotrophic biomass. This approach favours filamentous growth on hydrolysis products.
- **(Model 2)** Two soluble substrates, S_S and S_H, are adsorbed and converted to the adsorbed substrates, X_{ADS} and $X_{ADS,H}$, respectively. However, no reverse mechanisms to this process are incorporated in this model. The adsorbed substrates are utilized for the growth of heterotrophic biomass. Although formation of an intermediate readily biodegradable substrate from hydrolysis is considered, it does not appear in the liquid phase and remains on the floc surface. This approach disfavours the filamentous growth on hydrolysis products.
- **(Model 3)** The S_S substrate utilization is carried out through adsorption and subsequent utilization of the adsorbed substrate ($X_{ADS,S}$). The S_H substrate is utilized directly for the growth of heterotrophic biomass or it is adsorbed and subsequently converted to secondary hydrolysis soluble products (S_R) prior to the utilization. This intermediate component is introduced to allow for accumulation of soluble substrate (in the form of S_R) in the liquid phase when no electron acceptor is available. It is assumed that the S_H hydrolysis ceases under such conditions, whereas the $X_{ADS,H}$ hydrolysis continues at a reduced rate.

- **(Model 4)** The soluble substrate, S_S, is always removed at the first step by simple adsorption or internal accumulation. Then the adsorbed/ accumulated substrate, X_{ADS}, can be used for the growth of heterotrophic biomass either directly or through intermediate storage.

The adsorption kinetics can be mathematically expressed by the equation of modified Langmuir isotherm (Ekama and Marais 1979), which relates COD in the liquid phase with COD adsorbed in the sludge flocs by the heterotrophic biomass. This equation was used by Novak *et al.* (1995) for modeling adsorption of two fractions of soluble substrate (S_S and S_H), whereas Beccari *et al.* (2002) modified this equation by using a Monod term $S_S/(K_S+S_S)$ instead of S_S (Table 3.5). In terms of stoichiometry, the models of Novak *et al.* (1995) are represented by a simple transformation of the soluble substrates (S_S or S_H) to the adsorbed substrate, X_{ADS}, whereas the model of Beccari *et al.* (2002) assumes a demand for small amount of energy ($Y_{ADS} = 0.95$ mg COD/mg COD). The maximum fraction of S_S adsorbed, $f_{max,ADS}$, is similar in both models, i.e. 0.25 (Novak *et al.* 1995) and 0.2 (Beccari *et al.* 2002), but the maximum adsorption rate constant (k_{ADS}) cannot be compared directly due to different dimensions of this parameter in both models. Novak *et al.* (1995) introduced two different values of k_{ADS}, i.e. 0.6 and 0.1 m³/(g COD·d), for S_S and S_H, respectively. Beccari *et al.* (2002) estimated the k_{ADS} values within the range 0.26–0.84 g COD/(g COD·d) depending on the substrate used in the experiment (acetate, ethanol, glutamic acid, raw wastewater, filtered wastewater) with the highest value for the raw wastewater. At high concentrations of S_S (>10 mg COD/dm³), the adsorption process occurs much faster in the model of Novak *et al.* (1995) with comparison to the model of Beccari *et al.* (2002). The process rates predicted by both models become comparable only when the soluble substrate is almost depleted, i.e. the S_S concentration is in order of 1 mg COD/dm³.

Table 3.5 Stoichiometry and kinetics of the adsorption process

Stoichiometry				
S_O	S_S[1]	X_{ADS}[2]	Process rate	Reference
	−1	1	$k_{ADS} S_S X_H \left(f_{max,ADS} - \dfrac{X_{ADS}}{X_H} \right)$	Ekama and Marais (1979), Novak *et al.* (1995)
−(1−Y_{ADS})	−1	Y_{ADS}	$k_{ADS} \dfrac{S_S}{K_S + S_S} X_H \left(f_{max,ADS} - \dfrac{X_{ADS}}{X_H} \right)$	Beccari *et al.* (2002)

[1] The same kinetics and stoichiometry for S_H in the model of Novak *et al.* (1995)
[2] The same kinetics and stoichiometry for $X_{ADS,H}$ in the model of Novak *et al.* (1995)

In several simulation studies, the adsorption process has been reported without providing details on the process kinetics and stoichiometry. Ginestet *et al.* (2002) claimed that the addition of adsorption kinetics as proposed by Ekama and Marais (1979) to the surface limitation model (used in ASM1) led to better simulation results of oxygen uptake than the other hydrolysis models (pure surface limitation or first order). Larrea *et al.* (2002) introduced adsorption in a proposed model concept for carbonaceous substrate removal in high rate processes. The model assumed that a colloidal filterable fraction of X_S, X_{SF}, undergoes rapid adsorption on the floc and so is immediately added to non-filterable fraction of X_S.

3.5 HYDROLYSIS OF SLOWLY BIODEGRADABLE ORGANIC COMPOUNDS

The strict definition of hydrolysis is the breakdown of a polymer into smaller units by addition of water (Brock and Madigan 1991). In wastewater treatment modelling, however, the term "*hydrolysis*" is understood as a conversion of slowly biodegradable substrate into the readily biodegradable form (Henze *et al.* 1987). Therefore, hydrolysis summarizes all mechanisms that refer to the breakdown of complex organic compounds by means of extracellular enzymes and their conversion into smaller products that can subsequently be taken up and degraded by microorganisms. Because identification and description of these reactions in wastewater is impossible, they are conveniently collected into an overall single hydrolysis mechanism (Insel *et al.* 2002). A hydrolysis step is adopted to reflect the cumulative effects of other complex reactions ranging from physical entrapment and adsorption to storage prior to biochemical oxidation and synthesis (Orhon *et al.* 1999). With this broader definition of hydrolysis, other processes such as chemical dissolution and mass transport processes also have to be considered when evaluating hydrolysis rates (Morgenroth *et al.* 2002b). The hydrolysis process is slower than heterotrophic growth and it usually becomes the rate-limiting step for the biodegradation of organic matter. Moreover, hydrolysis also plays an important role in balancing the donor/acceptor ratio for denitrification and anaerobic phosphate release (Insel *et al.* 2003).

The hydrolysis processes can be grouped into two categories (Morgenroth *et al.* 2002b):

- Hydrolysis of primary substrate where organic substrate present in the original wastewater is broken down;
- Hydrolysis of secondary substrate that refers to the breakdown of substrate that has been produced by the bacteria (e.g. hydrolysis of internal storage products, of substances released by the bacteria during normal metabolism, or of particles produced during decay of bacteria).

Only the first issue is addressed in this section and more information about the hydrolysis of secondary substrate can be found in Section 3.2 and Section 3.3. Morgenroth *et al.* (2002b) presented a comprehensive review of hydrolysis of primary substrate including its mechanisms, broad range of mathematical models to describe this process and experimental techniques to determine mechanisms of hydrolysis and rate constants. In terms of the stochiometry, an overview of different approaches is presented in Table 3.6 and the main conclusions can be summarized as follows (Morgenroth *et al.* 2002b):

- In early models, hydrolysis was not considered as a separate process, but it was assumed that one group of microorganisms grows directly on both readily and slowly biodegradable substrate (Model 1);
- Ekama and Marais (1979) proposed in their model (Model 2) a two biomass system, in which slowly biodegradable organic compounds are adsorbed and two groups of microorganisms grow separately on soluble (readily biodegradable) or adsorbed (slowly biodegradable) substrate, respectively. This hypothesis was later supported by some experimental evidences;
- The Activated Sludge Models developed by the IWA (formerly IAWPRC and IAWQ) task groups (Henze *et al.* 2000), were based on the one-step hydrolysis approach (Model 3), in which slowly biodegradable organic compounds are converted into readily biodegradable substrate;
- More complicated hydrolysis stoichiometries incorporate multiple slowly biodegradable fractions in parallel hydrolysis (Model 4) or sequential hydrolysis (Model 5). The parallel hydrolysis provides greater flexibility in model calibration by considering the utilization of different fractions independent of each other, whereas the sequential hydrolysis should be considered when the accumulation of intermediate hydrolysis products is essential;
- In all of the reviewed model stoichiometries, hydrolysis involves no energy utilization and thus there is no utilization of electron acceptor associated with this process;
- The current understanding of hydrolysis is insufficient to judge which of the proposed model stoichiometries is most appropriate. Hence, a model structure should be selected based on the information concerning the required detail of simulation and the availability of data for model calibration.

Morgenroth *et al.* (2002b) also provided an overview of the kinetic rate expressions for hydrolysis (Table 3.7), which turned out to be largely independent of the model stoichiometry. Simple expressions have been proposed as zero order, first order or saturation type kinetics with a Monod term (Models 0–IV), whereas more complex expressions incorporate a surface-limited term (Model V) or multiple surface-limited and Monod terms with correction factors for different electron acceptor conditions (Model VI). Under specific conditions (very high or very low substrate to microorganism ratios), the saturation or surface-limited rate expressions can be approximated with the simpler models. Orhon *et al.* (1999) evaluated the applicability of Models I, IV, V to either municipal wastewater and a number of industrial wastewaters. The authors concluded that the surface-limited rate expression (Model V) proved to be the appropriate model for the description of hydrolysis of slowly biodegradable COD in both types of wastewater. Mino *et al.* (1995b) also observed that the hydrolysis of artificial substrate (starch) in the activated sludge process followed the surface-limited reaction kinetics. Moreover, since the slowly biodegradable COD fraction is composed of organic matter of different nature and complexity, some researchers (e.g. Sollfrank and Gujer, 1991; Orhon *et al.* 1999) argued that it might be difficult and sometimes misleading to characterize the entire fraction by a single hydrolysis rate. They observed that a dual (parallel) hydrolysis mechanism, which was described according to a simplified Model 4 in Table 3.6 for rapidly hydrolysable COD (S_H) and slowly hydrolysable COD (X_S), reduced the discrepancy between measured and predicted oxygen uptake rate (OUR) profiles with comparison to a single hydrolysis model.

In general, the accepted procedure for the experimental assessment of k_{hyd} and K_X under aerobic conditions involves model-based evaluation and curve fitting of OUR profiles in laboratory scale batch or semi-continuous reactors (Insel *et al.* 2003). Such an experimental approach have been extensively discussed in literature. For the estimation of parameters describing the surface saturation type hydrolysis, Ekama *et al.* (1986) proposed to use OUR measurements in a completely mixed reactor fed continuously under a daily cyclic square wave feeding pattern (Figure 3.5). In the second phase, after an immediate depletion of the readily biodegradable substrate, the accumulated slowly biodegradable substrate continues to be used at the same rate for a time period. For the batch experiments, the authors recommended that the value of k_{hyd} could be estimated at a high X_S/X_H ratio or performing simulations for curve fitting of OUR profiles. Sollfrank and Gujer (1991) determined the first order hydrolysis rate constant from the slope of the OUR_H plot vs. the concentration of degradable matter in a batch experiment. For a time $t > t_1$ (t_1 denotes the time when a nearly linear relation is reached), it was assumed that hydrolysis was the limiting process in the degradation of filtered wastewater. The rate constant was determined from the slope after the net respiration rate becomes proportional to the concentration of degradable matter by the following equation:

$$OUR_{H,net}(t \geq t_1) = (1 - Y_H) k_{hyd} X_S(t) \qquad (3.22)$$

Table 3.6 Review of hydrolysis and growth stoichiometries (Morgenroth *et al.* 2002b)

Processes	$X_{S,1}$	$X_{S,2}$	$X_{S,3}$	$X_{S,ADS}$	S_S	S_O	$X_{H,1}$	$X_{H,2}$
Model 1: Direct growth on both soluble and particulate organic matter								
Growth on slowly biodegradable COD ($X_{S,1}$)	$\dfrac{-1}{Y_H}$					$\dfrac{-(1-Y_H)}{Y_H}$	1	
Growth on readily biodegradable COD (S_S)					$\dfrac{-1}{Y_H}$	$\dfrac{-(1-Y_H)}{Y_H}$	1	
Model 2: Two biomass system								
Adsorption of hydrolysable COD ($X_{S,1}$)	-1			1				
Direct growth on adsorbed COD ($X_{S,ADS}$)				$\dfrac{-1}{Y_H}$		$\dfrac{-(1-Y_H)}{Y_H}$	1	
Growth on readily biodegradable COD (S_S)					$\dfrac{-1}{Y_H}$	$\dfrac{-(1-Y_H)}{Y_H}$		1
Model 3: One step hydrolysis								
Hydrolysis of hydrolysable COD ($X_{S,1}$)	-1				1			
Growth on readily biodegradable COD (S_S)					$\dfrac{-1}{Y_H}$	$\dfrac{-(1-Y_H)}{Y_H}$		

Table 3.6 (cont.) Review of hydrolysis and growth stoichiometries (Morgenroth et al. 2002b)

Processes	$X_{S,1}$	$X_{S,2}$	$X_{S,3}$	$X_{S,ADS}$	S_S	S_O	$X_{H,1}$	$X_{H,2}$
Model 4: Parallel hydrolysis								
Hydrolysis of slowly hydrolysable COD ($X_{S,1}$)	-1				1			
Hydrolysis of intermediate hydrolysable COD ($X_{S,2}$)		-1			1			
Hydrolysis of rapidly hydrolysable COD ($X_{S,3}$)			-1		1			
Growth on readily biodegradable COD (S_S)					$\dfrac{-1}{Y_H}$	$\dfrac{-(1-Y_H)}{Y_H}$	1	
Model 5: Sequential hydrolysis								
Hydrolysis of slowly hydrolysable COD ($X_{S,1}$)	-1	1						
Hydrolysis of intermediate hydrolysable COD ($X_{S,2}$)		-1	1					
Hydrolysis of rapidly hydrolysable COD ($X_{S,3}$)			-1		1			
Growth on readily biodegradable COD (S_S)					$\dfrac{-1}{Y_H}$	$\dfrac{-(1-Y_H)}{Y_H}$	1	

where

k_{hyd} - Specific hydrolysis rate constant, T^{-1}

$OUR_{H,net}$ - Net heterotrophic oxygen uptake rate, $M(O_2)L^{-3}T^{-1}$

X_S - Concentration of slowly biodegradable substrates, $M(COD)L^{-3}$

Table 3.7 Review of kinetic rate expressions for hydrolysis (Morgenroth *et al.* 2002b)

No.	Type of expression	Process rate
0	Zero order	$k_{hyd,0}$
I	First order with respect to X_S	$k_{hyd,I} X_S$
II	First order with respect to X_H	$k_{hyd,II} X_H$
III	First order with respect to X_S and X_H	$k_{hyd,III} X_S X_H$
IV	First order with respect to X_H with a Monod term (saturation type kinetics) with respect to X_S	$k_{hyd,IV} \dfrac{X_S}{K_{X,IV} + X_S} X_H$
V	First order with respect to X_H with a surface-limited term for the entrapment of X_S	$k_{hyd,V} \dfrac{X_S/X_H}{K_{X,V} + X_S/X_H} X_H$
VI	First order with respect to X_H with multiple surface-limited and Monod terms with correction factors for different electron acceptor conditions	$k_{hyd,VI} \left(\dfrac{X_S/X_H}{K_{X,VI} + X_S/X_H} \dfrac{S_O}{K_{O,hyd} + S_O} X_H \right.$ $\left. + \eta_{NO3,hyd} \dfrac{X_S/X_H}{K_{X,VI} + X_S/X_H} \dfrac{K_{O,hyd}}{K_{O,hyd} + S_O} X_H \right)$

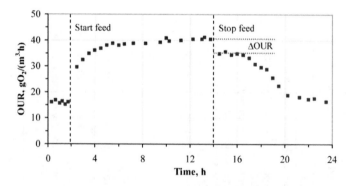

Figure 3.5 OUR measurements in a completely mixed reactor fed continuously under a daily cyclic square wave feeding pattern (Ekama *et al.* 1986).

Kappeler and Gujer (1992) suggested iterative curve fitting of the amount of slowly biodegradable substrate and the first order hydrolysis rate constant. Spanjers and Vanrolleghem (1995) also estimated the amount of slowly biodegradable organic matter from respiration rate measurements. However, they divided the slowly biodegradable organic matter into two fractions and curve fitting (so called "*optimization*" technique) allowed to estimate only the rapidly hydrolysable organic fraction, for which the process kinetics was described by the first order expression (Model I in Table 3.7). In addition, batch respirometric tests were proposed by adjusting the appropriate wastewater/biomass mixtures with different (either high or low) initial ratios between wastewater and biomass (S/X_V) (Orhon et al. 1999; Sperandio and Paul 2000). Insel et al. (2002, 2003) demonstrated that the model-based evaluation and curve fitting allows to generate not a unique set, but a relatively large combinations of different pairs of k_{hyd} and K_X coefficients equally applicable to the experimental data. Mogenroth et al. (2002) concluded that there is a large uncertainty dealing with model parameters estimated from respirograms.

The basic assumption of those experiments using respiration rate measurements to quantify hydrolysis is that hydrolysis determines respiration rates when the readily biodegradable substrate is depleted. However, in the presence of storage of internal polymers (see: Section 3.3) the interpretation of OUR profiles becomes confusing and extracellular hydrolysis of slowly biodegradable COD and intracellular degradation of storage polymers cannot be distinguished from the OUR profile (Goel et al. 1998). Goel et al. (1999) proposed an experimental approach to separate hydrolysis from storage by performing and analyzing two parallel OUR measurements: one with filtered wastewater (including soluble COD) and the other with non-filtered wastewater (including total COD).

The rate of hydrolysis depends upon the magnitude of two kinetic coefficients including k_{hyd} and the entrapment saturation constant for hydrolysis, K_X. Numerical values of these coefficients are not universal and reveal a significant variation, especially for industrial wastewaters (Insel et al. 2002). A literature review performed by Ginestet et al. (2002) revealed high values ($1.5-25$ d^{-1}) of the first order hydrolysis constants estimated from respirometric experiments shorter than 24 h with raw and settled wastewater. Lower values ($0.12-0.20$ d^{-1}) were associated with long-term experiments (over the period of several days) with the particulate matter in wastewater. The same authors investigated the biodegradability of physico-chemical fractions (settleable, unsettleable-coagulable and unsettleable-uncoagulable). The estimated first order constants, $k_{hyd,S}$, were 9 (± 2) d^{-1} for raw wastewater, 12 (± 3) d^{-1} for settled wastewater and 16 (± 5) d^{-1} for coagulated wastewater. The substrate observed in short-term respirometric experiments was classified as a "*readily hydrolysable*" COD fraction and could be modelled by a first order reaction. The hydrolysis of "*slowly hydrolysable*" COD fraction was correctly modelled with a two-step process (adsorption followed by surface-limited hydrolysis). For this fraction, the hydrolysis rate constants, $k_{hyd,x}$, and entrapment saturation coefficients, $K_{X,X}$, ranged from 0.25 to 1.05 d^{-1} and from 0.33 to 0.95 g COD/g COD, respectively. In the study of Orhon *et al.* (1999), the results for municipal wastewater yielded average $k_{hyd,S}$ and $K_{X,S}$ values of 3.1 d^{-1} and 0.2 g COD/g COD, respectively, associated with the hydrolysis of S_H and much lower values of $k_{hyd,X} = 1.2$ d^{-1} and $K_{X,X} = 0.5$ g COD/g COD associated with hydrolysis of X_S. It was observed, however, that the discrepancy between measured and predicted OURs was reduced considerably by shifting from a single hydrolysis model to a dual hydrolysis model.

A subject of ongoing debate has been whether hydrolysis rates are influenced by electron acceptor conditions (Morgenroth *et al.* 2002b). In general, this issue appears contradictory and requires further research (Henze *et al.* 1995a; Barker and Dold 1997a; Ekama and Wentzel 1999a). On the one hand, there is experimental evidence that the hydrolysis rate depends on the available electron acceptors. For example, Henze and Mladenovski (1991) found a significant reduction of hydrolysis rates under anoxic and anaerobic conditions compared to aerobic conditions. Also indirect evidence suggests that the rate under anaerobic conditions should not be high. If the anaerobic hydrolysis rate was significant, it would make a substantial contribution to EBPR which is contrary to experimental observations strongly relating EBPR to the influent readily biodegradable COD (RBCOD) (Ekama and Wentzel 1999a). On the other hand, some studies on enzymatic activity (Barker and Dold 1997a; Goel *et al.* 1999) indicated that anaerobic and anoxic hydrolysis rates were comparable with those under aerobic conditions.

Some complex activated sludge models incorporate two reduction factors to provide flexibility in accounting for reduced rates under anaerobic and anoxic conditions. In ASM1 (Henze *et al.* 1987), the anoxic hydrolysis reduction factor, $\eta_{NO3,hyd}$, was set to 0.4, whereas the anaerobic hydrolysis was not considered and the anaerobic hydrolysis reduction factor, η_{fe}, was assumed zero. In ASM2 (Henze *et al.* 1995a), the respective values of $\eta_{NO3,hyd}$ and η_{fe} defaults were 0.6 and 0.1. The latter coefficient was increased to 0.4 in ASM2d (Henze *et al.* 1999) and the value of $\eta_{NO3,hyd}$ remained unchanged. Barker and Dold (1997a) introduced in their model two (anaerobic and anoxic) solubilisation factors which were equivalent to the ASM hydrolysis reduction factors. Their values were set to 0.5 and 1.0, respectively. The latter value indicates that the hydrolysis rate under anoxic conditions is not reduced compared to aerobic conditions, but modifications are possible. In addition, two "*efficiency factors*" were included to allow for the possibility of "*COD loss*" during the hydrolysis process. There was no difference between hydrolysis rates under aerobic and anoxic conditions in ASM3 (Gujer *et al.* 1999). Rieger *et al.* (2001) extended ASM3 with a bio-P module and assumed that also the anaerobic hydrolysis rate did not include any reduction factors.

In the activated sludge systems performing COD and nitrogen removal and modelled according to the ASM1 concept, the $\eta_{NO3,hyd}$ coefficient can be determined along with $\eta_{NO3,H}$ in two parallel (aerobic and anoxic) batch experiments (see: Section 3.1.4).

3.6 FERMENTATION (CONVERSION OF "*COMPLEX*" READILY BIODEGRADABLE SUBSTRATE TO VFA)

The anaerobic metabolism of phosphate accumulating organisms (PAOs) requires volatile fatty acids (VFAs) which are part of the readily biodegradable organic fraction in wastewater (see: Section 2.3). The VFAs are present in the influent wastewater, but can be also generated inside the BNRAS systems. The conversion of "*complex*" readily biodegradable substrate (S_F) to VFA (S_A) under anaerobic conditions is commonly termed "*fermentation*" (Henze *et al.* 1995a). In practice, however, there are compounds (e.g. ethanol) which are known to be fermentation products but reported not to be utilized by PAOs (Satoh *et al.* 2000). Thus these authors proposed a modified conceptual model for anaerobic COD metabolisms that assumes the presence of soluble substrate, S_A', which is not utilized by PAOs either directly or via fermentation. The S_A' type of substrate becomes, however, available for "*ordinary*" heterotrophs in the presence of oxygen or nitrate.

Until now, the process of fermentation occurring in the BNRAS systems has not been well characterized and little is known about its kinetics (Henze *et al.* 1999). There is also little or no experimental evidence for the modelled fermentation process (Meijer 2004). In general, it is assumed that fermentation is the principal linkage between PAOs and "*ordinary*" heterotrophs in these systems. Although fermentation is likely a growth process, only a relatively small fraction (20-25%) of the influent biodegradable COD (BCOD) is fermented at a low process yield, i.e. approximately 0.10 g COD/g COD (Ekama and Wentzel 1999a; Hu *et al.* 2007). The authors concluded that the growth resulted from fermentation is of minor significance compared to the growth under anoxic/aerobic conditions and its incorporation in the model is not essential for improving model prediction capability.

Different approaches to modelling fermentation in terms of stoichiometry and process kinetics are presented in Table 3.8. Wentzel *et al.* (1985) incorporated fermentation in the enhanced culture model for completeness even though the influent COD to the system was in the form of VFA (acetate). The established kinetics of this process is first order with respect to the concentrations of S_F

and "*ordinary*" heterotrophs. Since the process is supposed to occur only under anaerobic conditions, switching functions for oxygen and nitrate are added to the process rate expression. In terms of stoichiometry, the process is a simple transformation yielding one unit of S_A per each unit of S_F converted. A similar approach was adopted later in ASM2 (Henze *et al.* 1995a) and ASM2d (Henze *et al.* 1999). In these models, growth of "*ordinary*" heterotrophs is not considered. The authors justified this approach by the fact that the growth would require both more complex kinetics and a greater number of model parameters which are difficult to obtain. The process rate expression is first order with respect to the concentration of "*ordinary*" heterotrophs with multiple Monod terms for S_F and alkalinity (S_{ALK}) as well as switching functions for oxygen ($K_{O,H}$) and nitrate ($K_{NO,H}$). Since fermentation releases negatively charged fermentation products, the change in alkalinity is predicted from the charge mass balance equation.

Table 3.8 Stoichiometry and kinetics of the fermentation process

Stoichiometry				
S_F	S_A	X_H	Process rate	Reference
-1	1		$q_{fe} S_F \dfrac{K_{O,H}}{K_{O,H} + S_O} \dfrac{K_{NO,H}}{K_{NO,H} + S_{NO}} X_H$	Wentzel *et al.* (1985)
-1	1		$q_{fe} \dfrac{S_F}{K_{SF,H} + S_F} \dfrac{S_{ALK}}{K_{ALK,H} + S_{ALK}} \cdot$ $\dfrac{K_{O,H}}{K_{O,H} + S_O} \dfrac{K_{NO,H}}{K_{NO,H} + S_{NO}} X_H$	Henze *et al.* (1995a, 1999)
-1	1		$q_{fe} \dfrac{S_F/(X_H + X_{PAO})}{K_{SF,H} + S_F/(X_H + X_{PAO})} (X_H + X_{PAO})$	Johansson *et al.* (1996)
-1	$(1 - Y_{H,ana})$ $\cdot Y_{AC}$	$Y_{H,ana}$	$q_{fe} \dfrac{S_F}{K_{SF,H} + S_F} \dfrac{K_{O,H}}{K_{O,H} + S_O} \dfrac{K_{NO,H}}{K_{NO,H} + S_{NO}} X_H$	Barker and Dold (1997a)

In the model of Barker and Dold (1997a), fermentation is assumed to be a growth process with an empirical factor for modelling "*COD loss*" resulting from fermentation. The process rate is modelled using a first order expression with respect to the concentration of "*ordinary*" heterotrophs with a Monod term only for S_F and switching functions for oxygen and nitrate. In terms of stoichiometry, the process yields $Y_{H,ana}$ units of the biomass of "*ordinary*" heterotrophs and $(1-Y_{H,ana})$ units of fermentation products. Further, it is assumed that only a portion (Y_{AC}) of the fermentation products is S_A and the reminding COD is lost from the system. This assumption is in accordance with results of the previous studies of Barker and Dold

(1995 and 1996a) on COD and nitrogen mass balances in activated sludge systems. These results suggested that there was a significant "*loss*" of COD in activated sludge systems incorporating anaerobic zones. The authors proposed two potential mechanisms justifying this COD loss, i.e. the generation of gas that evolves during the actual fermentation process or the production of volatile compounds that are released from the system under aerated conditions.

Johansson *et al.* (1996) assumed in their model of the anaerobic reactor that fermentation is accomplished both by "*ordinary*" heterotrophs and PAOs. The process rate expression, first order with respect to the sum of both biomass concentrations, was a surface-limited reaction $\dfrac{S_F/(X_H + X_{PAO})}{K_{SF,H} + S_F/(X_H + X_{PAO})}$ without other Monod terms and switching functions.

Johansson *et al.* (1996) reported that typical values for the fermentation rate constant, q_{fe}, remain within the range $1-3$ d^{-1}, although higher values have been also encountered, e.g. $q_{fe} = 5$ d^{-1} (Furumai *et al.* 1999). The authors investigated the process rates an two Swedish WWTPs using anaerobic batch test simulations. At the Malmö WWTP, the estimated rate constant was 1.5 d^{-1} for "*normal*" primary settled wastewater, whereas the one at the Helsingborg WWTP was only 0.25 d^{-1}. In the latter case, however, the constant was increased to 3 d^{-1} for the experiments with hydrolyzed wastewater, defined as wastewater containing primary sludge hydrolyzed and washed out in the primary clarifier. The higher fermentation rate with the hydrolyzed wastewater can be justified by the fact that readily biodegradable organic matter is pre-treated and maybe easier to ferment in the anaerobic reactor. It should be noted that the q_{fe} constant has a different dimension in the model of Wentzel *et al.* (1985) and thus its value was set to 0.04 $dm^3/(mg\ COD{\cdot}d)$.

In the Swiss studies (Koch *et al.* 2001b; Rieger *et al.* 2001), it was assumed based on model sensitivity analysis and experimental observations that fermentation was not the rate-limiting process for the anaerobic substrate storage by PAOs and release of phosphate in typical municipal wastewater in Switzerland. Therefore, fermentation was not incorporated in the EAWAG bio-P module (Rieger *et al.* 2001) coupled with ASM3 (Gujer *et al.* 1999). Meijer (2004) suggested, however, that fermentaion could not be considered in modeling EBPR only in situations where the biodegradable fraction of wastewater primarily consists of slowly biodegradable matter. This is typically encountered in sewer systems with a short hydraulic retention time (HRT), e.g. under Swiss conditions. Otherwise, wastewater will contain a relatively high amount of soluble substrate due to hydrolysis occurring in sewer systems with a long HRT, e.g. under Dutch conditions. In such cases, Meijer (2004) recommended that fermentation should be modelled and proposed to fit the anaerobic phosphate release by adjusting the rates of fermentation and anaerobic hydrolysis.

3.7 NITRIFICATION

Nitrification is a two-step biological oxidation of ammonia to nitrate. During the first step, ammonia is oxidized to nitrite by a group of bacteria collectively called the ammonia-oxidizing bacteria (AOB), represented by *Nitrosomonas*. In the second step, the oxidation of nitrite to nitrate is performed by a group of bacteria collectively called the nitrite-oxidizing bacteria (NOB), traditionally represented by *Nitrobacter*. The recent findings using the the fluorescence in situ hybridization (FISH) and laser scanning microscope techniques suggest, however, that the second step is rather performed by the *Nitrospirae* and *Nitrosospira* species (e.g. Parker and Wanner, 2007). The yield coefficient of nitrifying organisms is known to be very small and their growth rate is slow (Table 3.9). The microorganisms are sensitive to a number of environmental factors, such as pH, dissolved oxygen concentration, temperature and inhibitory chemicals. The effect of these factors on the process performance, design and operation as well as the general process fundamentals are well recognized and have been summarized in detail in the literature (e.g. Metcalf and Eddy 1991; Sedlak 1991; USEPA 1993; Grady *et al.* 1999; Metcalf and Eddy 2003). Recently, Ruel *et al.* (2005) discussed the process conditions that can be controlled to favour the AOB over NOB for nitrite accumulation. A selective inhibition by free ammonia and nitrous acid is considered to be the main reason for this accumulation. High pH and temperature are another factors that indirectly increase the free ammonia concentration by affecting the NH_3/NH_4^+ equilibrium. In addition, at temperatures over 15 °C nitrite oxidizers grow slower than ammonia oxidizers. A limited oxygen concentration is another known factor that promotes nitrite accumulation.

Table 3.9 Typical values of kinetic and stoichiometric coefficients for nitrifying bacteria

Coefficient	Symbol	Unit	*Nitrosomonas* (a, b, c)	*Nitrobacter* (a, b, c)	Overall (b, c)
Yield coefficient	Y_A	mg VSS/ mg N	0.03–0.13	0.02–0.08	0.1–0.3
Max. specific growth rate	$\mu_{A,max}$	d^{-1}	0.46–2.2	0.28–3.0	0.3–3.0
Ammonia saturation constant	$K_{NH,A}$	mg N/dm^3	0.06–5.6	0.06–8.4	0.2–5.0
Oxygen saturation constant	$K_{O,A}$	mg O_2/dm^3	0.3–1.3	0.25	0.5–1.0
Decay coefficient	b_A	d^{-1}	0.03–0.06	0.03–0.06	0.03–0.06

(a) Sharma and Ahlert (1977), (b) Metcalf and Eddy (1991), (c) Henze *et al.* (1995b)

Even though nitrification is generally considered to be autotrophic conversion, many heterotrophic microorganisms have recently been found to be able to nitrify (Zhao *et al.* 1999). The authors noted that compared to autotrophic nitrifiers, heterotrophic nitrifiers generally tend to grow more rapidly with higher yield, require lower DO concentration and tolerate a more acidic environment. Littleton *et al.* (2003) evaluated the importance of this processes in three BNRAS systems and concluded that heterotrophic bacteria were not substantial contributors to nitrification at the plants studied.

3.7.1 Modelling nitrification as a one-step conversion

For simplicity, the oxidation of ammonia to nitrate has generally been treated as a single-step oxidation reaction based on the assumption that the first step is typically rate-limiting (Henze *et al.* 1987; Gee *et al.* 1990; Chandran and Smets 2000a,b; Gernaey *et al.* 2004). Downing *et al.* (1964) used a simple Monod-type to model ammonia oxidation to nitrate as a one-step conversion by nitrifying bacteria in the activated sludge process (see: Section 1.2.3). In the 1970's, several dynamic models of the one-step nitrification in activated sludge systems were proposed (e.g. Lijklema 1973; Gujer and Jenkins 1975; Gujer 1977; Murphy *et al.* 1977). At the same time, the research group at the University of Cape Town started to develop complex activated sludge models (see: Section 1.2.3), in which nitrification was integrated with degradation of organic matter, denitrification and oxygen consumption in cascades of completely mixed reactors, first for steady state (Marais and Ekama 1976) and later for dynamic behaviour (Ekama and Marais 1979; Dold *et al.* 1980). The nitrification process was modelled according to the concept of Downing *et al.* (1964). Stenstrom and Poduska (1980) proposed an extension of the expression for autotrophic growth with a single Monod term with respect to the substrate (ammonia) limitation by including the possibility of growth limitation under low DO concentrations. The double-substrate limiting kinetic expression explained the variation in the reported results on nitrification rates at different DO concentrations. The DO saturation coefficient for autotrophic organisms, $K_{O,A}$, ranges from 0.2 to 1.0 mg O_2/dm^3 (Sedlak 1991) with a typical value of 0.5 mg O_2/dm^3 (Henze *et al.* 1987). In a more recent review, Bratby and Parker (2009) found the $K_{O,A}$ values in the range $0.25-2.0$ mg O_2/dm^3.

All the most common complex activated sludge models adopted that approach. Those models vary in the number of limiting terms in the Monod-type expression, which is especially apparent in the ASM series (Henze *et al.* 2000). A double saturation function, as originally proposed by Stenstrom and Poduska (1980), was used in ASM1 to account for the dependency of $\mu_{A,max}$ upon the concentrations of both ammonia and DO. In ASM3, a limiting term with respect to alkalinity was added. The number of limiting terms increased to four in the models incorporating EBPR (ASM2, ASM2d) by adding one more limiting term with respect to the inorganic soluble phosphorus concentration (as a nutrient for autotrophic growth). The same number of limiting terms (4) was included in other two complex models (van Veldhuizen et al. 1999a; Rieger et al. 2001), whereas the same two terms as in ASM1 were taken into account in the model of Barker and Dold (1997a).

Although the Monod-type expression is the most widely accepted approach to describe the autotrophic growth kinetics, several authors confirmed that the rate of nitrification was independent of the mixing regime and initial substrate concentration (Hall and Murphy 1980). Charley et al. (1980) summarized the nitrification studies which revealed zero order kinetics over a broad range of ammonia concentrations (1.6-673 mg N/dm^3). Moreover, some experimental evidence suggests that autotrophic growth kinetics follow an inhibitory pattern at high influent concentrations of ammonia. Under such conditions, a substrate inhibition function, such as the Haldane function (Equation 3.5), seems to fit the experimental data better than the Monod equation (Rozich and Castens 1986). Gee et al. (1990) found that the oxidation of ammonia to nitrite in batch tests with initial ammonia concentrations ranging from 100 to 1000 mg N/dm^3 was indeed well represented by the Haldane function. However, the oxidation of nitrite to nitrate was not satisfactorily described by that equation. The authors observed that the simultaneous presence of both nitrite and ammonia led to the inhibition of nitrate oxidation.

Significant variations exist in literature with respect to the reported kinetic parameters describing the nitrification process due to the wide array of biokinetic estimation techniques employed (Chandran and Smets 2000a). Moreover, an additional constraint of high importance in the analysis of experimental data in nitrifying systems is lack of an accurate and easy measurement method of autotrophic biomass concentration. Several techniques for determining the fraction of autotrophs in biomass, $f_{X,A}$, are available in the literature (e.g. Srinath et al. 1976, Hall and Murphy 1980; Metcalf and Eddy 1991; USEPA 1993; Albertson and Stensel 1994; Sinkjaer et al. 1994; Copp and Murphy 1995; Melcer et al. 2003; Manser et al. 2005). Below are discussed two kinetic parameters ($\mu_{A,max}$, $K_{NH,A}$) and one stoichiometric parameter (Y_A). The meaning and reported values of b_A were earlier discussed in Section 3.2.

Maximum specific growth rate constant for autotrophs, $\mu_{A,max}$. This is the most critical parameter for design and performance of nitrifying activated sludge systems (Henze et al. 1987; Metcalf and Eddy 1991; Sedlak 1991; USEPA 1993; Grady et al. 1999; Melcer et al. 2003). The $\mu_{A,max}$ parameter is used to determine the minimum SRT at which the system can be operated without losing nitrifying biomass. Typical values of $\mu_{A,max}$ at 20°C were found to vary in wide ranges: 0.34

to 0.65 d^{-1} (Henze *et al.* 1987); 0.46-2.2 d^{-1} (Sedlak 1991); 0.25-1.23 d^{-1} (Copp and Murphy 1995); and 0.55-1.62 d^{-1} (Cinar *et al.* 1998). These values, however, still remain lower than the upper range (= 3.0 d^{-1}) reported by Metcalf and Eddy (1991). The wide range of $\mu_{A,max}$ may be attributed to two factors, namely, wastewater characteristics (presence of inhibitory substances in the wastewater) and/ or differences in experimental techniques and methods of analysis (Metcalf and Eddy 2003). Koch *et al.* (2000b) found relatively high variation in $\mu_{A,max}$ values (0.9-2.0 d^{-1}) at three Swiss municipal wastewater treatment plants (WWTPs). The higher values were attributed to the favorable process conditions due to an increased air flow to the plant resulting in an improved CO_2 stripping and increased pH. The default $\mu_{A,max}$ value in ASM1 was 0.8 d^{-1} but in the later models of the ASM series (ASM2, ASM2d and ASM3) that value increased to 1.0 d^{-1} (Henze *et al.* 2000). Barker and Dold (1997a) and Rieger *et al.* (2001) provided ranges rather than a single value for $\mu_{A,max}$, i.e. 0.2-1.0 d^{-1} and 0.9-1.8 d^{-1}, respectively.

On the contrary, Melcer *et al.* (2003) hypothesized that *"there may well be consistency in nitrifier growth kinetics from plant to plant"*. The limited data from four WWTPs across North America revealed that the estimated $\mu_{A,max}$ values ranged from 0.75 to 1.17 d^{-1}. In the follow-up study, Jones *et al.* (2005) re-analyzed data from several studies outside of the WERF project by correcting $\mu_{A,max}$ estimates to 20 °C using a temperature dependency coefficient of 1.072 and using the autotrophic (nitrifier) decay rate constant, b_A, equal to 0.17 d^{-1} at 20 °C. In all the cases, the $\mu_{A,max}$ estimates remained in the narrow range 0.85-1.05 d^{-1} which seemed to confirm the limited variability of this crucial coefficient.

Another important contribution of the work of Melcer *et al.* (2003) was an attempt towards the development of reliable methods for estimating $\mu_{A,max}$ values. Two general approaches were identified including bioassay methods (low or high F/M tests, washout method) and model simulation by fitting the dynamic response of ammonia and/or nitrate+nitrite concentrations from experimental observations. Comparative studies presented by Melcer *et al.* (2003) revealed that the results obtained from the three bioassay methods were consistent and the maximum difference between the estimated $\mu_{A,max}$ values did not exceed 17% at all the studied plants. Model-based estimation of $\mu_{A,max}$ yielded very similar estimates to those derived from the bioassay tests only in the case of a well-characterized activated sludge system simulation. The simulations may yield results deviating from those in the bioassay tests in the systems in which factors affecting nitrification kinetics are not known, measured or sufficiently controlled. Hence, the authors did not consider simulation of full-scale systems to be an accurate method for estimating $\mu_{A,max}$.

A few procedures are available which do not require determination of the initial autotrophic biomass concentration. Henze *et al.* (1987) recommended a procedure that is based on a dynamic test on a completely mixed continuous reactor under study (provided it is barely nitrifying). During the test, the SRT should be increased to achieve a high degree of nitrification. The oxidized nitrogen (NO_3–$N + NO_2$–N) concentrations, S_{NO}, in the reactor effluent should be measured over time and the natural logarithm of these concentrations should be plotted vs. time. The slope of this plot will be:

$$\text{slope} = \mu_{A,max} - \frac{1}{SRT} - b_A \qquad (3.23)$$

where

$\mu_{A,max}$ - Maximum growth rate constant of autotrophic microorganisms, T^{-1}
b_A - Specific decay rate constant for autotrophic microorganisms, T^{-1}

A simpler procedure involves a batch test started with a small amount of biomass collected from a completely mixed reactor, preferably barely nitrifying (Grady et al. 1999). The S_{NO} concentrations in the batch reactor should be measured over time and the plot of the natural logarithm of S_{NO} vs. time results in the following slope:

$$\text{slope} = \mu_{A,max} - b_A \qquad (3.24)$$

Orhon et al. (1994b) and subsequently Sozen et al. (1996) determined the value of $\mu_{A,max}$ by monitoring the S_{NO} concentration in batch reactors. The final equation was presented in the following form (Sozen et al. 1996):

$$(S_{NO} - S_{NO,0}) \frac{Y_A}{X_{A,0}} \frac{\mu_{A,max} - b_A}{\mu_{A,max}} + 1 = e^{(\mu_{A,max} - b_A)t} \qquad (3.25)$$

where
S_{NO} - Concentration of oxidized nitrogen, $M(N)L^{-3}$
$S_{NO,0}$ - Initial concentration of oxidized nitrogen, $M(N)L^{-3}$
Y_A - "*True*" growth yield coefficient for autotrophic microorganisms, $M(COD)$
           $M(N)^{-1}$
$X_{A,0}$ - Initial concentration of autotrophic organisms, $M(COD)L^{-3}$

Equation 3.25 can be rearranged to the following linear form:

$$\ln\left[(S_{NO} - S_{NO,0})k_1 + 1\right] = (\mu_{A,max} - b_A)t \qquad (3.26)$$

where k_1 is a constant for the given experimental conditions and equal to:

$$k_1 = \frac{Y_A}{X_{A,0}} \frac{\mu_{A,max} - b_A}{\mu_{A,max}} \tag{3.27}$$

The value of $\mu_{A,max}-b_A$ is then calculated assigning different values for k_1 to get the highest correlation coefficient.

Respirometric methods have also been used to determine the specific growth rate constant of autotrophic organisms. Novak *et al.* (1994) and Nowak *et al.* (1994) demonstrated a similar approach to the estimation procedure of $\mu_{A,max}$ (batch tests) but used different techniques to determine $X_{A,0}$. Novak *et al.* (1994) estimated the value of $\mu_{A,max}$ using a combined technique of mathematical modelling and batch tests. This technique assumes that the proportions of autotrophic and heterotrophic microorganisms in the activated sludge biomass can be estimated from simulation of a continuously operating system. Nowak *et al.* (1994) determined the value of $\mu_{A,max}$ based on two measurements of respiration rates in batch tests. In the first one, the total OUR, OUR_T, is measured after the addition of an ammonia salt. The result of this measurement includes $OUR_{A,max}$ and the endogenous heterotrophic OUR, $OUR_{H,end}$. The $OUR_{H,end}$ has to be determined simultaneously in another test with the addition of a nitrification inhibitor, e.g. allythiourea (ATU). Finally, the $OUR_{A,max}$ is calculated by subtracting the $OUR_{H,end}$ from the previously measured OUR_T. The value of $\mu_{A,max}$ was then calculated from the following relationship:

$$\mu_{A,max} = \frac{OUR_{A,max}}{X_A} \frac{Y_A}{i_{OC,N}} \tag{3.28}$$

where
$OUR_{A,max}$ - Maximum oxygen uptake rate of autotrophic microorganisms, $M(O_2)L^{-3}T^{-1}$
$i_{OC,N}$ - Specific oxygen consumption for nitrification, $M(O_2)M(N)^{-1}$
X_A - Concentration of autotrophic organisms, $M(COD)L^{-3}$

The theoretical value of $i_{OC,N}$ is 4.33 mg O_2/mg N but in practice it typically ranges from 4.1 to 4.3 mg O_2/mg NO_3-N (Nowak *et al.* 1994). Furthermore, since it is impossible to determine X_A and Y_A by relatively simple methods, the X_A/Y_A ratio can be determined based on a mass balance of autotrophic biomass in the activated sludge system (Nowak *et al.* 1994):

$$V \frac{dX_A}{dt} = Y_A L_{N,ox} - E_A V X_A \tag{3.29}$$

where

$L_{N,ox}$ - Load of oxidized nitrogen produced, $M(N)T^{-1}$

E_A - *"Decrease rate of autotrophic organisms"*, which is a sum of the b_A coefficient and the excess sludge removal rate, T^{-1}

The integration and rearrangement of Equation 3.29 leads to the following form:

$$\frac{X_A^{n+1}}{Y_A} = \frac{L_{N,ox}^n}{VE_A^n}\left(1-e^{-E_A^n\Delta t}\right) + \frac{X_A^n}{Y_A}e^{-E_A^n\Delta t} \qquad (3.30)$$

The first value of X_A/Y_A can be estimated from the average conditions. The next values of this ratio have to be determined on each consecutive day. The excess sludge removal rate, which is part of E_A, can be calculated from operating data. In order to estimate daily values for b_A, additional investigations have to be carried out or the literature data should be assumed.

Ammonia saturation coefficient for autotrophs, $K_{NH,A}$. This coefficient is not affected by inhibitory compounds as strongly as $\mu_{A,max}$ (Grady *et al.* 1999). In the literature review of Copp and Murphy (1995), the $K_{NH,A}$ values ranged from 0.06 to 5.6 mg N/dm^3. This range is consistent with the data listed in Table 3.9. The default values used in the series of ASMs were identical and equal to 1.0 mg N/dm^3 (Henze *et al.* 2000). In general, lower values of this coefficient are encountered in pilot plants due to lower diffusion limitation in the flocs (higher turbulence and smaller flocs) compared to full-scale systems (Henze *et al.* 1995a). Recently, however, Satoh *et al.* (2000) reported the use of an extremely high $K_{NH,A}$ value (= 8 mg N/dm^3) for simulation of a pilot-scale plant consisting of two parallel anaerobic-aerobic activated sludge systems. The authors could not provide a rational explanation for such a high value of $K_{NH,A}$. One potential reason was that $K_{NH,A}$ accounted for all limitations of the actual growth conditions of autotrophic organisms.

In general, determination of $K_{NH,A}$ in batch culture should be avoided since depletion of growth is frequently due to factors other than substrate limitation (Prosser 1990). Henze *et al.* (1987) recommended the infinite-dilution procedure of Williamson and McCarthy (1975) for determining $K_{NH,A}$. This method, developed for a general substrate, was adopted by Sharma and Ahlert (1977) to determine a combination of $K_{NH,A}$ and the maximum specific nitrification rate, $r_{N,max}$. The final (simplified) equation was presented in the following form:

$$r_{N,max}\frac{S_{NH}}{S_{NH}+K_{NH,A}} = \frac{S_{NH,in}Q_{in}}{VX_A} \qquad (3.31)$$

where

$r_{N,max}$ - Maximum specific nitrification rate, $M(N)L^{-3}T^{-1}$

$S_{NH,in}$ - Influent concentration of ammonia (and ammonium) nitrogen, $M(N)L^{-3}$

A respirometric method developed originally by Cech *et al.* (1985) and modified by Drtil *et al.* (1993) may be directly applied to determine the value of $K_{NH,A}$. This method was, in turn, recommended by Grady *et al.* (1999). Substrate (an ammonia salt) is injected at 10 min intervals to a batch reactor (respirometer). After each dose of substrate, the nitrification rate, r_N, increases. It is assumed that the $i_{OC,N}$ coefficient remains constant and thus the corresponding couples, r_N and S_{NH}, can be calculated from the following relationship:

$$r_N(t) = \frac{OUR_T(t) - OUR_{H,end}}{i_{OC,N}} \tag{3.32}$$

where

$OUR_{H,end}$ - Endogenous heterotrophic oxygen uptake rate, $M(O_2)L^{-3}T^{-1}$

OUR_T - Total (autotrophic and endogenous heterotrophic) oxygen uptake rate, $M(O_2)L^{-3}T^{-1}$

The total ammonia amount in the respirometer is a sum of ammonia injected and ammonia remaining after previous injections:

$$S_{NH}(t_i) = S_{NH,I}(t_i) + S_{NH,R}(t_i) \tag{3.33}$$

where

$S_{NH,I}$ - Concentration of injected ammonia in the respirometer after the ammonia injection, $M(N)L^{-3}$

$S_{NH,R}$ - Residual ammonia concentration in the respirometer after the previous ammonia injections, $M(N)L^{-3}$

The concentration of ammonia remaining after previous injections can be calculated by rearranging Equation 3.32:

$$S_{NH,R}(t_i) = S_{NH}(t_{i-1}) - \frac{(OUR_T(t_{i-1}) - OUR_{H,end})(t_i - t_{i-1})}{i_{OC,N}} \tag{3.34}$$

Yield coefficient for autotrophic biomass, Y_A. Values of the autotrophic yield coefficient, Y_A, can vary in a relatively broad range (Table 3.9). In another review, Hulsbeek et al. (2002) reported the values ranging from 0.07 to 0.28 mg COD/mg N but the value of 0.24 mg COD/mg N has been established as a good approximation for Y_A (Vanrolleghem et al. 1999). This value was used in most of the common complex activated sludge models, apart from the model of Barker and Dold (1997a) in which $Y_A = 0.15$ mg COD/mg N. It should be noted that variations in Y_A are thought to affect the autotrophic biomass concentration (X_A) rather than values of the kinetic coefficients ($\mu_{A,max}$, $K_{NH,A}$) (Prosser 1990).

In order to estimate the actual value of Y_A, Vanrolleghem et al. (1999) proposed a respirometric batch test in which a pulse of ammonia is added to a nitrifying activated sludge sample and the cumulative OUR_A is measured:

$$Y_A = \frac{4.57 S_{NH,0} - \int OUR_A(t)\,dt}{S_{NH,0}} \qquad (3.35)$$

where
OUR_A - Autotrophic oxygen uptake rate, $M(O_2)L^{-3}T^{-1}$
$S_{NH,0}$ - Initial concentration of ammonia (and ammonium), $M(N)L^{-3}$

3.7.2 Modelling nitrification as a two-step conversion

The approach of modelling nitrification as a one-step conversion is not usually acceptable for elevated nitrogen conditions, temperatures or in cases of inhibition. This is because nitrite is an environmentally important intermediate product of both nitrification and denitrification. Typical examples of such situations include anaerobic digester supernatant effluents, food processing facilities, concentrated animal feeding operations, and industrial facilities that manufacture fertilizers, chemicals, pharmaceuticals, and other products (Hiatt and Grady 2008). Henze et al. (1995a) noted that modelling the intermediate nitrite production and consumption would be relatively easy in the context of nitrification, however, nitrite is also produced and consumed in the denitrification process. Therefore, considering nitrite as a state variable in the nitrification model, but neglecting it in the denitrification model, would be inconsistent and result in erroneous model predictions. An example of the stoichiometric matrix and process rates for growth of AOB and NOB in the two-step nitrification process is presented in Table 3.10 and Table 3.11, respectively. Table 3.12 contains the ranges of parameter values reported for both mainstream and sidestream treatment systems.

Knowles *et al.* (1965) presented the first two-step nitrification model for a mixed bacteria culture in water from the Thames Estuary. The first dynamic model for nitrification in the activated sludge process considering ammonia and nitrite oxidation as two separate steps of the overall conversion was developed by Poduska and Andrews (1975). The model consisted of five mass balance equations (for ammonia, nitrite, nitrate, *Nitrosomonas* and *Nitrobacter*) and was tested against the data obtained from a laboratory scale system. The validated model was a tool to investigate design criteria and operating techniques by which the efficiency of nitrification might be improved. Beck (1981) applied the model of Poduska and Andrews (with a few minor revisions) to predict process performance (nitrification recovery) under assumed scenarios for the future operating conditions over the period of 40 days at the studied full-scale plant (Norwich Sewage Works, UK). The author claimed that the model was "*possibly overly complex in assuming a two-step, as opposed to one-step conversion process for nitrification*". According to the classification of Chandran and Smets (2000b), available two-step nitrification models are based on substrate depletion (and product formation) profiles (e.g. Gee *et al.* 1990), a combination of substrate/product profiles and OURs (e.g. Ossenbruggen *et al.* 1996) as well as sole oxygen uptake measurements (e.g. Brouwer *et al.* 1998; Chandran and Smets 2000a,b). Chandran and Smets (2000a) concluded that one-step nitrification models are only appropriate as long as ammonia to nitrite oxidation is the sole rate-limiting step in the entire oxidation path of ammonia to nitrate. There is a number of factors affecting ammonia and nitrite oxidation to be considered in mathematical models. These factors are discussed in Table 3.13 and Table 3.14, respectively.

Table 3.10 Stoichiometric matrix for growth of AOB and NOB in the two-step nitrification process

Process	S_O	S_{NH}	S_{NO2}	S_{NO3}	X_{AOB}	X_{NOB}	S_{ALK}
	gO_2/m^3	gN/m^3	gN/m^3	gN/m^3	$gCOD/m^3$	$gCOD/m^3$	mole HCO_3^-/m^3
Growth of X_{AOB}	$-\dfrac{(3.43-Y_{AOB})}{Y_{AOB}}$	$-\dfrac{1}{Y_{AOB}}-i_{N,BM}$	$\dfrac{1}{Y_{AOB}}$		1		$\upsilon_{ALK,AOB}{}^*$
Growth of X_{NOB}	$-\dfrac{(1.14-Y_{NOB})}{Y_{NOB}}$	$-i_{N,BM}$	$-\dfrac{1}{Y_{NOB}}$	$\dfrac{1}{Y_{NOB}}$		1	

$$^* -\upsilon_{ALK,AOB} = -\frac{1}{7Y_{AOB}} - \frac{i_{N,BM}}{14}$$

Table 3.11 Process rates for growth of AOB and NOB in the two-step nitrification process

Process	Process rate gCOD/(m³·d)
Growth of X_{AOB}	$\mu_{AOB} \dfrac{S_O}{K_{O,AOB}+S_O} \dfrac{S_{NH}}{K_{NH,AOB}+S_{NH}} \dfrac{S_{ALK}}{K_{ALK,AOB}+S_{ALK}} X_{AOB}$
Growth of X_{NOB}	$\mu_{NOB} \dfrac{S_O}{K_{O,NOB}+S_O} \dfrac{S_{NO2}}{K_{NO2,NOB}+S_{NO2}} \dfrac{S_{ALK}}{K_{ALK,NOB}+S_{ALK}} \dfrac{S_{NH}}{K_{NH,NOB}+S_{NH}} X_{NOB}$

Table 3.12 Parameter values used in two-step nitrification models for mainstream and sidestream treatment systems

Parameter	Unit	Sin et al. (2008b)* (T = 20–35 °C)	Hiatt and Grady (2008)
AOB			
Y_{AOB}	mg COD/mg N	0.11–0.21	0.18
μ_{AOB}	d⁻¹	0.5–2.1	0.78
b_{AOB}	d⁻¹	0.15	0.096
$b_{AOB,AE}$	d⁻¹	0.071–0.3	–
$b_{AOB,AX}$	d⁻¹	0.015–0.15	–
$K_{O,AOB}$	mg O₂/dm³	0.5–3.0	0.6
$K_{NH,AOB}$	mg N/dm³	0.14–5.0	–
$K_{NH3,AOB}$	mg N/dm³	0.468–0.75	0.0075
$K_{HCO3,AOB}$	mmol/dm³	0.008–4.2	–
$K_{INH3,AOB}$ (Haldane eq.)	mg N/dm³	10–3000	1
$K_{IHNO2,AOB}$	mg N/dm³	0.22–2.8	0.1
$K_{PH,AOB}$	–	5.5–9.5	–
NOB			
Y_{NOB}	mg COD/mg N	0.03–0.09	0.06
μ_{NOB}	d⁻¹	0.9–1.8	0.78
b_{NOB}	d⁻¹	0.15	0.096
$b_{NOB,AE}$	d⁻¹	0.08–0.22	–
$b_{NOB,AX}$	d⁻¹	0.022–0.1	–
$K_{O,NOB}$	mg O₂/dm³	0.3–1.1	1.2
$K_{NO2,NOB}$	mg N/dm³	0.05–3.0	–
$K_{HNO2,NOB}$	mg N/dm³	0.0015–0.27	0.0001
$K_{HCO3,NOB}$	mmol/dm³	0.008–4.2	–
$K_{INH3,NOB}$	mg N/dm³	0.1–20	0.2
$K_{IHNO2,NOB}$	mg N/dm³	0.27–2.8	0.04
$K_{PH,NOB}$	–	5.5–9.5	–

* values reported by Anthonisen et al. (1976), Hellinga et al. (1999), Volcke (2006), Moussa et al. (2005), Kampschreur et al. (2007), Wett and Rauch (2003), Jones et al. (2007), van Hulle et al. (2007), Sin and Vanrolleghem (2006), Kaelin et al. (2009)

Table 3.13 Factors affecting ammonium oxidation considered in mathematical models applied to mainstream and sidestream treatment systems (adopted from Sin *et al.* (2008b) and Hiatt and Grady (2008))

Factor	Effect on growth of AOB
True substrate (total ammonia (TAN) vs. free ammonia (FA))	Most models assume that TAN is the substrate for growth of AOB and the Monod term sufficiently describes the substrate limitation. However, it has been reported in several studies that FA is actually the true substrate for AOB. In such a case, substrate inhibition should be considered (in particular, with regard to the sidestream processes).
Free ammonia inhibition	In practice, the inhibitory effects of FA may only be significant in some sidestream processes. A simple switching function in combination with the Monod term or Haldane equation. The respective inhibition coefficients were set to high values, which limits the inhibitory effect of FA (Table 3.12).
Nitrous acid inhibition	Nitrite concentrations do not reach high levels in the mainstream processes which justifies the exclusion of the inhibitory effects of nitrous acid. A simple switching function is used for the sidestream processes.
pH effect	The pH effect can be ignored in modelling the mainstream processes as pH variations are insignificant. In the sidestream processes, three approaches may be considered: - direct effect by considering an inhibition function with respect to μ_{AOB} only or μ_{AOB} and $K_{NH,A}$; - indirect effect by considering pH dependent equilibrium of weak acid/base reactions (the effects of pH were accounted for entirely through the FA concentration, which directly follows the true AOB mechanism), bicarbonate limitation (sufficient alkalinity to buffer acidification) and CO_2 stripping kinetics (Hellinga *et al.* 1999; Wett and Rauch 2003). - combination of both approaches.
Inorganic carbon limitation	In addition to buffer acidification, bicarbonate is the substrate for AOB. In the mainstream processes, influent alkalinity is high enough to exclude the possibility of this kind of limitation (unless low BOD and bicarbonate concentrations). Inorganic carbon could become an important limiting factor in high-rate nitrification systems with low ratios of influent biacarbonate to TAN. The effect can be modelled with a Monod term or exponential term (Wett and Rauch, 2003).

Table 3.14 Factors affecting nitrite oxidation considered in mathematical models applied to mainstream and sidestream treatment systems (adopted from Sin *et al.* (2008b) and Hiatt and Grady (2008))

Factor	Effect on growth of NOB
True substrate (total nitrite (TNO_2) vs. free nitrous acid (HNO_2)	Some models consider HNO_2 as the substrate source for growth of NOB. In such cases, a kind of inhibition should be considered. When TNO_2 is used as the substrate source, the substrate limitation can be sufficiently modelled using the Monod term.
Free nitrous acid inhibition	The HNO_2 inhibition could follow different kinetics, such as a simple switching function, Haldane kinetics and mixed inhibition kinetics. However, no comparison of these terms has been made on experimental observations.
Free ammonia (FA) inhibition	The effect of FA inhibition on NOB is more pronounced compared to AOB, which has been reflected by considerably lower values of the respective inhibition coefficients (Table 3.12).
pH effect	The pH effect can be considered indirectly through weak/acid base equilibrium reactions between TNO_2–HNO_2 and TAN-FA.
Inorganic carbon limitation	Similar to the effects on growth of AOB (Table 3.12).

3.8 DENITRIFICATION

Denitrification, known also as dissimilative nitrate reduction or anoxic respiration of nitrate, is the biological conversion of nitrate (NO_3-N) to nitrogen gas (N_2). The entire conversion involves four steps with three intermediates including nitrite (NO_2-N), nitric oxide (NO) and nitrous oxide (N_2O). A common assumption is that the conversions of the last two intermediates occur faster than the former, hence the NO and N_2O kinetics can be neglected (Sin *et al.* 2008b). However, the emission of N_2O is may be considered in some cases due to environmental concerns (a greenhouse gas). Wild *et al.* (1995) hypothesized that an increased N_2O production results from high NO_2-N concentrations.

The denitrification process is accomplished by a variety of facultative heterotrophic microorganisms which can utilize nitrate (and/or nitrite) instead of oxygen as the final electron acceptor (e.g. Metcalf and Eddy 1991). At this moment it is not clear, however, whether all the four steps are accomplished by one group of denitrifying microorganisms resulting in a complete reduction to N_2 or by the activity of more than one group when different steps carried out by different organisms (Sin *et al.* 2008b). For example, Wilderer *et al.* (1987) hypothesized that some microorganisms are capable of reducing nitrate, but only to nitrite. Furthermore, it is unclear whether the conversions are accomplished sequentially (NO_3-N $\rightarrow$ NO_2-N $\rightarrow$ gaseous forms) or in parallel (NO_3-N $\rightarrow$ gaseous forms, NO_2-N $\rightarrow$ gaseous forms), nor what is the electron acceptor (NO_3-N vs. NO_2-N) preference of the denitrifying microorganisms.

Sufficient amounts of organic carbon (electron donor) must also be ensured to provide energy for the denitrification process. The energy sources can be categorized as internal (present in the influent wastewater), endogenous (self-generated within the system as a result of organism decay) and external (not present in wastewater). The latter (external) sources are added to enhance the denitrification process and improve the overall efficiency of N removal within the existing capacities of activated sludge systems. There is a number of effective, commercially available organic compounds (such as methanol, ethanol, acetic acid, sodium acetate and glucose) which can be categorized as the "*conventional*" carbon sources. Among them, methanol has been most commonly used and best documented. However, due to high costs of commercial compounds and acclimation periods (usually) required, the effective use of internal carbon sources for denitrification is preferred. Also various industrial by-products or waste materials have recently received more attention as the "*alternative*" external carbon sources for denitrification. In particular, food industry effluents appear to be good candidates for this purpose due to their high C/N ratios and high content of readily biodegradable organic fraction (Makinia *et al.* 2009).

In the denitrification process, nitrate and nitrite serve as electron acceptors in the respiratory electron transport chain in a similar manner as oxygen in aerobic growth of heterotrophs (Zhao *et al.* 1999). The needed corrections for the anoxic stoichiometry ($Y_{H,NO}$) and kinetics ($\eta_{NO3,H}$ and $\eta_{NO3,hyd}$) were discussed in Section 3.1.4. In extending the nitrification-denitrification models with EBPR the role of PAOs in the denitrification process has also to be taken into account (Ekama and Wentzel 1999b). This issue is addressed in Section 3.9.3.

Van Haandel *et al.* (1981) reviewed early studies, in which the denitrification kinetics were described based on batch test results. In those studies, the process rate was a zero order reaction with respect to the nitrate concentration, and was only a function of the organic fraction of biomass (volatile solids concentration):

$$\frac{dS_{NO}}{dt} = -r_{DN} X_V \qquad (3.36)$$

where

r_{DN} - Specific denitrification rate, $M(N)M^{-1}T^{-1}$

Van Haandel *et al.* (1981) found that the denitrification kinetics were adequately described by Equation 3.36 under constant load and flow conditions, but the r_{DN} coefficient had different values depending upon the position of the denitrification zone in the process configuration (4-stage Bardenpho). Next, the authors incorporated denitrification in the general (aerobic) model of Dold *et al.* (1980). The denitrification behaviour was modelled using the same expressions as proposed by Dold *et al.* for aerobic conditions. The utilization rate of RBCOD was described by a Monod-type expression with respect to the RBCOD concentration, whereas the utilization rate of SBCOD was described by a surface-limited type expression. The only difference was that the rate of SBCOD utilization under anoxic conditions needed to be reduced to 0.38 of the rate under aerobic conditions. The model predicted near linear two-phase denitrification behaviour in the primary anoxic reactor. The first phase was attributed to utilization of RBCOD and adsorbed SBCOD, and the second phase arose from utilization of adsorbed SBCOD. A single near linear phase was predicted in the secondary anoxic reactor which was attributed to utilization of adsorbed SBCOD generated from organism decay.

In the study of Shah and Coulman (1978) and Beccari *et al.* (1983), the denitrification rate was described by a Monod-type expression with two growth-limiting substrates (organic carbon and nitrate or nitrite). In the first case, batch tests were performed with glucose as an organic carbon source and the results revealed high values of the saturation coefficients for both glucose, $K_{S,H} = 17.4$ mg (glucose)/dm^3, and nitrate, $K_{NO,H} = 83$ mg N/dm^3. In contrast, Beccari *et al.* (1983) observed in batch tests run under non-carbon limiting conditions that the nitrate utilization rates were zero order even with nitrate concentrations below 5 mg N/dm^3. Thus the authors suggested a low value attributed to $K_{NO,H}$ (<1 mg N/dm^3), although the reviewed literature data revealed a considerable dispersion even for only methanol ($K_{NO,H} = 0.08\text{-}52$ mg N/dm^3). USEPA (1993) also reported very low values of $K_{NO,H}$ ranging from about 0.1 to 0.2 mg N/dm^3. It was noted that with these values nitrate concentrations higher than 1-2 mg N/dm^3 have almost no effect on denitrification rates. The default value of $K_{NO,H}$ in the IWA ASM series is 0.5 mg N/dm^3 (Henze *et al.* 2000). The same value is used in the other complex activated sludge models outlined in Section 2.4.1, except for the model of Barker and Dold (1997a) and Hu *et al.* (2007) in which $K_{NO,H} = 0.1$ and 0.2 mg N/dm^3, respectively.

In denitrification systems, a DO concentration is the critical parameter as the presence of oxygen in the anoxic reactor suppresses the enzyme formation and activity needed for denitrification. In order to reflect this effect, Metcalf and Eddy (1991) and Sedlak (1991) incorporated the term (1-DO) in the denitrification rate expression which indicates that the denitrification rate decreases linearly to zero when the DO concentration reaches 1.0 mg O_2/dm^3. Above this value, it was assumed for practical purposes that denitrification can be ignored. Henze *et al.* (1987) proposed in ASM1 a concept based on switching functions to reflect inhibition of denitrification due to the presence of oxygen. In general, switching functions serve a mathematical purpose to turn process rates on/off when concentrations of specific components are above/below some threshold magnitude (Barker and Dold 1997a). These authors demonstrated how such functions can facilitate interpretation of the complex process model in the matrix format. Henze *et al.* (1987) incorporated a switching function $K_{O,H}/(K_{O,H}+S_O)$ in the expression for anoxic growth of heterotrophs and this approach was commonly accepted in the complex activated sludge models. It should be noted, however, that this function has no strict kinetic meaning and its aim is only to start and stop aerobic and anoxic processes (Grady 1989). The function is mathematically continuous which prevents problems with numerical instability of models describing rate equations that are turned on and off discontinuously (Henze *et al.* 1987). The typical range of $K_{O,H}$ is 0.01-0.2 mg O_2/dm^3 (Hulsbeek *et al.* 2002), but the values even up to 2.0 mg O_2/dm^3 were reported (Grady *et al.* 1999). The default value of $K_{O,H}$ in the IWA ASM series is 0.2 mg O_2/dm^3 (Henze *et al.* 2000). The same value occurs in the other complex activated sludge models, except for the models of Barker and Dold (1997a) and Hu *et al.* (2007) in which an extremely low value of 0.002 mg O_2/dm^3 is used. Higher values of $K_{O,H}$ (0.5-1.0 mg O_2/dm^3) have also been reported in the literature (Xu and Hultman 1996; Meijer *et al.* 2001; Petersen *et al.* 2002; Plosz *et al.* 2003). Meijer *et al.* (2001) used $K_{O,H}$ as a black box calibration parameter to correct hydrodynamic and physical shortcomings of the model, but higher $K_{O,H}$ also could account for the phenomenon of aerobic denitrification.

The effect of oxygen penetration across the liquid surface of bench scale batch reactors proved to be highly dependent on the denitrification rate. At low rates, the effect was considerable but it became insignificant when rapid nitrate consumption occurred, i.e. > 30 mg N/(g•h) (Jobbagy *et al.* 2000). In later studies of this group (Plosz *et al.* 2003), it was demonstrated using ASM1-based simulation model that oxygen entering the reactors through the liquid surface may affect denitrification both metabolically (resulting in the utilization of substrate which cannot then be utilized in denitrification) and kinetically (resulting in an inhibitory effect of the DO concentration on the denitrification rate). When the initial exogenous substrate concentration in a bioreactor was high (485 mg COD/dm^3), a high consumption rate of the substrate was attained and the DO concentration was kept low enough to minimize the effect of oxygen penetration through the liquid surface.

Oh and Silverstein (1999) investigated the effect of DO (over a range from 0.09 to 5.6 mg O_2/dm^3) on denitrification by activated sludge in a bench scale

SBR. Even the DO concentration as low as 0.09 mg O_2/dm^3 was found to inhibit denitrification, resulting in a rate decrease of 35% compared to the *"true"* anoxic conditions. When the DO concentration was 2.0 mg O_2/dm^3, the denitrification rate was reduced by 85%. Some denitrification activity (4% of the *"true"* anoxic rate) was observed with the DO levels as high as 5.6 mg O_2/dm^3. The authors concluded that aerobic denitrification is more persistent in activated sludge than in water and soil. There are two theories (biological and physical) explaining the phenomenon of aerobic denitrification in the activated sludge process (von Munch *et al.* 1996; Pochana *et al.* 1999). According to the physical (traditional *"engineering"*) theory, aerobic denitrification occurs in the anoxic microzones inside microbial flocs as a consequence of DO concentration gradients within the flocs due to oxygen diffusion limitations. The DO gradient is controlled by several factors including bulk DO concentration, floc size (turbulence level), organic substrate loading (or SRT) and aeration on/off cycle (Zhao *et al.* 1999). In contrast, the biological theory assumes that some heterotrophic microorganisms can denitrify in an aerobic environment. Compared to anoxic denitrifying bacteria, however, aerobic ones appear to have slower denitrification rates and prefer certain organic substrates, such as methanol (Zhao *et al.* 1999b).

Aerobic denitrification is part of the phenomenon called simultaneous nitrification and denitrification (SND), which was reported in the literature for various systems (von Munch *et al.* 1996). In the aeration tanks of many full-scale BNR plants, total nitrogen losses of up to 30% due to concurrent oxygen and nitrate/nitrite utilization have often been encountered (Zhao *et al.* 1999b). These authors found in their own studies in a 2-stage anaerobic-aerobic system that the nitrogen loss due to SND in the aeration tank contributed to 10-50% of the overall nitrogen removal. The highest process efficiency was observed under low DO concentrations (0.3-0.8 mg O_2/dm^3). Pochana and Keller (1999) found that even up to 95% of total nitrogen was removed through SND from SBRs installed in an industrial (abattoir) WWTP.

Pochana *et al.* (1999) developed a dynamic microbial floc model to evaluate the phenomena that occur due to internal floc effects, such as SND. The simulation results supported the hypothesis that SND is a physical phenomenon resulting from oxygen diffusion limitations within flocs. Von Munch *et al.* (1996) attempted to model the SND phenomenon by a switching function consistent with the ASM1 representation. In order to fit model predictions to the experimental data, it was needed to apply two switching functions in terms of the time from the beginning of aeration period (t_{aer}):

$$r_{DN} = r_{DN,max} \frac{K_{O,H,I}}{K_{O,H,I} + S_O} \qquad \text{for } 0 \leq t_{aer} < 1.5 \text{ h} \qquad (3.37)$$

$$r_{DN} = r_{DN,max} \frac{K_{O,H,2}}{K_{O,H,2} + S_O} \qquad \text{for } 1.5 \le t_{aer} < 3 \text{ h} \qquad (3.38)$$

where

$K_{O,H}$ — Oxygen saturation coefficient for heterotrophic organisms, $M(O_2)L^{-3}$

$r_{DN,max}$ — Maximum specific denitrification rate, $M(N)L^{-3}T^{-1}$

S_O — Concentration of dissolved oxygen, $M(O_2)L^{-3}$

t_{aer} — Time from the beginning of aeration period, T

In the first half of the aeration period, the $K_{O,H}$ coefficient was higher (0.67 mg O_2/dm^3) and it decreased to 0.4 mg O_2/dm^3 in the second half, whereas $r_{DN,max}$ had the same value of 3.8 mg $N/(dm^3 \cdot h)$ in both phases. These results implied that aerobic denitrification was less inhibited in the initial phase of the aeration period.

The nitrate utilization rate (NUR) is commonly accepted as the sole parameter that characterizes the denitrification kinetics, and nitrite is not even measured. Experimental evidence indicates, however, that on some occasions nitrite is reduced at a considerably lower rate causing nitrite accumulation (Wilderer *et al.* 1987). Based on the literature review, these authors listed three potential mechanisms responsible for the nitrite accumulation including repression of the synthesis of nitrite reductase, selection and enrichment in favour of microorganisms capable of reducing nitrate to nitrite, and capability of *Nitrobacter* to catalyze the reverse reaction to nitrite oxidation under anoxic conditions. In the developed model, two groups of microorganisms were considered: *"true"* denitrifiers capable of both nitrate and nitrite reduction, and nitrate reducers (facultative anaerobes) capable only of reducing nitrate to nitrite. Conversions of nitrate, nitrite and organic carbon were assumed to be the results of three processes: nitrate reduction by *"true"* denitrifiers, nitrate reduction by nitrate reducers and nitrite reduction by *"true"* denitrifiers. Model predictions were tested against the experimental data from a laboratory scale SBR. The observed and predicted nitrite utilization rates progressively decreased, whereas the nitrate utilization rates remained almost unaffected. The authors concluded that this fact indicated enhanced proliferation of facultative anaerobes at the expense of *"true"* denitrifiers.

Wild *et al.* (1995) and von Schulthess and Gujer (1996) proposed another approach to describe the conversions of denitrification intermediates. For this purpose, the authors formulated two models and coupled them with ASM1. The first model included the reduction of nitrate, nitrite and nitrous oxide (Table 3.15). Non-competitive inhibition (switching functions) of these processes by oxygen and nitrite was incorporated in the process rate expression. The other model (Table 3.16) was extended by adding the synthesis and decay of denitrification enzymes to correct predictions of nitrate, nitrite and N_2O concentrations as well as explain observations of the delay in denitrification after growth under aerobic conditions.

Table 3.15 Stoichiometric matrix and process rate equations in Model 1 (Wild *et al.* 1995)

Process	S_{NO3}	S_{NO2}	S_{N2O}	S_{N2}	Process rate (ρ_j)
NO$_3$-N reduction	-1	1			$r_1 \dfrac{S_{NO3}}{K_{NO3}+S_{NO3}} \dfrac{K'_{NO2,1}}{K'_{NO2,1}+S_{NO2}} \dfrac{K'_{O,1}}{K'_{O,1}+S_O} X_H$
NO$_2$-N reduction		-1	1		$r_2 \dfrac{S_{NO2}}{K_{NO2}+S_{NO2}} \dfrac{K'_{NO2,2}}{K'_{NO2,2}+S_{NO2}} \dfrac{K'_{O,2}}{K'_{O,2}+S_O} X_H$
N$_2$O reduction			-1	1	$r_3 \dfrac{S_{N2O}}{K_{N2O}+S_{N2O}} \dfrac{K'_{NO2,3}}{K'_{NO2,3}+S_{NO2}} \dfrac{K'_{O,3}}{K'_{O,3}+S_O} X_H$

Table 3.16 Additional stoichiometric matrix and process rate equations in Model 2 (Wild *et al.* 1995)

Process	E_{Sat}	Process rate (ρ_j)
Enzyme-synthesis	1	$r_4 \left[\dfrac{S_{NO3}}{K_{NO3,4}+S_{NO3}} \dfrac{S_{NO3}}{S_{NO2}+S_{NO3}} + \dfrac{S_{NO2}}{K_{NO2,4}+S_{NO2}} \dfrac{S_{NO2}}{S_{NO2}+S_{NO3}} \right]$ $\cdot \dfrac{K'_{O,4}}{K'_{O,4}+S_O}(1-E_{Sat})$
Enzyme-decay	-1	$r_5 E_{Sat}$

In all the most common, complex activated sludge models (outlined in Section 2.4.1), denitrification is assumed to be a one-step process. Hiatt and Grady (2008) and Sin *et al.* (2008b) reviewed several two-step denitrification models that have been proposed in recent years. These models do not consider the gaseous forms of nitrogen (NO and N_2O) as the model components and thus, with this approach, it is impossible to estimate the potential greenhouse gas emissions from denitrification systems. Hiatt and Grady (2008) developed a comprehensive model, termed the Activated Sludge Model for Nitrogen (ASMN), in which four anoxic processes were incorporated for all the intermediate nitrogen conversions. The authors noted that *"incorporation of emissions of nitric oxide (NO) and nitrous oxide (N₂O) and consideration of both inadvertent and intentional accumulation of nitrite necessitated more definite denitrification modelling than provided by ASM1, particularly for wastewaters with high nitrogen concentrations*". The ranges of kinetic and stoichiometric coefficients used in the denitrification models for nitrate and nitrite reduction are listed in Table 3.17.

Table 3.17 Parameter values used in two-step denitrification models (nitrate and nitrite reduction) for mainstream and sidestream treatment systems (values at T = 20 °C)

Parameter	Unit	Sin *et al.* (2008b)*	Hiatt and Grady (2008)
$\mu_{H,max}$	d^{-1}	1.5–6.0	6.25
η_{NO3}	–	0.15–0.8	0.28
η_{NO2}	–	0.15–0.8	0.16
$\eta_{NO3,end}$	–	0.25–0.50	–
$\eta_{NO2,end}$	–	0.35–0.70	–
$K_{NO3,H}$	mg N/dm^3	0.14–0.5	0.2
$K_{NO2,H}$	mg N/dm^3	0.12–0.5	0.2
$Y_{NO3,H}$	mg COD/mg COD	0.41–0.65	0.54
$Y_{NO2,H}$	mg COD/mg COD	0.24–0.65	0.54

* values reported by Hellinga *et al.* (1999), Volcke (2006), Wett and Rauch (2003), Jones *et al.* (2007), Sin and Vanrolleghem (2006), Kaelin *et al.* (2009)

3.9 ENHANCED BIOLOGICAL PHOSPHATE REMOVAL (EBPR)

3.9.1 Mechanism of the EBPR process

In conventional activated sludge systems, biological phosphate removal from wastewater is coupled stoichiometrically to microbial growth when phosphorus is utilized for cell maintenance and synthesis. Consequently, the amount of phosphorus incorporated in the activated sludge biomass typically remains within the range 1.5–3.0 mg P/mg VSS (Wentzel and Ekama 1997; Grady *et al.* 1999) and 10–30% of the influent phosphorus load can be removed in such a way (Metcalf and Eddy 1991). The capability of storing large quantities of phosphorus in the form of poly-P chains, originally called "*luxury uptake*", is the key mechanism of the process termed as biological excess phosphorus removal (BEPR) (Wentzel and Ekama 1997) or enhanced biological phosphorus removal (EBPR) (Mino *et al.* 1998).

The EBPR process is accomplished by heterotrophic microorganisms collectively referred to as phosphate accumulating organisms (PAOs), termed also variously as bio-P organisms, phosphotrophs and poly-P organisms (Wentzel and Ekama 1997). The PAOs can incorporate up to 0.38 mg P/mg VSS and therefore the phosphorus content of the activated sludge biomass in EBPR systems can increase to values of approximately 0.06-0.15 mg P/mg VSS (Wentzel and Ekama 1997). According to Grady *et al.* (1999), a typical value is 0.06 mg P/mg VSS, but higher contents (0.08-0.12 mg P/mg VSS) have also been observed. The results of experimental observations suggest that the stored polyphosphate is portioned between a low- and a high-molecular-weight fraction, and only the former fraction can be released again after being taken up (Mino *et al.* 1998). Barker and Dold (1997a) assumed the value of 94% for the releasable fraction which was in accordance with the literature data.

The principal mechanism for attaining the EBPR process is a continual circulation of the activated sludge biomass through an alternating anaerobic-aerobic (or anoxic) phases or zones and provision of volatile fatty acids (VFAs), also termed short-chain fatty acids (SCFAs), during the anaerobic phase. Such a configuration of activated sludge systems favours the growth of PAOs over "*ordinary*" heterotrophic organisms. In contrast to the latter group of microorganisms, PAOs are capable of taking up some carbon sources under anaerobic conditions and hydrolyzing intracellular stored polyphosphate in order to supply energy for the anaerobic uptake of the carbon sources. This process results in releasing orthophosphate into solution. In the early stages of the EBPR research, polyphosphate was considered to be the sole energy source for PAOs under anaerobic conditions. Now it is generally accepted that the utilization of stored glycogen (glycolysis) provides PAOs with reducing equivalents and additional energy during anaerobic substrate uptake (Mino *et al.* 1998). In the subsequent aerobic (or anoxic) phase, PAOs take up excessive amounts of orthophosphate to recover the intracellular polyphosphate level by oxidizing the stored PHA. Simultaneously, they grow and replenish the glycogen pool using PHA as both carbon and energy sources. Net phosphorus removal is achieved by wasting P-rich sludge after the aerobic (or anoxic) phase (Pijuan *et al.* 2004).

Wentzel and Ekama (1997) categorized the parameters that influence directly EBPR into two groups: wastewater and system characteristics (SRT, RAS and internal recycles, anaerobic mass fraction, series or single anaerobic reactor configuration). Indirect effects result from nitrification and denitrification, which may affect the nitrate discharged to the anaerobic reactor. The nitrate effect, in turn, depends on the temperature, maximum specific growth rate of autotrophic organisms, denitrification design of the plant, and the influent wastewater ratio of TKN/COD. Temperature also may have a direct influence on EBPR. This has not been established, but the effect seems to be relatively small, provided the aerobic sludge age is sufficient to sustain PAOs.

In a review paper, Mulkerrins *et al.* (2004) discussed some of the key wastewater composition parameters, which influence the EBPR process, such as COD, VFAs and cations (K^+ and Mg^{2+}) concentrations, phosphorus load and pH. In addition, it appears that careful consideration must be given to operation of the system at the correct F/M (food to microorganism) ratio, HRT, SRT, temperature and DO concentration in the aerobic zone.

3.9.2 Carbon sources and storage products

During the anaerobic phase, PAOs take up VFAs, primarily acetate and propionate which are most common VFAs in domestic wastewater (Pijuan *et al.* 2004). Moser-Engeler *et al.* (1998) and Meijer *et al.* (2001) found that acetate and propionate constituted the main part (approximately 60% and 90%, respectively) of the COD of all VFAs in the fermentation products. Von Muench (cited in Chen *et al.* 2004) summarized data from four full-scale systems with prefermenters in Canada and Australia. Acetate and propionate constituted 49-71% and 24-33%, respectively, of total influent VFAs (by weight). The propionate to acetate carbon molar ratios were in the range 0.41–0.82 in these systems. Lopez-Vazquez *et al.* (2008) found the propionate to acetate ratios (by weight) ranging from 0.12 to 0.88 at seven Dutch WWTPs. Ekama and Wentzel (1999a) reported several studies in which fermentation of primary sludge generated significant amounts of propionate (up to 50% of VFAs).

The majority of experimental studies on EBPR have been carried out using acetate or glucose as the sole carbon source, and acetate in combination with other substrates (Chen *et al.* 2004; Pijuan *et al.* 2004). Propionate as a carbon source has not been commonly studied (Abu-ghararah and Randall 1991; Pijuan *et al.* 2004; Chen *et al.* 2004), although as stated above propionate is also an important component of wastewater. The EBPR process also occurred with other organic substrates, such a mixture of peptone and glucose, or even only glucose (Carucci *et al.* 1999) as well as carboxylic acids and amino acids (Mino *et al.* 1998). Ubukata (2005) found that the organic compounds present in real wastewater contribute to a different phosphate release than acetate.

The carbon sources are stored under anaerobic conditions in the form of organic polymers collectively called poly-β-hydroxyalkanoate (PHA), such as poly-β-hydroxybutyrate (PHB) or poly-β-hydroxyvalerate (PHV) (Barker and Dold 1997a). When acetate is the only carbon source in the anaerobic phase, PHB is the major component of PHA formed (Mino *et al.* 1998). Since many laboratory scale systems were supplied with acetate this may explain the observations that PHB was recognized as the most commonly reported anaerobic intracellular storage product in the early stage of the EBPR research (USEPA 1987a). When propionate becomes the predominant carbon source, most PHA produced appears as PHV (Satoh et al. 1992; Lemos et al. 1998; Randall and Liu 2002; Chen et al. 2004; Pijuan et al. 2004).

The anaerobic storage of PHA is accompanied by degradation of polyphosphate and consequent release of orthophosphate. The stoichiometric coefficient Y_{PO4} indicates the ratio of ortophosphate released to COD utilized. In the literature review by Freitas et al. (2004), the observed values of Y_{PO4} ranged from 0.4 to 0.7 mg P/mg COD, but for modelling, the values within the range 0.3–0.43 mg P/mg COD were used (Johansson et al. 1996). A wide range (i.e. 0–0.88 mol P/mol C) of molar ratios of phosphorus released per acetate taken up (P/Hac) were reported by Mino et al. (1998). A possible explanation for the lower value of this range is that if the systems are operated to allow the presence of nitrate in the anaerobic zone, denitrification of some of the acetate would reduce the amount of substrate available to PAOs. Brdjanovic et al. (2000) justified relatively low P/Hac ratios obtained in batch tests by either long SRTs at the studied plant or the presence of glycogen accumulating organisms (see below). Moreover, the presence of other VFAs, such as propionate or butyrate, in wastewater will result in a different value for the P/Hac ratio (Meijer 2004). The ratio is also a function of pH. Smolders et al. (1994a) and Filipe et al. (2001b) developed linear relationships between the external pH and the P/Hac ratio:

- Smolders et al. (1994a):

$$Y_{P/Hac} = 0.19pH_{ext} - 0.85 \text{ (P-mole/C-mole)} \tag{3.39}$$

- Filipe et al. (2001b):

$$Y_{P/Hac} = 0.16pH_{ext} - 0.55 \text{ (P-mole/C-mole)} \tag{3.40}$$

3.9.3 Anoxic growth of PAO

Initially, it was thought that PAOs could not grow and accumulate phosphorus under anoxic conditions (e.g. Clayton et al. 1991). Later it has been demonstrated in several studies that PAOs can do so and anoxic phosphate uptake has been observed in laboratory scale systems and in full-scale WWTPs (Mino et al. 1998; Filipe and Daigger 1999). Experimental evidence has suggested that two different populations of PAOs exist in EBPR systems (Kerrn-Jespersen and Henze 1993; Barker and Dold 1996b; Meinhold et al. 1999). Some PAOs have the ability to use nitrate and oxygen as the terminal electron acceptors, while the reminding ones only use oxygen. Taking this observation into account Hu *et al.* (2002b) divided PAOs into two groups:

- Aerobic PAOs (APAOs) which can use oxygen as electron acceptor only;
- Denitrifying PAOs (DPAOs) which can use both oxygen and nitrate as electron acceptors.

The reported fraction of DPAOs among the entire phosphate accumulating population (i.e., PAOs + DPAOs) in full-scale EBPR plants ranged from 0 to 0.8 (Zeng *et al.* 2003b; Lopez-Vazquez *et al.* 2008). Kuba *et al.* (1996b), Wachtmeister *et al.* (1997) and Meinhold *et al.* (1999) developed experimental methods to measure the fraction of DPAOs based on the comparison of anoxic and aerobic acetate uptake rates. In practice, the denitrifying capability of PAOs is important for two reasons (Mino *et al.* 1998):

- In the mathematical modelling of the EBPR process, behaviour of phosphate and nitrogenous compounds like ammonia, nitrate and nitrite can be predicted only by introducing denitrifying PAOs into the model.
- The available amount of COD in the wastewater is a crucial limiting factor for both EBPR and denitrification.

The use of nitrate rather than oxygen for metabolism of PAOs appears to be advantageous for several reasons. The supply of organic substrates in wastewater, needed for both EBPR and denitrification is normally limited. Hence, improved nutrient removal is expected if the same organics can be used for both purposes. This "*double use*" of wastewater organics will also result in a reduced sludge production, and the use of nitrate rather than oxygen as electron acceptor for at least a portion of the phosphate uptake will reduce aeration demand (Copp and Dold 1998; Filipe and Daigger 1998; Meinhold *et al.* 1999). In order to maximize anoxic phosphate uptake and hence utilize as much influent readily biodegradable COD for denitrification as possible, several two-sludge systems based on the activity of DPAOs have been proposed in the recent years (Wanner *et al.* 1992; Bortone *et al.* 1994; Bortone *et al.* 1996; Kuba *et al.* 1996b; Sorm *et al.* 1997; Bortone *et al.* 1999; Hu *et al.* 2000; Shoji *et al.* 2003). In these systems, the nitrification process has been removed from the suspended medium activated sludge, and transferred to an external fixed medium system, so that the slow growing autotrophic organisms no longer have to be sustained in the suspended medium activated sludge part of the system (see: Section 5.5). This allows the anoxic zone to be significantly enlarged at the expense of the aerobic zone to stimulate DPAOs (Hu *et al.* 2002b).

Hu *et al.* (2002c) evaluated the advantages and disadvantages of anoxic phosphate uptake EBPR both in conventional BNRAS and external nitrification BNRAS systems.

The authors concluded that that anoxic phosphate uptake does not add a significant advantage to the BNR system. This statement was justified by the observation that the available experimental data exhibited a variable anoxic phosphate uptake ranging from 0 to 62% of the total phosphate uptake, while the contribution of the DPAOs to denitrification only varied from 0 to about 25%. Moreover, as noted earlier by Ekama and Wentzel (1999a), the P removal with anoxic-aerobic phosphate uptake was significantly reduced (to approximately 65-75%) in comparison with the P removal induced predominantly by aerobic phosphate uptake.

Filipe and Daigger (1999) demonstrated that, although DPAOs can grow both under anoxic and aerobic conditions, these microorganisms have a significant disadvantage in comparison with APAOs due to the lower efficiency of utilization of stored PHA under anoxic conditions and the lower aerobic growth rate resulting from the reduced amount of the stored PHA due to the partial utilization in the previous (anoxic) phase. Based on the results of previous experimental investigations, Hu et al. (2002b) noted that the growth yield of PAOs and the stoichiometric coefficient for phosphate uptake per PHA utilized under anoxic conditions should be reduced to approximately 70% and 80%, respectively, of those under aerobic conditions. This means that DPAOs show a significantly lower EBPR performance and use influent readily biodegradable COD less "efficiently" compared with APAOs. In practice, a positive aspect of the less efficient energy generation is a lower sludge production (Zeng et al. 2003b).

Although anoxic phosphate uptake has been demonstrated in many laboratory scale and full-scale plants, sometimes anoxic activity of PAOs is very low or even not observed in the EBPR systems operating with alternating anaerobic/anoxic/aerobic conditions (Ekama and Wentzel 1999a). The factors that could stimulate or inhibit anoxic phosphate uptake are still unknown (Spagni et al. 2002). Barker and Dold (1996b) suggested that lack of anoxic phosphate uptake could possibly be related to the low influent TKN/COD ratios and, consequently, to low nitrate loads recycled to the anoxic reactor. Also Hu et al. (2002b) indicated that the main factor affecting the occurrence of DPAOs and associated anoxic phosphate uptake was the nitrate load to the main anoxic reactor. This means that if the nitrate load is large enough or exceeds the denitrification potential of "ordinary" heterotrophs in the anoxic reactor, then DPAOs are stimulated in the system, and vice versa. The other potentially stimulating factors include the small aerobic mass fraction, sequence of reactors (nitrification and then denitrification) and reduced frequency of sludge alternation between the aerobic and anoxic states (Hu et al. 2002b; Hu et al. 2002c; Hu et al. 2003a). This variability of factors would suggest that the value of reduction factor for PAO anoxic activity (η_{PAO}) may be not universal in model applications but rather system- and time-specific (Hu et al. 2002b). Also the results of other studies (Wachtmeister et al. 1997; Filipe and Daigger 1999; Meinhold et al. 1999; Zeng et al. 2003b) have suggested that the exposure of sludge to aerobic and anoxic conditions strongly affects accumulation of DPAOs in the system.

3.9.4 Approaches to modelling the EBPR process

A lot of research studies have indeed been dedicated to investigate the metabolism of PAOs, but details of their biochemical pathways are still *"only hypotheses without confirmation"* (Pijuan *et al.* 2004). The first step of the EBPR process is the anaerobic uptake of a favourable substrate by the PAOs and its storage as PHA. This biochemical conversion requires energy, which is provided by polyphosphate and reducing power for synthesis PHA (Filipe and Daigger 1998). Proposed explanations of the source of the reducing power have been presented in two biochemical models referred to as *"the Comeau-Wentzel model"* (Comeau *et al.* 1986; Wentzel *et al.* 1986) and *"the Mino model"* (Mino *et al.* 1987).

In the Comeau-Wentzel model (Figure 3.7), the reducing power is generated from a parallel metabolism of acetate. Some of the acetate taken up by the cells is directed through the TCA (tricarboxylic acid) cycle. Reducing power is generated and used to transform the remaining acetate to PHB (Filipe and Daigger 1998). An alternative idea has been proposed in the Mino model (Figure 3.8), This model is based on the observation that a significant decrease in the carbohydrate content of PAOs is observed during the uptake of acetate. In the Mino model, reducing power is considered to be derived from degradation of intracellularly stored glycogen to generate glucose that is directed through the Embden-Mayerhof-Parnas (EMP) pathway (Filipe and Daigger 1998).

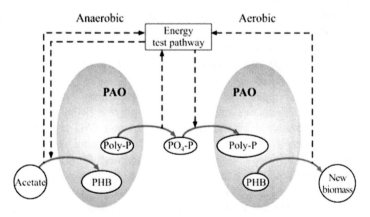

Figure 3.7 Schematic diagram showing the relationships in the Comeau-Wentzel model for the uptake and release of inorganic phosphate by PAOs.

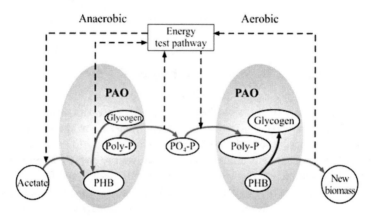

Figure 3.8 Schematic diagram showing the relationships in the Mino model for the uptake and release of inorganic phosphate by PAOs.

Both biochemical models are largely in agreement regarding the biochemical control mechanisms and have provided an explanation for the essential requirements for attaining EBPR, namely, an alternating anaerobic/aerobic sequence with the provision of VFAs during the anaerobic phase (Barker and Dold 1997a). Differences between the models result in different amounts of acetate removed, PHB produced and phosphorus released (Carucci *et al*. 1999). In the study of Filipe and Daigger (1998), the Mino model described the transformations occurring in the anaerobic phase more accurately. Mino *et al*. (1998) concluded that the reducing power required for anaerobic PHA synthesis is primarily supplied by the degradation of stored glycogen and therefore the Mino model is likely to be correct. However, in spite of several experimental evidences which strongly support the Mino model, the possibility of partial functioning of the TCA cycle cannot be totally excluded.

The biochemical aspects of EBPR have significantly contributed to the development of mathematical modeling of the process (Mino *et al*. 1998). A comprehensive structured kinetic model for enriched PAO cultures was first proposed by Wentzel *et al*. (1989b). These cultures were developed in the continuous-flow activated sludge systems (modified Bardenpho and UCT process configurations) fed with acetate as the only carbon source (Wentzel *et al*. 1988). Using a single set of kinetic and stoichiometric coefficients the model provided a reasonable description of the response in these systems and batch experiments with biomass drawn from these systems (Wentzel *et al*. 1989a). The EBPR model concept of Wentzel *et al*. (1989b) was used and further modified by Dold (1990) and Barker and Dold (1997a). The latter concept is graphically presented in Figure 3.9a. Under anaerobic conditions, PAOs store VFAs (S_A) in the form of cell internal storage material assumed to be PHA (X_{PHA}). The storage of PHA yields Y_{PHA} units (as COD) for each unit of VFAs (as COD) taken up. Monod terms allow the storage to be switched off when either polyphosphate or VFAs become limiting. The energy for this process is supplied from "*cleavage*" of "*releasable*" stored polyphosphate (X_{PP-LO}), which subsequently leads to release of soluble phosphate (S_{PO4}). The cleavage of X_{PP-LO} also provides the energy for anaerobic maintenance which is modelled as a separate process. Under aerobic (or anoxic) conditions, PAOs grow on X_{PHA}. There are four different process equations depending on whether S_{PO4} is limiting or not and whether ammonia or nitrate serves as the nitrogen source for cell synthesis. Switching functions enable the model to predict a change to P limitation or a change in N source. The S_{PO4} uptake and storage as polyphosphate (X_{PP-LO} and X_{PP-HI}) is directly coupled with growth of PAOs by a stoichiometric coefficient ($f_{P,UPT}$). Only a fraction (f_{PP}) of S_{PO4} is stored in the form that can be released subsequently ("*releasable*" polyphosphate, X_{PP-LO}), whereas the remaining fraction ($1-f_{PP}$) becomes fixed ("*fixed*" polyphosphate, P_{PP-HI}). Under anoxic conditions, the rate of PAO growth and polyphosphate storage is reduced by a correction factor (η_P). Moreover, the stoichiometric coefficient for anoxic uptake of S_{PO4} ($f_{P,UPT2}$) differs from that for aerobic conditions ($f_{P,UPT1}$). Decay of PAO cells is modelled according to the "*death-regeneration*" concept (see: Section 3.2) with portions, $f_{EP,P}$ and $f_{ES,P}$, of the dead cells adding to the particulate endogenous (non-biodegradable) cell residue (X_I) and soluble non-biodegradable residue (S_I), respectively. The remaining portion becomes part of the pool of particulate biodegradable COD (X_S). The effect of electron acceptor is accounted for. The associated lysis of cell internal components, i.e. X_{PHA} and both fractions of polyphosphate (X_{PP-LO}, X_{PP-HI}), leads to release of S_A and S_{PO4}, respectively.

The IAWQ Task Group on Mathematical Modelling for Design and Operation of Biological Wastewater Treatment Processes (Henze *et al.* 1995a) presented a different concept for modelling the EBPR process (based on the Comeau-Wentzel biochemical model), which was included in Activated Sludge Model No. 2 (ASM2). The concept was further extended by the same task group (Henze *et al.* 1999) to account for anoxic behaviour of PAOs in a modification of ASM2 called ASM2d (Figure 3.9b). In ASM2d, it is assumed that PAOs take up cell external fermentation products (S_A) and store them in the form of cell internal organic storage material (X_{PHA}). This process occurs primarily under anaerobic conditions but has also been observed under aerobic and anoxic conditions. Therefore, no inhibition terms for oxygen or nitrate are implemented in the kinetic expression.

The energy for storage of S_A is provided by release of phosphate (S_{PO4}) from polyphosphate (X_{PP}). Both processes, i.e. storage and release, are coupled by the stoichiometric parameter Y_{PO4}. Under aerobic (or anoxic) conditions, PAOs take up S_{PO4} to recover the X_{PP} levels. The energy for this process is gained from aerobic (or anoxic) respiration of X_{PHA}. Storage of X_{PP} becomes inhibited as the X_{PP}/X_{PAO} ratio approaches the maximum allowable value (K_{MAX}). Under anoxic conditions, the rate of X_{PP} storage is reduced relative to its value under aerobic conditions by a correction factor ($\eta_{NO3,PAO}$). In ASM2d, PAOs grow only at the expense of X_{PHA}. These organisms are known to grow at the expense of soluble substrates (e.g. S_A), but this possibility has been ignored because, in practice, it is unlikely that such substrates ever become available under aerobic or anoxic conditions in a BNR plant. Furthermore, the growth rate of PAOs in ASM2d is reduced under anoxic conditions relative to its value under aerobic conditions by the $\eta_{NO3,PAO}$ factor.

The EAWAG Bio-P module (Rieger *et al.* 2001), shown in Figure 3.9c, considers similar processes to those included in ASM2d. There are, however, several differences between the two model concepts. The module neglects fermentation of the readily biodegradable substrate assuming that there is no limitation of phosphate release due to the fermentation process in typical municipal wastewater. Besides neglecting fermentation, the main difference in comparison with ASM2d is the use of endogenous respiration instead of lysis for modelling disappearance of PAO biomass components (see: Section 3.2). Moreover, the disappearance under anoxic conditions is slower than under aerobic conditions, which is reflected by introducing anoxic reduction factors ($\eta_{NO,end,PAO}$, $\eta_{NO,lys,PP}$, $\eta_{NO,resp,PHA}$). Anaerobic decay is minimal and is therefore neglected. This assumption results from the observations made during batch experiments under starvation conditions (Siegrist *et al.* 1999).

The New UCTPHO model (Hu *et al.* 2007), shown in Figure 3.9e, uses the ASM2d concept with some modifications. Both ammonia (S_{NH}) and nitrate (S_{NO}) can be used by PAOs as N source and the aerobic growth of PAOs can occur under P-limiting (S_{PO4}) conditions. The storage of polyphosphate (X_{PP}) is not modelled as a separate process but as part of the growth of PAOs on PHA (X_{PHA}). Furthermore, there is no limit on the X_{PP} stored (no K_{MAX} coefficient). The lysis of X_{PHA} on aerobic/anoxic decay leads to the formation of the enmeshed slowly biodegradable substrate (X_S) but not VFAs (S_A) as it is described in ASM2d. Glycogen was not considered as a state variable assuming that "*it was unlikely that in "normal" BEPR operation glycogen would be limiting*".

The most detailed mathematical model for EBPR was developed by Smolders *et al.* (1994a, 1994b, 1995a), which included a stoichiometry of the EBPR metabolism with glycogen, PHA and polyphosphate as cell internal storage polymers, and the energy budget under aerobic and anaerobic conditions (Mino *et al*. 1998). This model was further extended by Kuba *et al*. (1996a) and Murnleitner *et al*. (1997) to describe the anoxic conversions for EBPR. Because the approach used by Smolders *et al*. was based on degradation and formation of all the relevant cell internal storage compounds, the model concept was called "*metabolic*" (Figure 3.9d). More details on the metabolic model development and its practical applications were presented in Section 1.2.3. In this model concept, it is assumed that VFAs (S_A) are taken up by PAOs (X_{PAO}) and stored as PHA (X_{PHA}) under anaerobic conditions. The energy for anaerobic uptake and storage of S_A (for stoichiometric calculations assumed to be acetate) is generated by degradation of cell internal glycogen (X_{GLY}) and polyphosphate (X_{PP}). In addition, X_{PP} is degraded to provide the energy for anaerobic maintenance. As a consequence of both processes, PAOs release large amounts of soluble phosphate (S_{PO4}) into the bulk liquid. In a subsequent aerobic (or anoxic) phase, X_{PHA} is oxidized to generate the energy for restoring X_{GLY} and X_{PP}. In order to restore X_{PP}, PAOs have to take up S_{PO4} from the bulk liquid. The remaining energy is used for growth of PAOs and maintenance of their cell structure. It is assumed that PAOs always have internal substrate (X_{PHA}) available to satisfy the maintenance requirements (van Veldhuizen *et al*. 1999a).

Among the outlined models, only the metabolic model includes the conversions of glycogen. In the other cases, glycogen is not introduced as a variable, because initially its role was not recognized in the metabolism of PAOs (see: the Comeau-Wentzel biochemical model in Figure 3.7). This aspect could also be considered too specific for a general mathematical model and relevant data available were very limited when the models were developed (Mino *et al*. 1998). Rieger *et al*. (2001) neglected an additional glycogen pool to keep the model structure simple. This assumption was adopted based on the results of measurements in pilot and full-scale experiments in Swiss municipal WWTPs. These experiments did not show the expected significant influence of glycogen on the EBPR process, even in situations with a low COD load, e.g. after weekends, where glycogen limitation could play a crucial role. Therefore, the authors considered a model including glycogen "*unnecessary complex*". This finding remains in contradiction to the statement of Mino *et al*. (1998), who emphasized that the inclusion of glycogen is necessary if shock loading conditions are modelled where glycogen can be the limiting substance. It should also be noted that glycogen and PHA have counteracting dynamics and the use of one lumped substrate pool damps the net dynamics of such a pool. Consequently, this reduces the sensitivity and accuracy of the modelled storage (Meijer 2004).

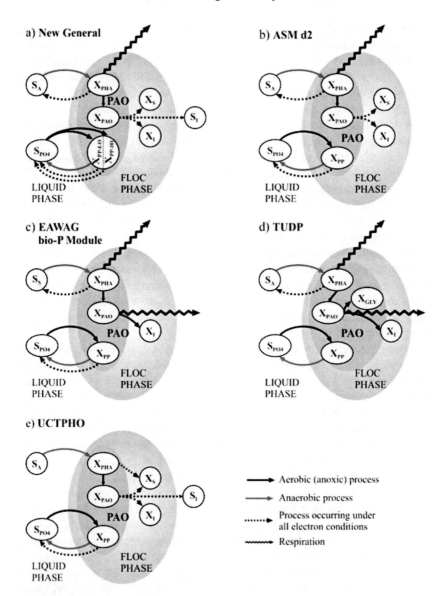

Figure 3.9 Comparison of the concepts for modelling the EBPR process.

3.9.5 Effect of GAO metabolism on EBPR

Occasional failures of EBPR processes due to unknown reasons have been reported even in well-designed laboratory scale reactors. In many of these cases, a particular group of microorganisms surviving and proliferating under the alternating anaerobic/aerobic conditions could be found (Zeng *et al.* 2003a). Cech and Hartman (1993) found such microorganisms in a glucose-fed SBR and named them "*G bacteria*", which are now more commonly referred to as glycogen accumulating organisms (GAOs) (Liu *et al.* 1996). The GAOs have the potential to directly compete with PAOs in EBPR systems as both groups of microorganisms are capable of taking up VFAs under anaerobic conditions and growing on the intracellular storage products under aerobic conditions. The difference is that GAOs are able to take up these compounds without involvement of anaerobic phosphate release and subsequent aerobic phosphate uptake. For this reason, a significant GAO population is undesirable due to the increase of VFA requirements that could otherwise be used by PAOs. Under anaerobic conditions GAOs hydrolyze glycogen in the process called glycolysis to gain energy and reducing equivalents to take up the VFAs and to synthesize PHA. In the subsequent aerobic zone, PHA is oxidized to gain energy for glycogen replenishment (from PHA) and for cell growth (Zeng *et al.* 2003a).

Wanner *et al.* (2000) claimed that the competition between PAOs and GAOs in full-scale WWTPs is less common that could be expected from previous laboratory studies. In the study of Saunders *et al.* (2003), GAOs were present in all of six full-scale WWTPs investigated in Australia, but only in one case GAOs made a significant contribution to the amount of VFAs taken up under anaerobic conditions. Also Lopez-Vazquez *et al.* (2008) found that GAOs did not occur in the range that might have affected the EBPR process performance at seven Dutch WWTPs under winter conditions (T = 9-15 °C).

The mechanism of the competition between PAOs and GAOs in taking up VFAs is not very well recognized at present. In several studies, however, certain environmental and operational conditions have been identified as potential factors to understand that competition including SRT, type of VFA (e.g. acetate or propionate) or other substrate (e.g. glucose) present in the influent wastewater, pH, temperature and the P/VFA ratio (Makinia 2006; Lopez-Vazquez *et al.* 2008; Lopez-Vazquez *et al.* 2009).

Mino *et al.* (1995a) proposed a preliminary conceptual model for the GAO metabolism without providing specific values for the kinetic and stoichiometric coefficients for this model (Figure 3.10). The authors concluded that for the purpose of modeling EBPR phenomena in activated sludge processes the classification of microorganisms should also incorporate GAOs. A detailed metabolic model for the anaerobic processes of GAOs was developed by Filipe *et al.* (2001c), whereas Zeng *et al.* (2002) proposed further amendments to the model stoichiometry. A complete anaerobic/aerobic metabolic GAO model was presented by Zeng *et al.* (2003a). According to these models, GAOs take up VFAs under anaerobic conditions and store them in the form of intracellular PHA. The required energy and reducing power are released in glycolysis of glycogen without involving phosphate release. Under the subsequent aerobic conditions, GAO use the stored PHA to grow and replenish the glycogen pool.

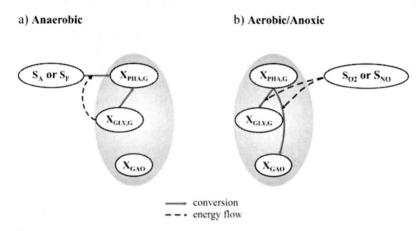

Figure 3.10 Conceptual model for the GAO metabolism (Mino *et al.* 1995a).

Several mathematical models describing the PAO-GAO competition in BNRAS systems were proposed and evaluated by Manga *et al.* (2001), Zeng *et al.* (2003b), Yagci *et al.* (2004), Whang *et al.* (2007) and Lopez-Vazquez *et al.* (2009). Manga *et al.* (2001) proposed an extension of ASM2 that includeed denitrification by PAOs (in a similar way to ASM2d), glycogen storage by PAO and metabolism of GAO. In comparison with ASM2, only four new particulate components were added: an additional to $X_{PHA,P}$ cell internal storage product of PAOs ($X_{GLY,P}$) assumed to be glycogen, GAO biomass (X_{GAO}) as well as two cell internal storage products of GAO ($X_{PHA,G}$ and $X_{GLY,G}$) assumed to be PHA and glycogen, respectively. The GAO metabolism model was described by eight processes as shown in Table 3.18 including storage of $X_{PHA,G}$, aerobic/anoxic growth of GAOs, aerobic/anoxic glycogen storage of GAOs and lysis of GAO and their two storage products.

Table 3.18 Stoichiometry and kinetics of the GAO behaviour (adapted from Manga et al. (2001))

Process	S_O	S_A	S_{NO}	S_{PO4}	X_I	X_S	X_{GAO}	$X_{PHA,G}$	$X_{GLY,G}$	Simplified process rate expression[1]
Storage of PHA by GAOs		$-Y_{SA,G}$						1	$-(1-Y_{SA,G})$	$q_{PHA,G}X_{GAO}$
Aerobic growth of GAOs	$-\dfrac{(1-Y_{GAO})}{Y_{GAO}}$			$-i_{P,BM}$			1	$-\dfrac{1}{Y_{GAO}}$		$\mu_{GAO}X_{GAO}$
Anoxic growth of GAOs			$-\dfrac{(1-Y_{GAO,NO})}{2.86 \cdot Y_{GAO,NO}}$	$-i_{P,BM}$			1	$-\dfrac{1}{Y_{GAO,NO}}$		$\mu_{GAO}\eta_{NO3,GAO}X_{GAO}$
Aerobic storage of glycogen by GAOs	$-\dfrac{(1-Y_{GLY,G})}{Y_{GLY,G}}$							$-\dfrac{1}{Y_{GLY,G}}$	1	$q_{GLY,G}X_{GAO}$
Anoxic storage of glycogen by GAOs			$-\dfrac{(1-Y_{GLY,GNO})}{2.86 \cdot Y_{GLY,GNO}}$					$-\dfrac{1}{Y_{GLY,GNO}}$	1	$q_{GLY,G}\eta_{NO3,GAO}X_{GAO}$
Lysis of GAOs				$\upsilon_{17,SPO4}$	f_p	$1-f_p$	-1			$b_{GAO}X_{GAO}$
Lysis of PHA		1						-1		$b_{PHA,G}X_{PHA,G}$
Lysis of glycogen		1							-1	$b_{GLY,G}X_{GLY,G}$

[1] Saturation terms are neglected.

3.10 BULKING SLUDGE (GROWTH OF *MICROTHRIX PARVICELLA*)

A widespread problem in the operation of activated sludge systems is the excessive abundance of filamentous microorganisms that tend to overgrow non-filamentous (floc-forming) microorganisms. This undesirable condition can result in two major operational impacts including the inability to separate solids from treated effluents in the secondary clarifiers ("*bulking sludge*") and/or forming a stable scum layer on the top of activated sludge reactors (or on the water surface at any location in wastewater treatment plants). Although the distribution of specific filamentous microorganisms considerably varies between geographical regions, several surveys have identified *Microthrix parvicella* as a dominating filament responsible for most bulking and scumming events in BNR activated sludge systems in Europe (Jardin 2006). These surveys have also shown that the abundance of *M. parvicella* follows a typical seasonal pattern with massive growth occurring in winter and early spring. Furthermore, *M. parvicella* has been identified as a specialized microorganism responsible for degradation of lipids, but the processes that allow this filament to successfully compete with other microorganisms in activated sludge systems still need to be clarified (Hug *et al.* 2006).

The competition between filamentous and non-filamentous (floc-forming) bacteria has been described by a few theories summarized by Martins *et al.* (2004) including diffusion selection, kinetic selection, storage (metabolic) selection and nitric oxide (NO) hypothesis. Models that can predict the bulking and scumming phenomena are in an early stage of development. In general, these models can be grouped into two categories: "*pure*" biokinetic models (considering the bacterial physiology and kinetics) and biokinetic models coupled with the morphology of bacteria. Existing models attempt to predict the development of filamentous and non-filamentous bacteria considering either these two general species or a group of species (e.g. floc formers, filaments, low DO filaments, low F/M filaments) competing for a single substrate or dual substrates (e.g. readily biodegradable COD vs. slowly biodegradable COD) (Martins *et al.* 2004).

3.10.1 Conceptual model for *M. parvicella* in activated sludge

Based on the results of their own studies, Andreasen and Nielsen (2000) and Nielsen *et al.* (2002) proposed a conceptual model explaining the proliferation of *M. parvicella* in activated sludge systems (Figure 3.11). The authors assumed that the microorganism is a specialized consumer of lipids, however, only the dissolved long-chain fatty acids (LCFA) can directly be taken up and stored by *M. parvicella*. The particulate lipids and LCFA can become available for consumption in two ways, i.e. thorough dissolution or adsorption and hydrolysis on the cell surfaces. The floc-forming bacteria compete with *M. parvicella* for both fractions of lipids. However, due to its filamentous structure and more hydrophobic surface *M. parvicella* is capable of adsorbing much of the substrate present in wastewater.

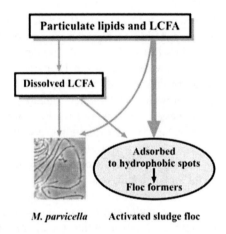

M. parvicella   Activated sludge floc

Figure 3.11 Conceptual model explaining the proliferation of *M. parvicella* in activated sludge systems (adopted from Andreasen and Nielsen (2000)).

In addition to the storage capabilities of *M. parvicella* and floc-formers, the lipid content of municipal wastewater is a crucial issue for the competition between these two groups of microorganisms. The occurrence of lipids in municipal wastewater and possible transformations in activated sludge were discussed in detail by Dueholm *et al.* (2001). Lipids can be characterized either as oils, greases, fats or LCFA. The reported amounts of lipids in municipal wastewater range from 9 to 40% of total COD (Dueholm *et al.* 2001; Morgenroth *et al.* 2002b; Nielsen *et al.* 2002). The main lipid constituents in wastewaters, comprising 80-85% of the total amount, are triacylglyerides (TAG), whereas free LCFA are the remaining portion. Both groups are generally hydrophobic and primarily exist in the forms adsorbed to particles or dissolved by detergents. Only a minor portion of LCFA is directly dissolved in wastewater. The fatty acid composition of TAG and LCFA is very similar with approximately 80% of the content in the form of palmitic, stearic, oleic and linoleic acids. However, only LCFA can be considered readily biodegradable by microorganisms in activated sludge with or without excessive growth of *M. parvicella*. The utilization rate of LCFA is comparable to acetate in terms of the measured oxygen or nitrate utilization rates. TAG cannot directly be utilized by the microorganisms, but must be hydrolysed first at a very slow rate.

3.10.2 Mathematical model structure for *M. parvicella* in activated sludge

Hug *et al.* (2006) proposed a matrix of stoichiometric coefficients (Petersen matrix) for an extension of ASM3 to describe the seasonal variations of the *M. parvicella* abundance in a full-scale WWTP. The details of the process rate expressions have not been provided except for the statement that *M. parvicella* growth was modelled according to ASM3 concept for heterotrophic growth. Spering *et al.* (2007, 2008) presented a full model structure ("*Microthrix module*") and described the new processes and components. Basically, five processes are considered in the newly developed module including hydrolysis of lipids, storage of lipids, growth and decay of *M. parvicella*, and decay of internally stored lipids. Apart from the hydrolysis, all the processes incorporated in the module have reduction factors for anoxic conditions. Moreover, the storage of lipids can be accomplished by both *M. parvicella* and other heterotrophs. Furthermore, the module contains four new model components: particulate lipids and LCFA (X_{LIP}), readily biodegradable dissolved lipids (S_{LIP}), *M. parvicella* biomass (X_{MIC}) and a cell internally stored lipids (X_{STOLIP}). The model concept used in the Microthrix module is illustrated in Figure 3.12.

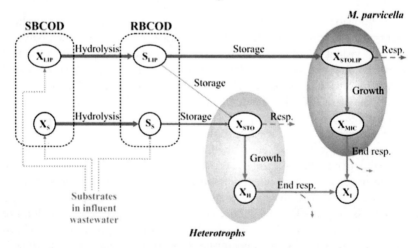

Figure 3.12 Schematic diagram illustrating the substrate flow in the *M. parvicella* module combined with ASM3 (Spering *et al.* 2008).

In order to model population dynamics, quantitative data and selective identification method are needed (Wilderer *et al.* 2002). In the developed model, a target model prediction is the *M. parvicella* abundance which can be calculated as the concentration ratio of *M. parvicella* solids ($X_{TS,MIC}$), which include both the biomass and stored lipids, to total suspended solids (X_{TS}) (Hug *et al.* 2006):

$$X_{TS,MIC} = i_{TS,XLIP}X_{STOLIP} + i_{TS,BM}X_{MIC} \tag{3.41}$$

$$\begin{aligned} X_{TS} = i_{TS,XS}X_S + i_{TS,STO}X_{STO} + i_{TS,XLIP}(X_{LIP} + X_{STOLIP}) \\ + i_{TS,BM}(X_H + X_A + X_{MIC}) + i_{TS,XI}X_I + X_{Precip} \end{aligned} \tag{3.42}$$

The rapid quantification of $X_{TS,MIC}$ in Equation 3.42 can be carried out using a method described by Hug *et al.* (2005). Instead of a real concentration, the outcome of that method provides a "*pixel concentration*" which represents the ratio of *M. parvicella* solids volume to the apparent total floc volume and the calculated values are comparable among themselves.

3.10.3 Kinetic and stoichiometric parameters, temperature effects

Rossetti *et al.* (2005) emphasized that very limited quantitative data on kinetic and stoichiometric coefficients exist in literature for *M. parvicella* as this filament is difficult to isolate and cultivate. The authors summarized the reported values with regard to the maximum growth rate constant, μ_{MIC}, substrate half-saturation coefficient, $K_{S,MIC}$ (not existing in ASM3), aerobic yield coefficient, $Y_{MIC,O2}$, and temperature correction factor (Arrhenius coefficient) for growth, θ_{MIC}. In most cases, there are substantial discrepancies between the values of these parameters obtained in different studies.

Relatively comprehensive data are available for the μ_{MIC} coefficient (Table 3.19). Batch tests with both pure and mixed cultures revealed the μ_{MIC} values may remain in a relatively narrow range 0.3-0.66 d^{-1}, however, the reported value (=1.44 d^{-1}) was considerably higher when estimated for a pure culture of *M. parvicella* in a chemostat. It should be noted, however, that results of batch experiments depend on the initial substrate to biomass (S/X) ratio, which directly affects the actual system behaviour and determines whether the catabolic or anabolic processes prevail.

Table 3.19 Summary of the maximum growth rate values of *M. parvicella* estimated in pure culture and mixed biomass (Rossetti *et al.* 2005)

Conditions	Culture	Limiting substrate	Growth indicator	$\mu_{MIC,max}$, d^{-1}	Reference
Batch (liquid medium)	Pure	Tween 80	Protein	0.38	Slijkhuis (1983)
Batch (liquid medium)	Pure	Complex carbon sources	TEFL	0.3–0.5	Tandoi *et al.* (1998)
Batch (agar plate)	Pure	R2N	TEFL	0.37–0.66	Rossetti *et al.* (2002)
Batch (agar plate)	Mixed	R2N	TEFL	0.66	Rossetti *et al.* (2002)
Chemostat	Pure	Tween 80	Dry weight	1.44	Slijkhuis (1983)

Literature values of the aerobic yield coefficient of *M. parvicella* ($Y_{MIC,O2}$) are not consistent and no experimental data for this parameter are available with regard to anoxic conditions. Rossetti *et al.* (2005) found the reported values of $Y_{MIC,O2}$ in the range 0.1–0.17 mg VSS/mg COD and concluded that such low values of the yield coefficient suggest that *M. parvicella* has a high energy requirement for growth. On the contrary, Slijkhuis (1983) determined the $Y_{MIC,O2}$ value equal to 1.41 mg biomass/mg oleic acid. This value was used in the later studies of Lebek (2003) and Hug *et al.* (2006). Lebek (2003) calculated its equivalent in the COD units (=0.72 mg $COD_{biomass}$/mg $COD_{oleic\ acid}$), which is slightly higher than the default value (=0.67 mg $COD_{biomass}$/mg $COD_{substrate}$) used for heterotrophs in ASM1 (Henze *et al.* 1987). It should be noted, however, that all those values refer in fact to the "*apparent*" (or net) yield ($Y_{MIC,net,O2}$) which incorporates both storage and growth. Hug *et al.* (2006) adapted the value determined by Slijkhuis (1983) to ASM3 model concept by using separate yield coefficients for these two phenomena. The calculated "*true*" $Y_{MIC,O2}$ = 0.67 mg $COD_{biomass}$/mg $COD_{stored\ substrate}$ is identical with the yield for heterotrophs in the calibrated ASM3 based on the experimental data from a few Swiss WWTPs (Koch *et al.* 2000b).

The reported values of $K_{S,MIC}$ ranged from 23 and 30 mg COD/dm^3 for Tween 80 and oleic acid as the substrates, respectively, to 3.9 mg COD/dm^3 for two strains (RN1 and 4B) cultivated and maintained on R2A medium (Rossetti *et al.* 2005). A high variability of the coefficients may be attributed to the cultivation conditions. In the latter case, the authors of that study (Rossetti *et al.* 2002) emphasized that the obtained low value of $K_{S,MIC}$, estimated in a pure culture study, refers to filaments growing in the liquid bulk under intrinsic conditions. Higher $K_{S,MIC}$

values, however, may be expected for filaments developing inside the flocs due to the diffusional resistance of the substrate transport from the liquid phase to the flocs (so called "*apparent*" K_S values). A high substrate affinity of *M. parvicella*, expressed by low $K_{S,MIC}$ values, is advantageous in terms of the kinetic competition with floc-formers in systems operating at low substrate concentrations.

Rossetti *et al.* (2002) investigated the effect of temperature on the growth rate of *M. parvicella* using the Arrhenius equation in the range 7–20 °C. The estimated temperature correction factors (θ) for strains 4B and RN1 were 1.114 and 1.105, respectively, and these values are comparable to the maximum growth rate constants of nitrifying bacteria in ASM3 ($\theta=1.11$). Hug *et al.* (2006) calibrated the pronounced seasonal variations of *M. parvicella* abundance by adjusting the temperature correction factor for the saturation constant for storage of lipids by heterotrophs ($K_{SLIP,H}$). The authors justified their choice by the fact that the solubility of lipids decreases in low temperatures and therefore reduces their availability for heterotrophs. *M. parvicella* is less affected as a specialized lipid consumer. However, the adjusted value of the θ coefficient was 0.67 which resulted in exceptionally high values of $K_{SLIP,H}$ at low temperatures (e.g. $K_{SLIP,H} = 245$ mg COD/dm^3 at 12 °C). That fact led the authors to the suggestion that the seasonal fluctuations of *M. parvicella* abundance result not from the solubility of lipids but from a change within the heterotrophic population towards the microorganisms that are incapable of utilizing lipids.

3.11 ANAEROBIC AMMONIUM OXIDATION (ANAMMOX)

3.11.1 Mechanism of the Anammox process

Broda (1977) published a theoretical paper describing, based on thermodynamic calculations, the potential existence of chemolithotrophic bacteria capable of oxidizing ammonia to dinitrogen with nitrate, carbon dioxide or oxygen as oxidant. Mulder *et al.* (1995) observed unexplainable nitrogen losses in a denitrifying fluidized bed reactor treating the effluent from a laboratory methanogenic reactor. The authors were first who attributed those losses to a new anaerobic process, named **AN**aerobic **AMM**onium **OX**idation (Anammox), in which ammonium was used as electron donor for denitrification. Van de Graaf *et al.* (1995) demonstrated by inhibition experiments that Anammox is indeed a microbiologically mediated process which involves the oxidation of ammonia with nitrite as the electron acceptor to yield gaseous nitrogen:

$$NH_4^+ + NO_2^- \rightarrow N_2 + 2\,H_2O \left[\Delta G^{\circ\prime} - 357\,kJ\,mol^{-1} \right] \qquad (3.43)$$

Equation 3.43 implies that the name "*anaerobic*" ammonium oxidation should rather be termed "*anoxic*" ammonium oxidation since nitrite is present as electron acceptor (anoxic conditions). Anammox bacteria do not, however, consume ammonia and nitrite in a ratio 1:1 as it could be expected from Equation 3.43. The observed NO_2-N/NH_4-N removal ratios in various Anammox reactors ranged from 0.5 to 4 depending on the substrate, operating conditions and reactor configuration (Ahn 2006). For example, Strous *et al.* (1999b) demonstrated that increasing the total nitrite nitrogen (TNO_2=HNO_2+NO_2^-), concentration changed the stoichiometry of the total ammonia (ammonium) nitrogen (TAN=NH_3+NH_4^+) and TNO_2 consumption from 1.3 mg TNO_2-N/mg TAN-N at 0.14 mg TNO_2-N/dm^3 to almost 4 mg TNO_2-N/mg TAN-N at 0.7 mg TNO_2-N/dm^3.

The role of the Anammox bacteria in the nitrogen cycle is presented in Figure 3.13. Even though the main product of the Anammox process is the dinitrogen gas, a small amount of the nitrogen feed is also converted to nitrate. Based on a mass balance over Anammox enrichment cultures in a SBR, Strous *et al.* (1998) derived the global stoichiometric equation, in which ammonia and nitrite are utilized in a ratio approx. 1:1.3 with the excess 0.3 mol of nitrite anaerobically oxidized to nitrate (the electrons derived from this oxidation are probably used for the fixation of CO_2):

$$NH_4^+ + 1.32\ NO_2^- + 0.066\ HCO_3^- + 0.13\ H^+$$
$$\rightarrow 1.02\ N_2 + 0.26\ NO_3^- + 0.066\ CH_2O_{0.5}N_{0.15} + 2.03\ H_2O$$

$$(3.44)$$

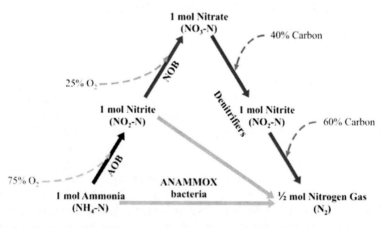

Figure 3.13 Role of the Anammox bacteria in the nitrogen cycle.

Bacteria carrying out the anaerobic ammonium oxidation had not been known earlier and were identified as lithotrophs of the order of *Planctomycete* (Strous *et al.* 1999a). Two of these bacteria were tentatively named "*Candidatus Brocadia anammoxidans*" (Strous *et al.* 1999a) and "*Candidatus Kuenenia stuttgartiensis*" (Schmid *et al.* 2000). The former bacterium was discovered in the Netherlands (Strous *et al.* 1999a), while the latter was found in several biofilm systems in Switzerland and Germany (Schmid *et al.* 2000; Egli *et al.* 2001; Helmer-Madhok *et al.* 2002). A third genus was soon observed in rotating biological contractors in the UK and the two species of that genus were called "*Candidatus Scalindua brodae*" and "*Candidatus Scalindua wagneri*" (Schmid *et al.* 2003). The different genera of the Anammox bacteria of rarely occur in the same WWTP or enriched culture. It is likely that they all occupy their own niche and environmental conditions select for only one of the genera (Schmid *et al.* 2003). The factors influencing the Anammox process are outlined in Table 3.20.

Table 3.20 Environmental conditions influencing the Anammox process in wastewater treatment systems

Factor	Effect on the Anammox bacteria
TAN and TNO$_2$ inhibition	"*C. Brocadia anammoxidans*" has a very high affinity for the substrates ammonia and nitrite. It is irreversibly inhibited by nitrite at concentrations in excess of 70 mg N/dm^3 for several days. "*C. Kuenenia stuttgartiensis*" has a higher, but still low, tolerance to nitrite (180 mg N/dm^3) (Schmidt *et al.* 2003).
Dissolved oxygen	The Anammox bacteria (*C. Brocadia anammoxidans*) are very sensitive to DO and even low DO concentrations (0.5% of air saturation) were found to inhibit the Anammox activity completely, but reversibly (Strous *et al.* 1997).
Phosphate	*C. Brocadia anammoxidans* is irreversibly inhibited by phosphate at concentrations in excess of 60 mg P/dm^3 for several days. *C. Kuenenia stuttgartiensis* has a higher, but still low, tolerance to phosphate (600 mg P/dm^3) (Schmidt *et al.* 2003).
Temperature	Both Anammox bacteria have a similar temperature (37 °C) optimum (Schmidt *et al.* 2003), although *C. Brocadia anammoxidans* was found to be active at temperatures between 10 and 43 °C (Strous *et al.* 1999b).
pH	The pH interval for the Anammox bacteria ranges from 6.7 to 8.5 (Jetten *et al.* 2001; van Hulle 2005) and both bacteria have a similar pH (8.0) optimum (Schmidt *et al.* 2003).
Organic compounds	Van de Graaf *et al.* (1996) showed that carbon sources such as acetate, glucose and pyruvate had a negative effect on the Anammox activity. Güven *et al.* (2005) also observed that the Anammox bacteria were very sensitive to the presence of some organic carbon sources, such as alcohols, especially methanol. On the other hand, however, propionate and potentially acetate were converted by the Anammox bacteria.

The Anammox process is primarily addressed to nitrogen removal from ammonium-rich wastewater containing little organic material. Several kinds of reactors allowing for high biomass retention have been investigated in a bench- and pilot-scale, including sequencing batch reactors (SBRs), air/gas lift reactors, fixed bed reactors, rotating biological contactors and upflow anaerobic sludge blanket (UASB) reactors (Ahn 2006). Preliminary experiments and design calculations suggested that Anammox reactors would be extremely compact with volumetric ammonium loading rates exceeding 15 kg $N/(m^3 \cdot d)$ (Schmidt *et al.* 2003). In practice, however, the observed rates were lower and highly variable. The reported values from several studies, summarized by Ahn (2006), ranged from 0.003 to 8.9 kg $N/(m^3 \cdot d)$. Strous *et al.* (1997, 1998) investigated the feasibility of using the Anammox process to remove ammonia from sludge digestion effluents (reject water). Different types of reactors were tested including a fixed-bed reactor, fluidised-bed reactor and sequencing batch reactor (SBR). In the latter study, the authors concluded that a SBR is the most suitable system for enrichment and cultivation of Anammox organisms.

Before feeding into the Anammox process, ammonium has to partially be converted to nitrite (55-60% of ammonium), but not to nitrate. Thus, the Anammox process needs to be combined with partial nitritation, which can be obtained in either one- or two-reactor systems. In the latter case, partial nitrification occurs in the first reactor and is followed by the Anammox process in the second (separate) one. In the two-reactor systems, streams with higher ammonium concentrations can be treated. Moreover, both steps can be controlled independently which results in a more flexible and stable operation (van Hulle 2005).

In the one-reactor systems, both processes, i.e. partial nitrification and Anammox, take place in the same reactor. With such systems generally a higher volumetric nitrogen removal rates can be achieved for low loaded streams and an important footprint reduction can be accomplished. Even though the use of CSTR is possible, the one-reactor system is typically a biofilm reactor where the ammonium oxidizers are active in the outer layers of the biofilm, consuming oxygen and producing a suitable amount of nitrite for the Anammox bacteria. These bacteria are active in the inner layers protected from oxygen (van Hulle 2005). Various names were used to describe the 1-reactor systems (Fux and Siegrist 2004): the OLAND process (oxygen limited autotrophic nitrification and denitrification) at Ghent University (Belgium), the CANON process (completely autotrophic nitrogen removal over nitrite) at Delft University of Technology (the Netherlands) and aerobic/anoxic deammonification at the University of Hanover (Germany). The main difference between CANON and the other two processes is related to the microorganisms responsible for nitrogen removal. CANON incorporates the Anammox process, whereas OLAND and deammonification make use of the denitrification activity of conventional aerobic nitrifiers (Jetten *et al.* 2001). Recent studies with FISH analyses confirmed that anaerobic ammonium oxidation in all reactors was performed by Anammox organisms, although a specific role for the aerobic ammonium oxidizers can not be excluded (van Hulle 2005).

3.11.2 Approaches to modelling the ANNAMOX process

Even though the Anammox microorganisms have now been studied since mid-1990's, knowledge on the model parameters of Anammox is still very limited and can primarily be gained from the studies of Strous *et al.* (1998, 1999b). Using a SBR, Strous *et al.* (1998) obtained persisting stable and strongly selective conditions which resulted in a high degree of enrichment (74% of the desired microorganism). These conditions allowed, for the first time, to determine the biomass yield and maximum specific growth rate of Anammox microorganisms. The biomass yield of anaerobic ammonium oxidation (= 0.07 mg of protein/mg of NH_4-N) was lower by 30% compared to nitrification. The maximum specific growth rate ($\mu_{AN,max,}$) of Anammox microorganisms in the absence of mass-transfer limitations was only 0.0027 h^{-1} (equivalent to 0.065 d^{-1}) at 32-33 °C. This value was still much higher than reported previously (0.001 h^{-1}) by van de Graaf *et al.* (1996) for a fluidized bed reactor with mass-transfer limitations. Strous *et al.* (1999b) compiled the physiological parameters of Anammox, including affinity and inhibition constants, and compared them with nitrification. The affinity constants for ammonia and nitrite were below 0.1 mg N/dm^3. The Luong inhibition model (Equation 3.6) fitted the experimental data best.

Only a few studies on modelling and simulation of the Anammox process have been reported in literature so far. Koch *et al.* (2000a) performed simulations of an aerobic rotating biological contactor (RBC) treating ammonium-rich wastewater (leachate). The Anammox process contributed to a nitrogen loss of up to 70% in that one-reactor (CANON-like) system. Values of the kinetic and stoichiometric parameters were adopted from Strous *et al.* (1998), derived from laboratory experiments or adjusted during the model calibration. The distribution of the populations within the biofilm as well as the ammonium, nitrite and nitrate degradation along the RBC were predicted at steady state and during short-term (24-hour) experiments under different operating conditions, such as variable rotation speed, air flow, oxygen partial pressure and ammonium load. Hao *et al.* (2002 a,b) performed a comprehensive theoretical simulation study on the behaviour of a CANON system. The sensitivity of kinetic constants and process parameters to the process performance was evaluated (Hao *et al.* 2002a). The model was subsequently used to evaluate the system for its temperature dependency and behaviour under variable inflows and DO concentrations (Hao *et al.* 2002b). However, no comparison with real data was performed.

The studies of Koch *et al.* (2000a) and Hao *et al.* (2002a,b) did not consider any start-up or long-term dynamic effects. Furthermore, total nitrogen concentrations amounted to about 150 mg N/dm^3, while in practice these concentrations could be several times higher. Taking these factors into account, Dapena-Mora *et al.* (2004) successfully isolated and enriched Anammox biomass from sludge of a municipal WWTP by using a SBR reactor. The feeding solution was a mineral medium with the total nitrogen concentrations of 900 mg N/dm^3 (close to practical situations). In order to interpret the experimental results, the start-up and dynamic operation of the reactor were predicted using ASM1 extended for the Anammox process (Tables 3.21 and 3.22). The simulations accurately predicted the measured concentrations of nitrogenous compounds and also allowed to estimate the evolution of Anammox and heterotrophic biomass in the reactor. It was revealed that heterotrophs still remained in the reactor after the start-up period and could protect the Anammox microorganisms from a negative effect of the oxygen.

Table 3.21 Stoichiometric matrix for the Anammox process (Dapena-Mora *et al.* 2004; van Hulle 2005)

Process	S_{NH} gN/m^3	S_{NO2} gN/m^3	S_{NO3} gN/m^3	S_{N2} gN/m^3	X_{AN} gCOD/m^3	X_S gCOD/m^3	X_I gCOD/m^3
Growth of X_{AN}	$-\dfrac{1}{Y_{AN}} - i_{N,BM}$	$-\dfrac{1}{Y_{AN}} - 1.52*$	1.52*	$\dfrac{2}{Y_{AN}}$	1		
Decay of X_{AN}	$i_{N,BM} - f_i i_{N,XI}$				-1	$(1-f_i)$	f_i

* different stoichiometric coefficients were used by Koch *et al.* (2000a) and Hao *et al.* (2002a)

Table 3.22 Process rates for the Anammox process (Dapena-Mora *et al.* 2004; van Hulle 2005)

Process	Process rate gCOD/(m^3×d)
Growth of X_{AN}	$\mu_{AN,max} \dfrac{K_{O,AN}}{K_{O,AN}+S_O} \dfrac{S_{NO2}}{K_{NO2,AN}+S_{NO2}} \dfrac{S_{NH}}{K_{NH,AN}+S_{NH}} X_{AN}$
Decay of X_{AN}	$b_{AN} X_{AN}$

Using the same model as Dapena-Mora *et al.* (2004), but with different kinetic parameters as shown in Table 3.23, van Hulle (2005) estimated the minimum start-up time of the Anammox process for different reactor configurations and operational conditions as function of temperature, HRT, initial Anammox biomass concentration and separator efficiency. The simulation results were presented in the form of calculated curves meant as a tool supporting design decisions.

Jones *et al.* (2007) applied an Anammox model to simulate a sidestream pilot plant (SBR) as a part of an integrated model developed for whole-plant simulations. While testing different control strategies, appropriate inhibitions (i.e. nitrite toxicity to the Anammox organisms) and limitations had to be incorporated. The pH control band had to be narrower (6.97–7.01) and the DO setpoint was kept at 0.3 mg O_2/dm^3 to prevent the accumulation of nitrite even at a very low level, i.e. over 5 mg NO_2–N/dm^3. The effect of nitrite toxicity was implemented in the model by increasing the decay rate by the product of the nitrite sensitivity constant ($K_{TNO2,AN}$) and the nitrite concentration.

The stoichiometric and kinetic coefficients used in the simulation studies outlined above were summarized and discussed by Sin *et al.* (2008b) (Table 3.23). The authors concluded that the approaches to modelling the Anammox process appeared to be consistent among different research groups except for the issue of nitrite toxicity (described only in one model). The parameter values were mostly similar for the yield and growth (referring to the studies of Strous *et al.* (1998, 1999b), while differences in the reported values of the affinity constants could be explained from the process characteristics, i.e. extent diffusion limitation in biofilm and suspended growth systems, and the effects of toxicity. It was also emphasized that the ammonia affinity constant lacks sensitivity in most sidestream treatment systems as residual ammonia concentrations are beyond the rate-limiting range.

A few important differences, not mentioned by Sin *et al.* (2008b), require a further explanation. Koch *et al.* (2000a) used higher values of $\mu_{AN,max}$, (0.08 d^{-1} at 20 °C) and the affinity and inhibition constants ($K_{NH4,AN}$, $K_{NO2,AN}$, $K_{IO,AN}$). Hao *et al.* (2002b) noted that this value for the growth rate of Anammox bacteria would probably be too high, if the temperature effect was considered. A more realistic value of this rate constant should be approximately 0.03 d^{-1} at 20 °C as the activation energy of Anammox organisms was found to be 70 kJ/mol (Strous *et al.* 1999b). The increased values for the affinity and inhibition constants can be attributed to diffusion limitation in the studied biofilm system. In the growth rate expression, the higher values of $\mu_{AN,max}$, and $K_{IO,AN}$. are compensated by the higher values of $K_{NH4,AN}$ and $K_{NO2,AN}$.

Table 3.23 Parameter values used in Anammox models (Sin *et al.* 2008b)

Parameter	Unit	Koch *et al.* (T = 20 °C)	Hao *et al* (T = 30 °C)	Dapena-Mora *et al.* (T = 35 °C)	Van Hulle (T = 20 °C)	Jones *et al.* (T = 20 °C)
Y_{AN}	mg COD/ mg N	0.15	0.159	0.159	0.159	0.114
$\mu_{AN,MAX}$	d^{-1}	0.08	0.072	0.08	0.019	0.1
b_{AN}	d^{-1}	–	–	0.0011	0.0025	–
$b_{AN,AE}$	d^{-1}	0.016	0.003	–	–	0.019
$b_{AN,AX}$	d^{-1}	0.008	0.0015	–	–	0.0095
$K_{NH4,AN}$	mg N/dm^3	21	0.07	0.3	0.3	2.0
$K_{NO2,AN}$	mg N/dm^3	2.0	0.05	0.2	0.2	1.0
$K_{HCO3,AN}$	mol/m^3	–	–	–	–	4.0
$K_{IO,AN}$	mg O$_2$/dm^3	0.4	0.01	0.01	0.01	0.01
$K_{INO2,AN}$	mg N/dm^3	–	–	–	–	1000
$K_{TNO2,AN}$	dm^3/ (mg N d)	–	–	–	–	0.016
$K_{PH,AN}$	–					5.5-9.5
Temperature correction factor, θ, for the kinetic parameters						
$\theta_{\mu AN}$	–	$e^{0.09(T-20)}$	–	–	$e^{0.096(T-20)}$	$1.1^{(T-20)}$
θ_{bAN}	–	–	–	–	$e^{0.096(T-20)}$	–
$\theta_{bAN,AE}$	–	$e^{0.09(T-20)}$	–	–	–	–
$\theta_{bAN,AX}$	–	$e^{0.09(T-20)}$	–	–	–	–

The decay process of Anammox bacteria was modelled according two concepts including endogenous respiration (Koch *et al.* 2000a; Hao *et al.* 2002a,b; Jones *et al.* 2007) as proposed in ASM3 and death-regeneration (Dapena-Mora *et al.* 2004; van Hulle, 2005) as proposed in ASM1. No decay characteristics of Anammox bacteria have been reported so far and this topic remains of high-priority for further research (Hulle 2005). Dapena-Mora *et al.* (2004) justified the preference of the latter concept by the fact that heterotrophs were shown to be active in a reactor performing Anammox. This activity could only be described by applying the death-regeneration concept since no biodegradable substrate had been added in the influent.

In the Anammox process, the nitrate formation from nitrite results from the electron equivalents used for autotrophic biomass growth. The stoichiometry in the models of Dapena-Mora *et al.* (2004) and van Hulle (2005) was directly derived from Equation 3.44 in which $0.26 \cdot 14$ mg of NO_3-N is formed per $0.066 \cdot 36.4$ gCOD$_{biomass}$ (or $0.066 \cdot 24.1$ g of biomass). Thus, the resulting value of the yield coefficient is 1.52 mg NO_3-N/gCOD$_{biomass}$ (see: Table 3.21). Koch *et al.* (2000a) and Hao *et al.* (2002a) assumed that 14 mg NO_3-N is theoretically yielded from 16 mg $COD_{biomass}$ (equivalent to $1/1.14$ mg $NO_3-N/$mg $COD_{biomass}$). This assumption leads to the following stoichiometric coefficients for the growth of Anammox bacteria (X_{AN}):

$$v_{SNO2,XAN,growth} = -\frac{1}{Y_{AN}} - \frac{1}{1.14} \tag{3.45}$$

$$v_{SNO3,XAN,growth} = 1.14 \tag{3.46}$$

The above rates of NO_2-N utilization and NO_3-N formation are lower by approximately 8 and 25%, respectively, when compared to the corresponding values in the model of Dapena-Mora *et al.* (2004) presented in Table 3.21.

Chapter 4

Organization of a simulation study

An adequate modeling process including model calibration is a key issue for the successful application of these models in practice. This step is usually time consuming and cost intensive since many experiments may be required to determine accurately model parameters. Two approaches are possible: process engineering versus sytem engineering. Accuracy of the available plant data should be first evaluated using a continuity check for flow rates and mass balance calculations for oxygen demand, solids, nitrogen and phosphorus. This procedure is outlined including also the use of error diagnostics and data reconciliation techniques. A more standardized use of dynamic simulations is essential since a number of model applications for optimization studies and design studies has been growing very rapidly. Therefore, systematic protocols (guidelines) for conducting simulation studies are outlined and compared. Most of these protocols emphasized the importance of several elements, such as data quality control, determination of mixing conditions (flow pattern) in the bioreactors and clarifiers, characterization of wastewater and biomass as well as parameter estimation in the biokinetic and settling models.

4.1 APPROACHES TO A SYSTEMATIC ORGANIZATION OF THE SIMULATION STUDY

The process of evaluating a model against experimental data is described in control theory as system identification, although this term has several levels of interpretation (Beck 1989). It may simply be understood as "*curve fitting*", but more generally, it is understood as calibration of the model against a certain measured data set or estimation of model parameter(s) to fit the data set. In more detail, parameter estimation (or parameter identification) is defined as the application of field or experimental data to determine the values of the parameters in a model such that differences between the model predictions and the measured values are minimized (Patry and Chapman 1989). Gernaey *et al.* (2004) defined model calibration as the estimation of model parameters to fit a certain experimental data set obtained from a studied WWTP.

The objective of calibration is to select appropriate values for the model parameters and this selection can be accomplished in three different ways (USEPA 1993):

- By deriving from theoretical considerations when the parameters are fundamental in nature,
- By using experiences with the particular model that have been acquired in similar applications,
- By adjusting model coefficients to match the results of bench, pilot and/ or full-scale experiments (a rational procedure has to be used to produce a range of responses that is sufficiently broad to allow for accurate determination of model coefficients).

Model calibration also involves evaluation of the adequacy of model structure. The structure may need to be modified if the model is unable to properly predict trends in system response and/or if model parameters cannot be adjusted sufficiently to obtain an acceptable fit (USEPA 1993).

Model calibration is followed by model validation. The aim of validation is to compare predictions by the calibrated model with measured data to identify similarities and differences between model predictions and measured results. The data set must be different from the data set used for calibration and sufficiently broad to test the calibrated model over the entire range of the potential application. The model is considered to be validated when the differences do not exceed acceptable tolerances (USEPA 1993).

In the case of modelling activated sludge systems, there are still few references that contain details on the practical calibration procedure of mechanistic models, although ASM1 has been widely applied over last 20 years. Petersen *et al.* (2002)

claimed that *"Even though more than a decade has passed since the publication of ASM1, a fully developed model calibration procedure has not yet been defined"*. An adequate modelling process including model calibration is a key issue for successful application of these models in practice. This step is usually time consuming and cost intensive since many experiments may be required to determine accurately model parameters (Vanhooren *et al.* 2003). However, the calibration of a site-specific model does not always receive the same level of scrutiny as does model development (Melcer 1999). Initial attempts were varying with regard to calibration procedures, scope of monitoring and time investment (e.g. Henze *et al.* 1987; Lesouef *et al.* 1992; Pedersen and Sinkjaer 1992; Siegrist and Tschui 1992; Stokes *et al.* 1993; USEPA 1993; de la Sota *et al.* 1994; Dupont and Sinkjaer 1994; Xu and Hultman 1996; Kristensen *et al.* 1998).

The quality of simulation studies can vary strongly on the objectives, expertise available and resources spent. The objectives of the study determine which units and how detailed a plant needs to be modeled (Langergraber *et al.* 2004). The objectives will also decide to which degree a model has to be calibrated, since the accuracy and reliability of the desired model predictions will depend strongly on the quality of model calibration (Petersen *et al.* 2002). For example, reasonably good descriptions can be often obtained with default parameter sets for typical municipal cases without significant industrial influences. However, if the calibrated model is going to be used for process performance evaluation or optimization, then a more accurate description of the actual processes would be necessary (Petersen *et al.* 2002).

Table 4.1 Procedure for calibration of the WWTP model (adapted from Coen *et al.* 1997)

Element of the procedure	Description
Hydraulics	Determination of the hydraulics of the settler, the aeration tank, etc.
Choice of the atomic models	Determination of the model for the description of biotransformation, sedimentation, sludge thickening
Characterization of the influent	Estimation of the COD fractions of the waste-water by means of respirometry and standard laboratory analyses
Characterization of the sludge	Determination of the different fractions of the sludge based on analysis of the model equations under steady state conditions
Calibration of the atomic models	Determination of the parameters for heterotrophic growth. Determination of the sedimentation parameters

Coen *et al.* (1997) proposed one of the first comprehensive procedures for calibrating a general model of WWTP from the process engineering perspective (Table 4.1). In the first step, the hydraulics for each component was determined followed by the choice of a model for description of the processes. Secondly, the influent characteristics (wastewater fractionation), initial conditions (biomass characterization) and parameters of the atomic models needed to be determined.

Petersen *et al.* (2002) gathered and summarized the set of information needed to achieve a successful model calibration:

1. Design data, e.g. reactor volume, pump flows and aeration capacities.
2. Operating data including:
   2.1. Flow rates, as averages or dynamic trajectories of influent, effluent, recycle and waste flows.
   2.2. pH, aeration and temperatures.
3. Characterization for the hydraulic model, e.g. the results of tracer tests.
4. Characterization for the settler model, e.g. zone settling velocities at different MLSS concentrations.
5. Characterization for the biokinetic model including:
   5.1. Wastewater concentrations of full-scale WWTP influent and effluent (as well as some intermediate streams between the WWTP unit processes), as averages or dynamic trajectories, e.g. TSS, COD, TKN, NH_4-N, NO_3-N, PO_4-P, etc.
   5.2. Sludge composition, e.g. TSS, VSS, COD, nitrogen and/or phosphorus content.
   5.3. Reaction kinetics, e.g. growth and decay rates.
   5.4. Reaction stoichiometry, e.g. yields.

Design (1) and operating (2) data are always needed for a model calibration. The hydraulics (3) can be determined most accurately by using tracer tests at the studied plant. The settling characteristics (4) can be characterized via online or laboratory scale settling tests. Finally, the biokinetic model (5) can be characterized via different information sources including different types of laboratory scale experiments (see: Section 4.3.3). After collecting the necessary information, the model should be calibrated at different levels. At the first level a model is typically calibrated at steady state. Routine data obtained from the full-scale WWTP are averaged, thereby assuming that this average represents a steady state, and the model is calibrated to fit to average effluent and sludge waste data.

Petersen *et al.* (2002) discussed the procedure of model calibration. In general, with a steady state model calibration, only parameters responsible for the long-term behaviour of the WWTP can be determined, such as Y_H, f_p, b_H and X_I in the influent (based on Nowak *et al.* 1999). These parameters are correlated to a certain degree, meaning that a modification of one parameter value can be compensated by a modification of another parameter value. In the study of Nowak *et al.* (1999) and Petersen *et al.* (2002), Y_H and f_p were remained fixed, whereas X_I in the influent and b_H were adjusted from the steady state data based on mass balances of the full-scale plant. A steady state calibration may be very useful for a preliminary estimation of the model parameters and for the determination of initial conditions prior to a dynamic model calibration (e.g. Pedersen and Sinkjaer 1992; Stokes *et al.* 1993; Dupont and Sinkjaer 1994; Xu and Hultman 1996; Kristensen *et al.* 1998; Makinia and Wells, 2000). In the study of Lesouef *et al.* (1992), two WWTP models were calibrated at steady state only and the calibrated models were further applied to simulate dynamic process scenarios. In general, however, this approach should be avoided. Otherwise some problems may be encountered since the real input variations are usually faster than the slow process dynamics on which the steady state calibration was focused (Petersen *et al.* 2002).

If the objective is focused on predicting dynamic situations, a model calibration using dynamic data will be needed since such data contain more information than steady state data, especially on fast dynamic behavior. The important point in model calibration based on dynamic data is to obtain a more reliable estimation of the maximum specific growth rates, which are the most important parameters in predicting dynamic situations (Petersen *et al.* 2002). However, for calibrating the effluent ammonia concentrations, Hulsbeek *et al.* (2002) recommended that the autotrophic decay rate should be adjusted rather than the autotrophic growth rate due to a larger uncertainty in the former parameter.

A more standardized use of dynamic simulations is essential since a number of model applications for optimization studies and design studies has been growing very rapidly (Hulsbeek *et al.* 2002). Independent studies in Belgium by BIOMATH (Coen *et al.* 1997; Petersen *et al.* 2002; Vanrolleghem *et al.* 2003), in Holland by STOWA (Hulsbeek *et al.* 2002), in North America by WEFR (Melcer *et al.* 2003) and in German speaking countries (Germany, Austria and Switzerland) by HSG (Langergraber *et al.* 2004) have attempted to provide systematic guidelines for calibrating a general model of a WWTP. Sin *et al.* (2005a) summarized and compared in detail these guidelines using a SWOT (strengths, weaknesses, opportunities and threats) analysis, which also allowed to identify relative advantages and disadvantages of each approach. In recent years, two more protocols were proposed by the Japanese Sewage Works Agency (Itokawa *et al.* 2008) and an international IWA Task Group (Gillot *et al.* 2009). All these reports are outlined in the following sections.

4.1.1 BIOMATH calibration protocol (Belgium)

The protocol developed by Petersen *et al.* (2002) was refined by Vanrolleghem *et al.* (2003) and referred to as the BIOMATH calibration protocol. The protocol is sophisticated and oriented at using scientifically more exact methods, such as the optimal experimental design (OED) technique or sensitivity analysis, rather than more experimental methods. For this reason, it may not be user friendly for new modellers (Sin *et al.* 2005a).

The BIOMATH protocol is composed of four main stages and 12 modules (Figure 4.1). However, the proposed procedure is not fixed but it should be considered as dynamic depending on each specific *"case study"*. The first stage is the definition of target(s) of the simulation study followed by decision making on the necessary information from the studied plant to achieve the target(s). The decision is usually based on the time and/or budget available for the study. Some of the eleven modules (1–11) can be omitted depending on the study objectives, assumptions made and problems experienced during the data collection. Although each module is regarded as an individual unit, they are interrelated to each other and these interactions have to be considered to predict the general behaviour of the system.

The next stage (stage II) comprises the collection of detailed information on the studied plant. A general description of the plant, consisting of three kinds of data (design, operational and measured), can be compiled from the operation databases, design documents and personal communication with the plant operators. The design data consist of the physical characteristics such as volumes, a number of compartments, pumping capacities, a number of aerators and connections installed. The operational data are primarily the actual working volumes, flow rates, energy consumption, flow distributions, and type and implementation of control strategies. The measured data comprise the conventional characterization of influent and effluent (e.g. COD, PO_4-P, NH_4-N, NO_3-N, etc.), on-line measurements, sludge blanket height in the clarifier, TSS concentration in the RAS and effluent as well as sludge composition (P and N content, VSS/SS ratio, etc.).

The data collected at this stage should be processed and evaluated for understanding of the capacity and behaviour of the system. For data quality assessment, outlier detection is applied here and interpolation of data can be done whenever it is appropriate and possible. Mass balances regarding the flow rate and the sludge (considering the N and P content) are crucial for the model calibration; particularly for the accurate estimation of the system SRT. The information gathered at this stage is then used to select and characterize three different sub-models for mass transfer, settling and biological processes (including influent wastewater characterization). The OED technique can be applied for designing both lab-scale and full-scale experiments in order to reduce the required experimental work and uncertainty of the parameter estimates. In the case of full-scale systems, however, OED applications need a high level of a priori information and expert knowledge.

The appropriate sub-models should be selected based upon the objectives of the study. In the mass transfer characterization; two main aspects have to be considered, i.e. the determination of the oxygen transfer efficiency (K_La) in all reactors under process conditions and the hydraulic properties of the studied plant. A tracer test is recommended to obtain information about the mixing characteristics and determine a number of equivalent tanks-in-series.

If the solid separation in the settler is working well, no detailed settling characterization is necessary and a point settler model can be used. If a considerable amount of sludge is maintained in the settler, a virtual reactor can follow the point settler. A detailed settling characterization is necessary if the settling performance and the biochemical reactions during settling, i.e. denitrification and phosphorus release, affect the overall system behaviour or a specific analysis of the plant is required, e.g. optimization of settling or improvement of suspended solid removal efficiency. This can be done via one-, two- or three-dimensional settling models. Selection of the appropriate model should be based on the measurement of concentration profiles and/or dynamic profiles of sludge blanket-height.

The determination of all biokinetic model components is an expensive and time-consuming process. Therefore, it is suggested that the reported values from previous applications should be used whenever possible. Procedures for the determination of key kinetic parameters occurring in ASM1, i.e. those related to heterotrophic and autotrophic activity, are outlined in the protocol. Batch tests for measuring PAO activity under anaerobic and aerobic (or anoxic) conditions are also proposed for activated sludge systems performing EBPR. The influent wastewater characterization has to be performed carefully as it is essential for the simulation of the activated sludge plant. In the protocol, an experimental methodology is proposed for determining the influent fractions of COD and nitrogen based on the combination of biological (respirometric) and physical-chemical methods.

The model calibration is carried out at the third stage. Each sub-model is first calibrated separately and then the three sub-models are combined in an overall model. A steady state calibration of the overall model is carried out with the data collected in stage II. The data are averaged assuming that this average represents a steady state disregarding the sudden process disturbances, and the model is calibrated to fit to the average effluent COD, N and P concentrations, oxygen consumption and sludge production. The main purpose of the steady-state calibration is to estimate the appropriate biomass composition, which is subsequently used in the dynamic model calibration as the initial condition in the reactors and/or settler. In the dynamic calibration step, the influent data obtained from a dynamic measurement campaign (intensive sampling) are used as inputs to the model. A larger amount of data and optimized sampling provide sufficient information on the understanding of the system behavior and allow a precise estimation of model parameters. Dynamic simulations can be used for fine-tuning the parameters or can serve as a validation step by itself (using an independent data set, which was not used in the calibration).

A decision has to be made concerning the parameters to be adjusted (e.g. kinetic or stoichiometric constants for the biokinetic model). In order to identify which single parameters or sets of parameter may be selected for the calibration, a sensitivity analysis can be performed in both steady state and dynamic calibration. The sensitivity functions allow to distinguish between less and more sensitive parameters influencing the measured data. This information should be taken into account in the calibration protocol to minimize the calibration efforts and optimize the overall calibration procedure towards the study objectives.

At the last stage, a final evaluation is performed within the context of the study objectives and decisions should be made upon eventual re-iteration of a number of the preceding modules. If the calibrated activated sludge model does not accurately predict the dynamic behavior of the studied plant, new solutions have to be investigated by re-evaluation of the plant data, sub-models or even the defined targets. In addition, modifications or alternatives of the calibrated model may be proposed. However, in such a case, it is recommended that the protocol should be adopted to the new model.

4.1.2 STOWA calibration protocol (Holland)

In the period 1995–2001, over 100 full-scale WWTPs in the Netherlands were examined using dynamic models implemented in the simulation software SIMBA (see: Section 5.7.4). Such a wide application of the simulation tool revealed the need for a better standardization and some form of quality control. In response to this need, the Dutch Foundation of Applied Water Research (STOWA) stimulated the development of a protocol for dynamic modelling of activated sludge systems. Experienced practitioners were interviewed for obtaining an overview on the existing approaches to the use of ASM1 for dynamic simulation of full-scale plants. A special attention was paid to the influent wastewater characterization, model structure and calibration. During the interviews, bottlenecks were identified and analyzed, whereas solutions for these bottlenecks were formulated and incorporated in two STOWA protocols.

The influent wastewater characterization was addressed in a standard methodology proposed by Roeleveld and Kruit (1998) and Roeleveld and van

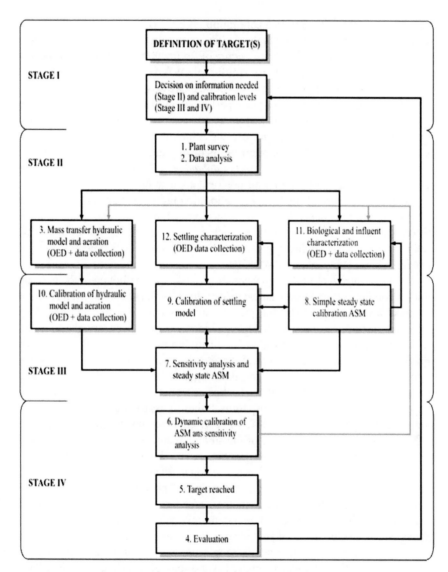

Figure 4.1 The BIOMATH model calibration protocol (Vanrolleghem *et al*. 2003).

Loosdrecht (2002). The STOWA calibration protocol (Hulsbeek *et al*. 2002) was seen as a guideline for improving the quality and controllability of the simulation studies by giving a standardized procedure of setting up and calibrating a specific model (ASM1) for the practical use at full-scale WWTPs. In both cases, practical feasibility was preferred rather than scientific exactness.

With the protocol, it was expected to perform a simulation study easier, develop a model as much as possible on historical data and well planned measurements, minimize time and costs, and increase the reliability of predictions. For Sin *et al*. (2005a), the STOWA calibration protocol appeared *"most straightforward, practical and easy to follow and implement"*. Therefore, these authors recommended it as most suitable for practical applications.

The main structure of the STOWA calibration protocol is presented in Figure 4.2. The structure is dependent on the objectives of the study and refers to three different kinds of simulation studies, such as a system choice during the basic design, optimization studies for existing WWTPs, and development of control strategies for existing and new WWTPs.

When the objectives of the study are clear (phase 1), a definition of the relevant process components and flows can be made (phase 2). In general, it is not required to model a complete WWTP but only those parts that fit within the described process dynamics, i.e. the activated sludge bioreactor and secondary clarifier(s) in most cases. The flow rates, composition of the flows as well as volumes of the process components are defined in phase 3. It is advised to initially generate the flows and compositions from available data (e.g. daily average concentrations, flow patterns). These data should be then checked using mass balances for flows, COD, suspended solids, nitrogen and phosphorus. If these mass balances are not correct or relevant data are missed, it is necessary to perform extra monitoring of the plant.

The hydraulic model structure of the studied plant (phase 4) is defined based on the actual process description. In this phase, such aspects as number of compartments, aeration configuration (including vertical and horizontal gradients), settling and control should be taken into account. The intended use of the model determines a modelling approach to the secondary clarifiers. The amount of suspended solids in the effluent should definitely be made equal to the measured values. Most clarifier models do not incorporate biological processes (see: Section 2.5.3). In general, however, denitrification at the bottom of the clarifier is even more important then the exact temporal variations of the sludge blanket. This denitrification can be checked by comparing the nitrate concentrations in the RAS line and effluent, and modelled by applying a virtual tank in the RAS line.

By using historical data and/or specific measurements, the important process flows can be characterised in terms of the required monitoring frequency and duration of the monitoring campaign (phase 5). These flows include influent, effluent, recirculation flows and internal flows. Under the Dutch conditions, if the model is used for a system choice, daily average concentrations of influent and effluent and the variations in the flow pattern are sufficient. If the model is used for optimization studies or developing control strategies, specific data from 4- or 2-hour composite samples are required. In these cases, other flows than the influent and effluent, e.g. recirculation flows, should also be sampled. The recommended duration of the measurement campaign depends on the accuracy of the required results. In general, 1-3 days is considered sufficient for normal cases, whereas a measurement campaign of 3 to 7 days and at least 7 days, respectively, is advised for optimization studies and developing control strategies.

A first attempt to calibration can be made after defining the model and implementing a preliminary set of data (phase 6). If the effluent quality is initially not properly predicted by the model, it is most likely that a structural error is present in the model. In such a case, the mass balance calculations should be repeated and some extra verification measurements should be carried out. A sensitivity analysis for several operating parameters, such as flow rates, set points, etc., can help in detecting where potential errors in the process have been made. The order of steps during calibration of the selected biokinetic model (ASM1) is presented in Figure 4.3. A stepwise procedure including sludge composition and production, nitrification and denitrification is recommended. The target variables and model parameters to be preferentially adjusted at each step are also identified.

After successful calibration, a validation of the model (phase 7) can be done prior to applying the model for the indented purpose in the actual study (phase 8). For the validation, it is recommended to collect sufficient data under different operating conditions with regard to process temperature or influent characterization (e.g. when primary clarifiers are taken off-line for maintenance).

4.1.3 WERF protocol (USA)

Among the existing guidelines, the WERF protocol (Melcer *et al.* 2003) is most extensively documented and summarizes a great number of full-scale model calibration experiences gathered in North America. The protocol is primarily focused on presenting experimental methods for the influent wastewater characterization, fractionation of active biomass and determination of key kinetic and stoichiometric coefficients in biokinetic models.

A generalized methodology of conducting a simulation study at a particular plant and guidance for model calibration are also provided. This generalized methodology should include the following steps: configuration of the plant in a simulator, data gathering (historical data, additional measurements, assumptions), calibration, verification and application for the intended purpose (Figure 4.4).

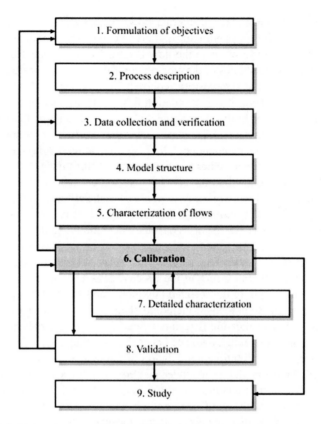

Figure 4.2 Main structure of the STOWA protocol for dynamic modelling of activated sludge systems (Hulsbeek *et al.* 2002).

As a guidance for model calibration, "*a tiered approach*" is proposed for practitioners to follow, depending on plant complexity, objective of the simulation study, available data and budget for additional data gathering and analysis. The overall objective of calibration is defined as minimization of the error between the data sets and model predictions. When evaluating the match of the model predictions against experimental data, it is crucial to observe all the important variables. It is recommended to fit to most of the measured variables reasonably, rather than fit "*perfectly*" to one selected (even though important) component concentration and poorly to others.

In addition to minimizing model error, two other objectives of the model calibration procedure should be observed: establishing the field of validity and

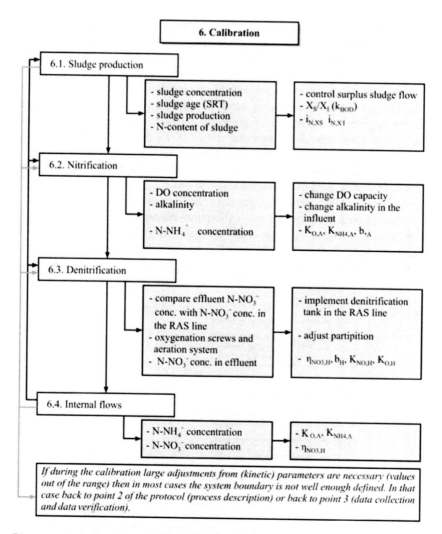

Figure 4.3 Order of steps during calibration of the biokinetic model (Hulsbeek *et al.* 2002).

expected accuracy of the model. The calibration should be used to establish the field of validity (or the "*design space*") of the model. If a particular model application is too far from the calibration conditions, simulation results might not be reliable, e.g. predicting the behaviour of a plant under storm flow conditions based on calibration performed on average or steady state data. Furthermore, an attempt should not be made to fit a variable that is influenced by a process not accounted for in the model (e.g. nitrite generation, chemical precipitation, GAO presence, inhibitory industrial component).

If the model is to be considered reliable for a particular application, the most important performance parameters must be within a certain accuracy provided that the experimental data are of good quality (see: Section 4.2). In general, these parameters should be matched within 5–20 percent overall in steady state simulations, while temporary deviations of 10–40 percent are not unusual during dynamic simulations, even from well-established models. Less important variables can vary even more in typical cases, without compromising the model's capability of truthfully predicting general process performance.

A generalized structure of the calibration procedure includes four steps as presented in Figure 4.4. Sin *et al.* (2005a) emphasized the significance of calibration level 2 (based on historical plant data) before attempting to plan dynamic measurement campaigns for data collection. In this context, the WERF protocol is expected to be attractive for consultants and inexperienced modellers.

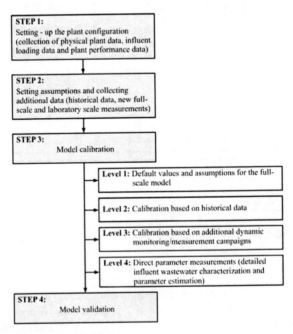

Figure 4.4 Main steps of the WERF protocol based on the summary of Sin *et al.* (2005a).

Calibration level 1 is related to cases when no process data are available. Simulation with "*uncalibrated*" model would still provide useful information that is not possible to obtain in any other way. However, the following three points should be realized:

- The literature data and previous case studies in the same geographical region provide information on a typical wastewater composition as well as default stoichiometric and kinetic coefficients;
- Qualitative comparisons are more reliable than quantitative predictions;
- Confidence in the predictions can be increased by simulating worst case scenarios (e.g. selecting model parameters on the low side of the default range or using peaking and other safety factors).

Historical data are incomplete for a full simulation study but can be used to obtain a rough calibration of the model (level 2). In almost all cases, the raw operating data require some adjustments, such as processing, filtering, cleaning and corrections, referred to as "*data conditioning*". If possible, the "*conditioned*" data set should be subjected to a simple data reconciliation. The aim of this step is to apply independent checks to verify the consistency of the data. These checks can be grouped into two categories: fundamental checks (based directly on a law of nature) and empirical checks (based on engineering knowledge). The first category comprises, for example, continuity checks for flow rates, mass balances on inert components and consistency of kinetic parameters in single sludge systems. The latter category comprises experience-based "*empirical*" knowledge of the studied plant, such as component ratios (e.g. particulate COD/VSS or VSS/TSS) and observed sludge production.

The calibration based on historical data alone is of poor quality and can be improved by providing data from additional measurements (level 3). The scope of testing depends on the availability of existing data and type of the studied plant but, in general, two principal categories of experiments can be identified:

- Additional composite and grab sampling to verify historical data and obtain more detailed information on influent wastewater characterization and some stoichiometric coefficients.
- Sampling during dynamic events (beyond the normal diurnal variation) to get better estimates of kinetic coefficients. There are two principal ways of collecting data under high-loading conditions, i.e. sampling high flow events that occur naturally under storm flow conditions or performing an artificial hydraulic stress tests. Sometimes disturbances resulted from planned maintenance (e.g. taking a bioreactor off-line) or even unintended equipment failure can make an opportunity to collect very useful, highly dynamic data.

A full-scale monitoring program may not provide all the required data if the studied plant is large and complex, many uncontrolled disturbances occur and mass balances are difficult to close or require expensive analyses. In such cases, bench-scale (lab-scale) experiments should be considered for direct parameter measurements (level 4). In comparison with the full-scale testing, the advantage of these experiments is that they are easily controllable and the number of samples is considerably reduced. A potential drawback is that these experiments may lead to the development of a unique microbial community which is different from the one present in the full-scale bioreactor. In many cases, however, bench-scale tests have successfully been applied to model calibration. This especially refers to the SBR procedure for influent wastewater characterization (see: Section 2.3) and rapid testing protocols for measuring the maximum specific growth rate of nitrifiers (μ_A) (see: Section 3.7).

4.1.4 HSG guideline (Austria, Germany, Switzerland)

In contrast to the STOWA protocol, the HSG (Hochschulgruppe) guideline (Langergraber *et al.* 2004) is not focused on the calibration stage only, but the process of the entire simulation study (Figure 4.5). The main goal of this guideline is to provide a standardized procedure for simulation studies with very high requirements on the simulation precision and accuracy, e.g. performance evaluation of existing WWTPs including also the plants with EBPR. The HSG guideline is primarily focused on "*careful data quality control and an extensive documentation*" and emphasizes four core points: explicit definition of the project objectives (phase 1), evaluation of plant data (phase 3), quality control of the data used for simulation (phase 3) and careful documentation (phase 7).

Even though trivial at the first glance, the definition of the project objectives is indeed of great significance as it will define the methods used and the quality demands required. The objectives should be set and quantified in cooperation with a client and a plant operator.

Data evaluation is essential for obtaining plausible simulation results. The proposed procedure for the evaluation of operating and performance data ensures that only validated and consistent measurement data are used for the simulation. The quality control of the data used for simulation should start with mass balance calculations for flux, COD, total N and total P. Furthermore, it is strongly recommended to check the analysis and measurement accuracy of the laboratory equipment used in the routine plant operation. Main sources of potential errors can be attributed to laboratory scales, micropipettes and photometers. The test kits commonly used for analyses are usually accurate enough provided that they are frequently checked with standards. In addition, an inspection of all sampling locations, auto samplers and flow rate measurements should be considered. The same procedure of data quality control as described for phase 3 (mass balances, reliability checks) should be repeated in phase 5 to evaluate the results of a monitoring campaign for model calibration.

The HSG guideline proposes to report the simulation study in a certain format and the written documentation in the final (7th) phase should include:

- Overall approach chosen in the study (including the used simulation software);
- All operational and performance data (including influent wastewater characterization);
- Final model layout (selected models with all modifications of the original model parameters);
- Results of the data evaluation,
- Calibration and validation results;
- Simulated scenarios and evaluation of the objectives.

By providing a standard format for reporting the overall simulation study, the HSG guideline can contribute to the improvement of evaluation and comparison of different calibration studies (Sin *et al.* 2005a). The same authors also noted that the HSG guideline does not impose specific experimental methods for model calibration. Consequently, the choice of the parameters to be adjusted (with the aid of sensitivity analysis) and methodology is left to more experienced model users. The beginners are not guided by the guideline to find adequate methods for parameter estimation and influent wastewater characterization.

According to the HSG guideline, the entire calibration procedure should be divided into a sequence of relatively independent steps, i.e. each step has only a minor effect on the results of the preceding steps. Therefore, the following general order is proposed:

- Model structure and hydraulics (number of tanks-in-series, size of internal unknown flows, etc.);
- Influent wastewater characterization;
- Parameters of the biokinetic model.

Due to the complex model structure of most biokinetic models an iterative approach is recommended. This includes the calibration of sludge production, effluent concentrations of NH_4-N and NO_3-N and, if implemented, PO_4-P. For some processes it could be necessary to carry out additional measurements in special points (e.g. PO_4-P at the end of the anaerobic zone for EBPR or NO_3-N in return sludge for denitrification in the secondary clarifier). The guideline emphasizes that mathematical criteria are of great advantage for the evaluation of the calibration quality, but no generally accepted and easy to use criteria actually exist. Therefore, the success of model calibration should also be judged through visual checks considering peaks and median values of the simulation results. Finally, the model needs to be validated with an independent set of performance data. For this purpose, an additional short monitoring campaign (e.g. 4 days) should be carried out under different process conditions, in terms of temperature or SRT, compared to the calibration period.

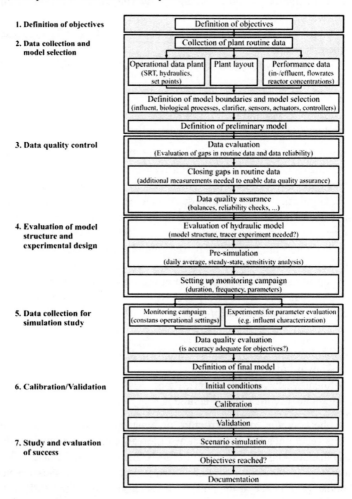

Figure 4.5 Flowchart of a simulation study according to the HSG guideline (Langergraber *et al.* 2004).

4.1.5 JS protocol (Japan)

The Japanese Sewage Works Agency (JS), a public corporation carrying out planning, design and construction of municipal WWTPs as well as R&D activities, established a committee in 2004 to discuss and evaluate an appropriate method for the practical use of activated sludge models. Two years later, the accomplishments of the committee were published as a report (Japan Sewage Works Agency 2006) and subsequently summarized by Itokawa et al. (2008). The important mission of the JS Protocol is to give a boost to the use of simulation studies in practical planning, design and operation. Consequently, it has to be easy to implement rather than scientifically-based, and feasible in terms of both cost and labour demands.

The JS Protocol is meant to use a specific biokinetic model, i.e. ASM2d (Henze et al. 2000), but most of the contents in the protocol are also applicable to other biokinetic models. An overall procedure of a simulation study (Figure 4.6) consists of seven stages which are basically set sequentially, but iterations are also possible if satisfactory results cannot be obtained at a certain stage. Actual simulations are carried out at the final stage, while six preceding stages are dedicated to the preparatory measures aiming at high quality of simulation results. Some important issues related to these measures are outlined below.

In stage 3, additional measurements are recommended for setting up a model of the plant (e.g. flow distributions, mixing conditions in bioreactors, air supply by aeration systems, solids accumulation in final clarifiers), the process analysis and model calibration (average daily influent and effluent concentrations and their diurnal fluctuation patterns, concentration profiles in bioreactors, DO concentrations in all aerobic compartments, sludge composition in terms of COD, N and P), and influent COD fractionation. With regard to the latter issue, three methods are proposed in the protocol including OUR-based (Kappeler and Gujer 1992), filtration-based (Roeleveld and van Loosdrecht 2002) and a method developed by the JS (based on a combination of OUR and ultimate BOD measurements).

Prior to getting into modelling and simulation stages, it is essential to analyze the process and make the full use of available data (stage 4). The recommended analyses include checks for the consistency of data sets subjected to calibration (representative for the historical operating data) and data sets used for flow rate and mass balance (P, N and COD) calculations (reflecting "*pseudo*" steady-state conditions). In the next stage ("*Process modelling*"), a model of the studied plant is constructed using either a commercial or self-made simulation software. Furthermore, it is emphasized that:

- The boundary conditions should be clearly defined with regard to the selection of appropriate lanes, unit processes and devices to be modelled;

- A simple tank-in-series model is allowed for bioreactors with an initial compartmentalization according to existing baffle walls. However, if some remarkable hydraulic features were identified in previous stages (e.g. short circuiting, back mixing and dead zones), they have to be incorporated (without providing a methodology in the protocol);
- Settling models are not specified in the protocol. The primary effluent wastewater characteristics are recommended as influent data (no modelling primary clarification). For final clarification, an ideal clarifier model is generally acceptable due to such advantages as simplicity, direct link between the WAS flow rate and SRT, and the possibility of incorporating biological reactions in the sludge blanket;
- Air supply in aerobic compartments can be modelled with either the conventional $K_L a$ expression or fixed DO concentration (provided that reliable DO data for each compartment are available).

Table 4.2 Outline of the "*calibration*" stage in the JS protocol

Step	Content
1. Definition of simulation method	Method of simulation, including steady or dynamic, duration, and initial condition, is defined
2. Selection of data sets	Data sets used for the fitting and its verification are fixed
3. Definition of input data	Input data for simulation for the fitting is defined
4. Sensitivity analysis (option)	Sensitivity of any uncertain input data on simulation results is evaluated by iterative simulation
5. Preparing outline of fitting	A general direction of fitting work is prepared by carefully evaluating the result of simulation with default values
6. Fitting	Change the value of selected input data so that simulation can represent the actual plant. A stepwise procedure is recommended, in the order of (i) solids production, (ii) conversion factors for N and P, (iii) nitrification, (iv) denitrification, and (v) EBPR.
7. Verification and evaluation	The result is verified with different data set(s). The reliability of simulation is evaluated.

The calibration stage (stage 6) is defined as "*an entire process in which prediction of the model is improved by any means*" and ensures the quality control of model predictions. It consists of seven steps which are outlined in Table 4.2, although the actual degree of efforts depends on the required quality of model predictions and available data. Since special experimental procedures for parameter estimation are considered too intensive for practical use of the models and, therefore, they have not been included in the protocol. Finally, the calibrated model is ready to be used for the defined objective, e.g. comparison of different scenarios (stage 7). Once a selected option is indeed implemented at an actual plant, it is recommended that the model predictions are re-evaluated based on a new set of experimental data.

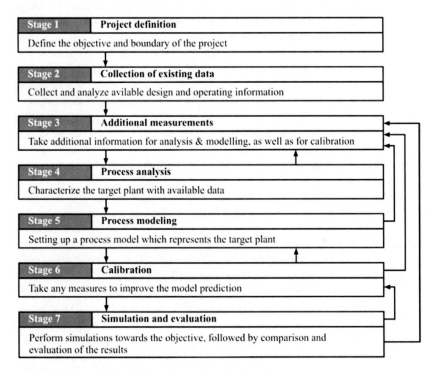

Figure 4.6 Overall procedure of the JS Protocol (Itokawa *et al.* 2008).

4.1.6 IWA Task Group protocol

An IWA Task Group on "*Good Modelling Practice (GMP): Guidelines for Use of Activated Sludge Models*" was established in 2005 in response to the growing concerns about a proper methodology of overall simulation studies and inter-comparability of the results. The main objective of the group was "*to enhance the quality and efficiency of activated sludge (AS) modelling projects, guidelines that include practical protocols on how to use AS models and specific aspects concerning the interaction between modellers and clients are required"*. A new protocol was developed (Figure 4.7) combining existing guidelines in the field of wastewater treatment field and emphasizing key elements in more general management procedures. The proposed protocol includes the following five main steps which were briefly described by Gillot *et al.* (2009).

Step 1: Project definition. The project objectives are established based on the requirements and availability of data. The required additional data have to be identified in order to fix the project budget. At this stage, expectations for the simulation results (including accuracy) and responsibilities are also defined.

Step 2: Data collection and reconciliation. Existing data are collected from historical records, whereas missing data must be gathered in additional measuring campaigns. The time for data collection and reconciliation should be carefully estimated as this step may consume over 30% of the overall time for a simulation study. Data quality should be validated using data series analysis, outlier detection, mass balances (for flows, P, inerts, etc.) and other checks (e.g. typical component ratios). Based on these data, a structure of the plant model should be defined including processes considered, flows and boundaries. This step ends up with an agreement between the modeller and the client concerning the used data and modelled processes.

Step 3: Plant model set-up. This step begins with selection of a simulation platform and sub-models of the general model. Input data, such as tank volumes, flows, influent concentrations, controllers, etc., are specified. Preliminary simulations with an initially built model are run to check mass balances, redefine boundaries and identify critical conditions. The functional model is then tested for sensible outputs. The model adequacy should be confirmed by the client.

Step 4: Calibration and validation. Model parameters are adjusted to obtain an agreement between measured data and model predictions. The final accuracy should meet the criteria established in step 1. Expert knowledge and guidance are required to select an appropriate set of parameters that can be *"fine tuned"* and how to handle the situations when the measured data and model predictions do not match reasonably. The model with the adjusted parameters should be validated using an independent data set.

Step 5: Simulation and result interpretation. With a validated model, various simulations are run to generate scenarios that can be analysed in terms of achieving the project objectives defined in step 1. Once the client accepts the simulation results, a document containing details of the simulation study is prepared. The content of the final report should be standardized to make the results comparable (e.g. for assessing their quality) and provide the adequate information for future model applications.

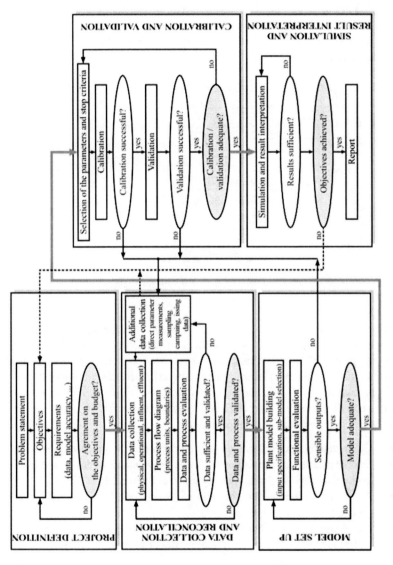

Figure 4.7 Overall procedure of the IWA Task Group protocol (Gillot *et al.* 2009).

4.1.7 Summary

In order to develop a credible model for both science (inter-comparability of the results) and decision making, any simulation study should be performed in accordance with a good, disciplined methodology. For example, Jakeman *et al.* (2006) proposed ten basic (iterative) steps in development and evaluation of environmental models (Figure 4.8).

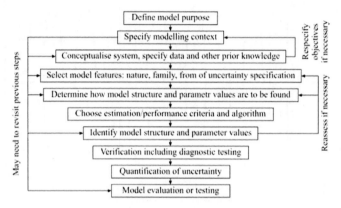

Figure 4.8 Iterative relationship between ten model building steps (Jakeman *et al.* 2006).

According to Jakeman *et al.* (2006), the main constituents of good model-development practice are a clear statement of modelling objectives, adequate setting out of model assumptions and their implications, and reporting of model results. As for the latter, the authors recommended that certain minimum standards should be maintained (subjected to "*a sceptical review by model users*") with regard to model structure, parameterisation, testing and sensitivity analysis. These standards include (but may not be limited to):

- Clear statement of the objectives and clients of the modelling exercise;
- Documentation of the nature (identity, provenance, quantity and quality) of the data used to drive, identify and test the model;
- A strong rationale for the choice of model families and features (encompassing alternatives);
- Justification of the methods and criteria employed in calibration;
- A thorough analysis and testing of model performance as resources allow and the application demands;
- A resultant statement of model utility, assumptions, accuracy, limitations, and the need and potential for improvement;

- Fully adequate reporting of all of the above, sufficient to allow informed criticism.

The calibration protocols, outlined in Sections 4.1.1–4.1.6, in principle follow the general procedure proposed by Jakeman *et al.* (2006) for environmental models. The above standards on reporting of model results should be considered in the final stage of a simulation study. Furthermore, the SWOT analysis of the four protocols (BIOMATH, STOWA, WERF, HSG) for a general model of a WWTP, performed by Sin *et al.* (2005a), revealed similarities and differences between the applied approaches (Table 4.3). Most of these protocols emphasized the importance of several elements, such as data quality control, determination of mixing pattern in the reactor, characterization of influent wastewater and activated sludge biomass as well as estimation of parameters in the biokinetic models. These issues are discussed in more detail in the following sections.

Table 4.3 Similarities and differences between systematic calibration protocols of BIOMATH, STOWA, WERF and HSG (based on Sin *et al.* 2005a)

Similarities	Differences
Effect of the definition of objectives and goals on the overall calibration procedure.	Design of the measurement campaign with respect to frequency, location, duration of measurements.
Significance of data collection, verification and reconciliation with respect to both design and operating data (e.g. SRT, flows, controllers, etc.) as well as additional measurements (e.g. intensive measurement campaigns).	Experimental methods for influent characterization and parameter estimation of the biokinetic models.
Validation using a data set obtained under different operating conditions than those of the calibration period.	Selection of parameter subsets for calibration and calibration methodology

4.2 DATA QUALITY CONTROL (COLLECTION, VERIFICATION AND RECONCILIATION)

In general, many historical data are available at WWTP including the design data (e.g. volumes of reactors), operating data (e.g. flow patterns, controller set points) and performance data (e.g. influent and effluent concentrations). However, much care needs to be invested into the screening (filtering) of existing data since they are often scarce, missing or of uncertain quality (Langergraber *et al.* 2004). Rieger *et al.* (2005) identified three main sources of potential error including flow measurements, sampling and analysis (Table 4.4). The HSG guideline recommends that the analysis and measurement accuracy of the laboratory, which performs the analysis of the routine plant data (laboratory scales, micropipettes and photometers), should also be checked. In the STOWA protocol and HSG guideline, it is advised to use the available data for generating the compositions and the flows and identifying relevant data that are missed. If these missing data are crucial for the simulation study (e.g. missing influent concentrations or P content of the excess sludge for P mass balance) they have to be obtained by expanding temporarily the routine sampling program of the plant (Langergraber *et al.* 2004).

Table 4.4 Sources of error during flow measurements, sampling and analysis (based on Rieger *et al.* 2005)

Flow measurement	Sampling	Analysis
Venturi or weir:	General:	Sample storage:
- Measurement of height	- Wrong location	- Biological
- Changing of	- Inhomogeneity	degradation,
cross-section	- Time	precipitation
- Calculation of flow		
- Miscellaneous (e.g.		
signal transmission		
or conversion)		
Electronic flow meter:	Autosampler:	Sample preparation:
- Air in pipe	- Kind of sampling	- Homogenity
- In- and outflow	- Settings	- Filtration
distances	- Cooling	- Dilution
- Miscellaneous	- Volume and pumping	- Digestion/fractionation
(e.g. signal	speed per single	
transmission or	sampling event	
conversion)	- Laying of inlet hose	
	- Inlet hose	
	Filtration unit for on-line	Lab analysis:
	sensors:	- Micropipettes
	- Inlet hose/filtrate outlet	- Laboratory scales
	- Laying of inlet hose/	- Photometer
	filtrate outlet	- Analysis
	- Coordination with	- Data storage
	sensor	
	- Membrane	
	- Pump rate	

The quality of the available (screened) plant data should be subjected to reconciliation using expert engineering knowledge, continuity checks (e.g. for flow rates) and mass balance calculations for oxygen demand, solids, nitrogen and phosphorus. Application of mass balances on plant historical data is not, however, straightforward because of the process dynamics and variability of the influent loading. Nowak *et al.* (1999) recommended that periods of at least 2–3 SRTs were considered to equalise short-term variations. If these mass balances do not hold, it is necessary to perform extra monitoring of the plant (Hulsbeek *et al.* 2002). Otherwise a calibration procedure can be laborious and lead to unjustified adaptation of the model parameters (Nowak *et al.* 1999; Meijer *et al.* 2002a; Puig *et al.* 2008). According to the HSG guideline, the mass balances for the biological stage should be calculated at least with an uncertainty of 5% for phosphorus, 10% for COD and 10% for TSS (including the secondary clarifiers). Meijer *et al.* (2002a) presented a detailed methodology for error diagnostics and data reconciliation using the free domain software "*Macrobal*" by Hellinga (1992). It detects gross errors (i.e. errors caused by non-random events, such as failures or wrong calibration of measuring devices), improves the data quality and provides accurate information concerning internal conversion rates.

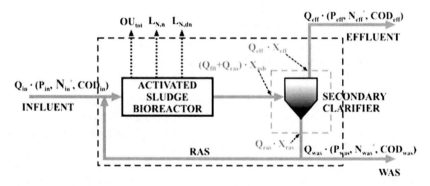

Figure 4.9 Schematic overview of the continuity checks and mass balances described in Equations 4.1–4.6 (* - refers also to TKN measurements in these sampling points).

The main conversions in an activated sludge system, shown in Figure 4.9, are incorporated in the oxygen demand (OD) mass balance:

$$OU_{tot} = Q_{in}\, COD_{in} - Q_{out}\, COD_{out} - 2.86\, L_{N,dn} + 4.57\, L_{N,n} - Q_{was}\, X_{was}\, i_{VT}\, i_{CV}$$

$$(4.1)$$

where
OU_{tot} - Total oxygen uptake in bioreactor, $M(O_2)T^{-1}$
COD_{in} - Influent COD, $M(COD)L^{-3}$
Q_{out} - Effluent flow rate from secondary clarifier, $L^3 T^{-1}$

COD_{out} - COD in secondary effluent, $M(COD)L^{-3}$
$L_{N,dn}$ - Total load of denitrified N, $M(N)T^{-1}$
$L_{N,n}$ - Total load of nitrified N, $M(N)T^{-1}$
Q_{was} - Waste activated sludge (WAS) flow rate, L^3T^{-1}
X_{was} - WAS concentration, ML^{-3}
i_{VT} - MLVSS/MLSS ratio, MM^{-1}
i_{CV} - COD/MLVSS ratio, $M(COD)M^{-1}$

The total loads of denitrified N ($L_{N,dn}$) and nitrified N ($L_{N,n}$) are part of the overall OD balance and can be calculated from two mass balance equations:

$$L_{N,dn} = Q_{in}N_{tot.,in} - Q_{out}N_{tot.,out} - Q_{was}N_{tot.,was} \qquad (4.2)$$

$$L_{N,n} = Q_{in}TKN_{in} - Q_{out}TKN_{out} - Q_{was}TKN_{was} \qquad (4.3)$$

where
$N_{tot,in}$ - Influent total N concentration, $M(N)L^{-3}$
$N_{tot,out}$ - Effluent total N concentration, $M(N)L^{-3}$
$N_{tot,was}$ - Total N concentration in WAS, $M(N)L^{-3}$
TKN_{in} - Total Kjeldahl N concentration in (primary) influent, $M(N)L^{-3}$
TKN_{out} - Total Kjeldahl N concentration in secondary effluent, $M(N)L^{-3}$
TKN_{was} - Total Kjeldahl N concentration in WAS, $M(N)L^{-3}$

Equation 4.1 is an open-type balance that cannot be closed by direct measurements as it contains components in the gas phase (oxygen utilized and N_2 produced). The calculated values of OU_{tot} and $L_{N,dn}$ could only be compared with predictions of the calibrated models which are used in a simulation study. In Equation 4.3, care should also be taken due to difficulties with accuracy and precision of TKN measurements (a number of organic compounds are refractory to Kjeldahl digestion) (Tan 1996).

The open-type balances should be checked with additional closed-type balances for which conversions are calculated from the measurable flows. Phosphorus is a particularly suitable parameter for the complete mass balance calculations since it is removed from the wastewater only in the solid or liquid phase (Nowak *et al.* 1999; Brdjanovic *et al.* 2000; Meijer *et al.* 2001; Langergraber *et al.* 2004; Makinia *et al.* 2005a; Puig *et al.* 2008):

$$Q_{in}P_{in} = Q_{out}P_{out} + Q_{was}P_{was} \qquad (4.4)$$

where
P_{in} - Total P concentration in (primary) influent, $M(P)L^{-3}$
P_{out} - Total P concentration in secondary effluent, $M(P)L^{-3}$
P_{was} - Total P concentration in WAS, $M(P)L^{-3}$

Table 4.5 Accuracy evaluation of measurements and operating parameters based on mass balance calculations for phosphorus and suspended solids (Makinia *et al.* 2006b)

			Average value (±*std*) during the study period			
			Reference line		Experimental line	
Parameter	Symbol	Unit	X_{asr}(lab)*	X_{asr} (online)**	X_{asr}(lab)*	X_{asr} (online)**
Measured:						
MLSS in the bioreactor	X_{asr}	g/m³	2916 (±*615*)	3067 (±*403*)	2898 (±*380*)	3243 (±*240*)
Influent flow rate	Q_{in}	m³/d	51.1 (±*10.0*)		51.6 (±*14.7*)	
Effluent flow rate	Q_{out}	m³/d	47.4 (±*10.9*)		47.9 (±*15.7*)	
Influent TP	P_{in}	g P/m³	8.6 (±*2.1*)		8.6 (±*2.1*)	
TP in WAS	P_{was}	g P/m³	99.5 (±*33.1*)		99.6 (±*19.3*)	
Effluent TSS	X_{out}	g/m³	7.1 (±*4.6*)		7.3 (±*6.7*)	
Effluent TP	P_{out}	g P/m³	1.42 (±*1.24*)		1.54 (±*1.28*)	
RAS flow rate	Q_{ras}	m³/d	54.1 (±*13.6*)		55.7 (±*11.1*)	
TSS in RAS	X_{ras}	g/m³	5131 (±*1123*)		5098 (±*1036*)	
WAS flow rate	Q_{was}	m³/d	3.83		3.83	
SRT	SRT	d	14.9	15.6	14.9	16.6
Balanced:						
TSS in RAS	X_{ras}	g/m³	5464	5748	5384	6026
WAS flow rate (continuity check)	Q_{was}	m³/d	3.72		3.72	
WAS flow rate (P balance)	Q_{was}	m³/d	3.69		3.75	
SRT	SRT	d	15.4	16.2	15.2	16.8

* Mass balance performed with the laboratory measurements of biomass concentration in the bioreactor.
** Mass balance performed with the on-line measurements of biomass concentration in the bioreactor.

Based on Equation 4.4, the WAS mass can be checked to evaluate the conventional calculations of SRT. For this purpose, the continuity check for flow rates and mass balances for suspended solids over the clarifier should also be checked. The appropriate set of these two equations can be written in the following form:

$$Q_{in} = Q_{out} + Q_{was} \tag{4.5}$$

$$(Q_{out} + Q_{ras} + Q_{was})X_{asr} = (Q_{ras} + Q_{was})X_{ras} + Q_{out}X_{out} \tag{4.6}$$

where
X_{asr} - Solids (MLSS) concentration in the bioreactor, ML^{-3}
X_{ras} - Solids concentration in RAS, ML^{-3}
X_{out} - Solids concentration in secondary effluent, ML^{-3}

Sample mass balance calculations, using Equations 4.1–4.6, for the Hanover-Gümmerwald pilot WWTP (Germany) were presented by Makinia *et al.* (2006b). The mass balance calculations for phosphorus revealed that the conventional SRT was calculated with sufficient accuracy (>96%) for both lines (Table 4.5). The Q_{was} flow rates balanced from the continuity equation also did not deviate substantially from the direct measurements. The good quality of the plant data was confirmed by comparing the calculated values of OU_{tot} ($OU_{tot,cal}$) with the corresponding model predictions, $OU_{tot,pre}$ (Table 4.6). These differences ranged from 2.3 to 4.0% depending on the model used.

Table 4.6 Accuracy evaluation of measurements and operating parameters based on mass balance calculations for nitrogen and oxygen demand (Makinia *et al.* 2006b)

Parameter	Symbol	Unit	Average value (±std) during the study period	
			Reference line	Experimental line
Measured:				
Influent COD	COD_{in}	g COD/m³	555 (±126.9)	555 (±126.9)
Effluent COD	COD_{out}	g COD/m³	39.6 (±5.8)	39.5 (±5.8)
Influent TN	$N_{tot,in}$	g N/m³	70.6 (±11.1)	70.6 (±11.1)
Effluent TN	$N_{tot,ut}$	g N/m³	13.0 (±3.2)	12.9 (2.6)
TN in WAS	$N_{tot,as}$	g N/m³	234.4 (±11.2)	231.1
Influent TKN	TKN_{in}	g N/m³	70.1 (±11.2)	70.1 (±11.2)
Effluent TKN	TKN_{out}	g N/m³	4.3 (±1.76)	4.6 (±1.75)
MLVSS/MLSS ratio	i_{VT}	–	0.74 (±0.032)	0.742 (±0.03)
COD/MLVSS ratio	i_{CV}	g COD/g	1.56 (±0.20)	1.57 (±0.14)
Calculated:				
Total oxygen uptake in the bioreactor	$OU_{tot,cal}$	kg O₂/d	19.47	19.68
Predicted:				
Total oxygen uptake in the bioreactor	$OU_{tot,pre}$	kg O₂/d		
– ASM2d			18.69	19.16
– ASM3P			18.81	19.23

The conventional calculations of SRT lead to much lower accuracy of this parameter, primarily due to difficulties with measuring the settled solids concentrations at the bottom of the clarifier and the return sludge transport pipe (Meijer *et al.* 2001). Indeed, Meijer *et al.* (2002a) found that measuring these concentrations based on grab sampling can result in measurement errors up to 40%. Since the assumed SRT is highly sensitive in the simulation model it is essential that this mass balance is scrutinized (Hulsbeek *et al.* 2002). For accurate simulation, the SRT should be known with accuracy higher than 95% (Meijer *et al.* 2001). Puig *et al.* (2008) proposed to calculate the SRT from three alternative formulae which are based on either TP leaving the process, TP entering the process, or COD particulate (COD_{TSS}) leaving the process via the WAS line (Table 4.7). Some of the formulae contain the particulate phosphorus in the sludge ($P_{TSS,R}$) which can be determined as the difference between the TP minus the soluble PO_4-P. The $P_{TSS,R}$ is best measured at the end of the aerobic zone where the phosphorus uptake is at its maximum and the excess sludge is retrieved.

Table 4.7 Different methods for calculation of the SRT (Puig *et al.* 2008)

Basis of calculation	Formula for SRT
Classical (the TSS leaving the process)	$$\dfrac{V_R\,MLSS}{\left(Q_{was}\,TSS_{was}\right)+\left(Q_{eff}\,TSS_{eff}\right)+\left(Q_{strs}\,TSS_{strs}\right)}$$
TP balance based on TP leaving the process	$$\dfrac{V_R\,P_{TSS,R}}{\left(Q_{was}\,P_{TSS,was}\right)+\left(Q_{eff}\,P_{TSS,eff}\right)+\left(Q_{strs}\,P_{TSS,strs}\right)}$$
The conserved TP balance based on TP entering the process	$$\dfrac{V_R\,P_{TSS,R}}{\left(Q_{in}\,P_{in}\right)-\left(Q_{was}\,PO_{4,was}\right)-\left(Q_{eff}\,PO_{4,eff}\right)-\left(Q_{str}\,PO_{4,strs}\right)}$$
The COD particulate leaving the process	$$\dfrac{V_R\,COD_{TSS,R}}{\left(Q_{was}\,COD_{TSS,was}\right)+\left(Q_{eff}\,COD_{TSS,eff}\right)+\left(Q_{strs}\,COD_{TSS,strs}\right)}$$

Subscripts: R – reactor, was – waste activated sludge, eff – secondary effluent, in – influent, strs – stripper stream

Using the four formulae in Table 4.7, the SRT was calculated based on the measured and balanced data for a full-scale WWTP (Figure 4.10). All the calculation methods led to a comparable SRT indicating that this was a correct value. However, the conventional SRT calculations based on TSS resulted in high standard deviations and the SRT calculated from COD measured in the WAS line only slightly improved by balancing. The lowest standard deviations were obtained by using the phosphorus balances which confirms that reconciliation is an effective technique for improving the quality of plant data.

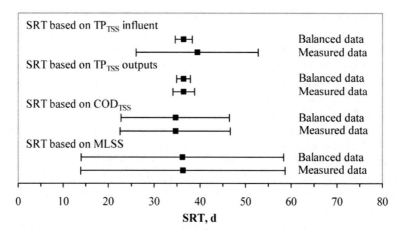

Figure 4.10 SRT calculation based on MLSS, COD_{TSS}, TP_{TSS} influent and outputs using measured or balanced data in the full-scale WWTP (Puig *et al.* 2008).

At WWTPs, the performance data are most often collected routinely at a daily or weekly sampling frequency. This sampling frequency may be insufficient, and therefore for more accurate modelling it may be required to run special monitoring (measuring) campaigns (Petersen *et al.* 2002). The monitoring campaign should start only after the mass balances could be closed without major gaps and all basic tests are carried out successfully (Langergraber *et al.* 2004). The biochemical behaviour of the activated sludge process spans a range of characteristic time constants from seconds (e.g. DO concentration) to days (e.g. biomass composition). It is therefore extremely difficult to implement specialized monitoring programs that enable the proper evaluation of models covering such a wide range of time constants (Lessard and Beck 1991).

A bottleneck for proper characterization of the different flows was the uncertainty about the required monitoring frequency and duration of the monitoring campaign. The sampling frequencies should be chosen in relation to the time constants of the process and influent variations. One of the important time constants of the process is the HRT. The best sampling frequency and test duration are about five times faster than the HRT and 3-4 times this key time constant, respectively (Petersen *et al.* 2002). According to the HSG guideline the monitoring campaign should include an intensive sampling program over a period of e.g. 10 days in series (to include a weekend and to get overlapping).

Generally, the monitoring campaign should be performed during dry weather conditions. In the STOWA protocol, the duration of the measurement campaign depends on the accuracy of the required results. In general 1-3 days is considered sufficient for a system choice study, whereas a measurement campaign of 3 to 7 days and at least 7 days, respectively, is advised for the models used in optimization studies and for control strategies. If the model is used for a system choice, daily average concentrations of influent and effluent and the variations in the flow pattern are sufficient. If the model is used for optimization, the development of control strategies specific data from 4-hour or 2-hour composite samples are required. In this case, other flows than the influent and effluent, such as recirculation flows, should also be sampled. Langergraber *et al.* (2004) recommended that the results of the monitoring campaign should be evaluated by means of mass balances and reliability checks. Due to the short monitoring duration, the storage terms of the mass balance often have to be taken into account and can not be neglected like during pseudo steady state situations.

Since measurements at full-scale WWTPs are relatively expensive these recommendations may not always be completely fulfilled (Petersen *et al.* 2002). Also Langergraber *et al.* (2004) pointed out that the budget available for the simulation study also plays a crucial role while planning the monitoring program. Furthermore, data from the full-scale installation alone may be insufficient for a dynamic model calibration since the reaction kinetics cannot readily be obtained from such data, except for specific designs like SBRs and alternating systems (Petersen *et al.* 2002). In parallel to the monitoring campaign, experiments for influent characterization and parameter estimation (e.g. respirometric measurements) have to be carried out (Langergraber *et al.* 2004). In general, a dynamic model calibration procedure at a full-scale WWTP should combine results derived from laboratory scale experiments (carried out using the process biomass and real wastewater) with data obtained from monitoring campaigns at the studied WWTP (Petersen *et al.* 2002).

4.3 MODEL CALIBRATION/VALIDATION PROCEDURES

Calibration and validation of activated sludge models is an inherent and most time consuming part of any type of model application. The calibration/validation procedure consists of several steps including dedicated lab-scale experiments and intensive measurement campaigns for the characterisation of the influent wastewater, the determination of the kinetics/stoichiometry of the biological processes, and the hydraulic (hydrodynamic) and settling behaviour of the system (van Huelle 2005).

4.3.1 Selection of a hydrodynamic mixing model

4.3.1.1 Estimation of the longitudinal dispersion coefficient, E_L, from tracer studies

Tracer studies involve finding the RTD curve, i.e. age distribution of fluid parcels moving through the reactor. Usually a tracer is introduced at the reactor inlet, and the tracer concentration is then measured at the outlet as a function of time. The tracer may be introduced instantaneously (an impulse signal) or it may be fed continuously (a step signal) as presented earlier in Figure 2.8. Tracers which can be used for this purpose include fluorescent dyes (e.g rhodamine WT), radioisotopes, bacteriophages, chemical salts (e.g. NaCl, LiCl) and floats (USGS 1986; Horan 1990). An ideal tracer should have the following properties (Horan 1990):

- Easily detectible at very low concentrations in the reactor outlet;
- Non-biodegradable within the reactor;
- Non-reactive in any way with the reactor contents (e.g. by binding to particulate material);
- Non-toxic;
- Invisible in the environment.

Although none tracer meets all these requirements, the dye tracers have important advantages, such as low detection and measurement limits, simplicity, and accuracy in concentration measurements (USGS 1986). The procedures of performing field tests in streams using such tracers were outlined in USGS (1982, 1986). Alternatively, a preliminary simulation can be used to estimate the required quantities of tracer, set up a sampling plan (locations, sampling frequencies) and evaluate the anticipated effect on the receiving water (Fall and Loaiza-Navia 2007). For calibrating the measurement instruments (fluorometers), dye standards of 100 (working solution), 25, 15, 10, 6, 4, 2, 1 $\mu g/dm^3$ can be prepared by a serial-dilution process. A preliminary estimate of the tracer (rhodamine WT 20%) dose to produce a desired peak concentration at sampling points was calculated from the following dosage formula for streams (USGS 1982):

$$V_t = 2.0 \cdot 10^{-3} \left(\frac{Q_{tot} L}{u} \right)^{0.93} C_{peak} \qquad (4.7)$$

where

V_t - Volume of tracer added to activated sludge reactor, L^3 (dm^3)
L - distance from injection to sampling point, L (km)
Q_{tot} - Total flowrate through activated sludge reactor, L^3T^{-1} (m^3/s)

C_{peak} - Peak concentration of tracer, ML^{-3} ($\mu g/dm^3$)
u   - mean velocity, LT^{-1} (m/s)

In activated sludge reactors, however, the actual dose should be more than doubled as the expected dispersion is significantly higher compared to streams (Makinia and Wells 2000). Chambers and Jones (1988) carried out tracer tests in 24 full-scale WWTPs, and found that the dispersion coefficient could be considered virtually constant in diffused-air systems. The estimated E_L value was approximately 245 m^2/h ($\pm 15\%$) for the following conditions: width 2-20 m, height 2.4-6 m, length 28-500 m, RAS recirculation ratio 0.7-1.5, hydraulic retention time 1.3-8 h. In the study of Iida (1988), tracer tests were carried out to evaluate mixing conditions in six full-scale aeration tanks with different volumes, geometries, flow rates and air supply intensity. The air was supplied by coarse porous plate diffusers set along one side of the tanks or longitudinally in rows. The estimated values of E_L ranged from 208 to 826 m^2/h with the mean = 476 m^2/h. Significantly higher dispersion in full-scale aeration tanks was observed in the tracer studies of Makinia and Wells (2005) (E_L = 1100-1500 m^2/h) and Zima et al. (2009) (E_L = 910-1690 m^2/h). In the latter case, due to variable mixing intensity in six compartments of the aerobic zone, two values of E_L were estimated and the highest difference between them reached 27%.

Using the results of tracer studies with the impulse signal, the curves of concentration versus time (or space) can be used to estimate the value of E_L by techniques outlined below.

Method of moments (French 1985). By definition, the E_L coefficient (when constant over time) is related to the rate of change of the variance of the tracer cloud:

$$E_L = 0.5 \frac{d(\sigma_x^2)}{dt} \tag{4.8}$$

where
σ_x^2 - Variance of a distribution curve about its mean in space, L^2

Equation 4.8 can be transformed to a form when tracer concentrations are measured at a specific point below the point of injection, as a function of time. This method, termed the method-of-moments, is based on assuming that the C vs. t curve is approximately Gaussian and that the velocity is approximately constant, such that Equation 4.8 can be transformed to:

$$E_L = \frac{u^2}{2} \left[\frac{\sigma_{t2}^2 - \sigma_{t1}^2}{\overline{t_2} - \overline{t_1}} \right] \tag{4.9}$$

where

σ_{t1}^2, σ_{t2}^2 - Variances of a distribution curve about its mean in time in the upstream and downstream sampling points, T^2

$\bar{t}_1$, $\bar{t}_2$ - Mean times of passage of the tracer cloud past the upstream and downstream sampling points, T

Laplace transforms. Another method involves establishing a relationship between σ_t^2 and E_L using Laplace transforms as a solution technique for various boundary conditions (Murphy and Timpany 1967). For a closed system ($E_L \cdot \partial C/\partial x = 0$ at inlet and outlet) and a constant E_L through the tank, the relationship between σ_t^2 and E_L can be presented in the following form:

$$\sigma_t^2 = 2\frac{E_L}{uL} - 2\left(\frac{E_L}{uL}\right)\left[1 - \exp\left(-\frac{uL}{E_L}\right)\right] \tag{4.10}$$

The variance, σ_t^2, for any experimental response curve can be calculated from a dimensionless plot of concentration and time:

$$\sigma_t^2 = \frac{\int_0^\infty \left(\frac{t}{\bar{t}} - 1\right)^2 C\,dt}{\int_0^\infty C\,dt} \tag{4.11}$$

where
$\bar{t}$ - Mean time for the C vs. t distribution, T

Combining and rearranging Equations 4.10 and 4.11, a value of E_L can be calculated from field data of C and t.

Numerical optimization. The two techniques outlined above (methods of moments, Laplace transforms) may not be appropriate when sludge recirculation occurs in activated sludge reactors. In many cases, in order to obtain relevant data for analysis, the recirculations (RAS and MLR) have to be turned off which does not happen under normal operating conditions. The technique flexible enough to account for this recirculation is a numerical solution of the 1-D ADE without an internal source term (Equation 2.13). Such a form of the equation can be used to estimate the value of E_L by a numerical optimization technique which involves minimizing the objective function, e.g. the mean square error test of convergence (Makinia and Wells 2005; Zima *et al.* 2009). In those studies, tracer studies with rhodamine WT 20% were performed at two full-scale activated sludge bioreactors located at the Rock Creek WWTP in Hillsboro, Oregon (USA) (Makinia and Wells 2005) and the Wschod WWTP in Gdansk (Poland) (Zima *et al.* 2009). The bioreactors had the following dimensions: length – 84 m, width – 15.6 m,

depth – 4.9 m (fully aerobic bioreactor, Rock Creek), and length – 252 m, width –
8.4 m, depth – 5.5 m (aerobic zone of the MUCT process, Wschod). At the Rock
Creek WWTP, 0.5 dm^3 of the rhodamine was injected at the reactor inlet and a
single sampling point was established at the reactor outlet (SP1) (Figure 4.11a).
At the Wschod WWTP, 1.0 dm^3 of the tracer was injected at the aerobic zone inlet
and three sampling point was established: middle of the zone (SP1), outlet of the
zone (SP2) and inlet to the zone (SP3) (to control recirculation of the rhodamine)
(Figure 4.11b). In that case, the method of moments and numerical optimization
generated comparable results with deviations normally not exceeding 14% for the
same measurement. However, secondary peaks of the tracer concentrations were
explicitly observed after 3.5-4.5 h from the beginning of the test, which resulted
from the recirculations (MLR and RAS) in the activated sludge system. These
peaks could be considered and accurately predicted by the optimized 1-D ADE.

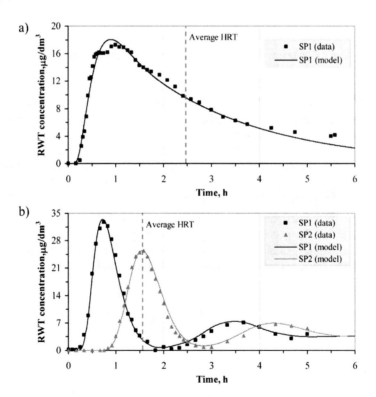

Figure 4.11 Measured vs. predicted concentrations of rhodamine WT during tracer
studies: (a) Rock Creek WWTP, Hillsboro, OR (USA) (adopted from Makinia and
Wells 2005), (b) Wschod WWTP, Gdansk (Poland) (adopted from Zima *et al.* 2009).

4.3.1.2 Estimation of the longitudinal dispersion coefficient, E_L, from empirical formulae

Alternatively, when data from a tracer test are not available for a studied reactor, the value of E_L can be approximated with one of the empirical formulae discussed below.

Murphy and Boyko (1970). Tanks of different width to depth ratios ranging from 0.87 to 2.04 were studied. Various combinations of width and depth were selected tentatively as the "*characteristic length*" and the most successful correlation ($r^2=0.885$ for 96 different test conditions) was obtained when the tank width was selected. Consequently, the following correlation relating the E_L coefficient, tank width and specific air flow rate per unit tank volume was proposed:

$$\frac{E_L}{W^2} = 3.118(q_A)^{0.346} \tag{4.12}$$

where
q_A - Air flow rate per unit reactor volume, T^{-1} $(m^3/(1000 \text{ m}^3 \cdot min))$
W - Reactor width, L (m)

It should be noted that USEPA (1993) recommended the use of Equation 4.12 as an acceptable approximation of the dispersion coefficient in reactors with both fine and coarse bubble diffused air systems.

Harremoes (1979). The tanks studied included a bench scale reactor with diffuser stones installed on the side and at the bottom of tanks, and two full-scale aeration tanks. The value of E_L was determined by the general flow pattern of the reactor as generated by the air supply. It was assumed that the coefficient was primarily a function of the characteristic velocity, defined as $(g \cdot q'_A)^{1/3}$, with corrections for the geometry of the tank. Multi-parameter regression analysis performed for the results of the pilot plant study (45 measurements) and the full-scale studies (26 measurements) gave the following result:

$$\frac{E_L}{(gq'_A)^{1/3} W} = 2.4 \cdot 10^{-3} \left(\frac{H}{W}\right)^{-0.68} Re_g^{0.26} \tag{4.13}$$

where
G - Gravity acceleration, LT^{-2}
H - Reactor depth, L
q'_A - Air flow rate per unit length of reactor, L^2T^{-1}
Re_g - Reynolds number associated with the aeration intensity

The Reynolds number associated with the aeration intensity (Re_g) was defined as:

$$Re_g = \frac{(gq'_A)^{1/3} H}{v_l} \qquad (4.14)$$

where
v_l - Kinematic viscosity of liquid, LT^{-2}

Fujie et al. (1983). Two full-scale tanks with diffusers (porous plastic tubes or porous ceramic plates) were studied. The E_L coefficient was related to spiral liquid circulation by applying random walk theory. The spiral liquid circulation was caused by the air supplied at right angles to the direction of liquid flow. During one circulation, the average travelling distance of the liquid element in the vertical cross section was $\xi(H + W)$, whereas the corresponding displacement of the liquid element per unit circulation was $\pm\lambda(H + W)$. After N_c circulations, the probability of the displacement was given by the Gaussian distribution and the variance of the displacement, σ_x^2, was calculated from the following formula:

$$\sigma_x^2 = N_c[\lambda(H + W)]^2 \qquad (4.15)$$

where
N_c - Number of circulations in the vertical cross-section,
λ - Non-dimensional correction factor which makes the displacement from the average flow during one circulation,
H - Depth of tank, L
W - Width of tank, L

The time t_N required for a liquid element of interest to circulate N_c times in the vertical cross section was expressed as:

$$t_N = \frac{2N_c\xi(H + W)}{\xi'u_{ls}} \qquad (4.16)$$

where
t_N - Time required for a liquid element of interest to circulate N times in the vertical cross section, T
ξ - Non-dimensional correction factor which makes the average travelling distance of the liquid element in the vertical cross-section,
ξ' - Non-dimensional correction factor which makes the average spiral circulation rate at liquid surface,
u_{ls} - Spiral circulation rate, LT^{-1}

The relationship between E_L and σ_x^2, developed from Equation 4.8, was given by:

$$\sigma_x^2 = 2E_L t_N \qquad (4.17)$$

Rearranging Equation 4.17 by substituting σ_x^2 and t_N from Equations 4.15 and 4.16, respectively, gives:

$$E_L = \lambda'^2 u_{ls}(H + W) \qquad (4.18)$$

where

$$\lambda'^2 = \frac{\lambda\xi'}{4\xi} \qquad (4.19)$$

In earlier studies, Kubota et al. (cited in Fujie et al. 1983) developed the following formula to measure u_{ls}:

$$u_{ls} = a_d \left[hu_g \left(\frac{h}{H} \right)^{1/2} \left(\frac{H}{W} \right)^{1/3} \right]^{m_d} \qquad (4.20)$$

where
a_d, m_d - Empirical constants dependent on the type of air diffuser,
H   - Diffuser depth, L
u_g - Superficial gas velocity, LT^{-1}

The constants a_d and m_d were determined in terms of the type of diffuser installed (Table 4.8). Values of λ'^2 were a function of the configuration of the reactor and the air supply rate per unit floor area of the reactor (superficial velocity), u_g. Based on literature data and results of their own studies, the authors developed the following empirical formula for λ'^2:

$$\lambda'^2 = 0.0115 \left(1 + \frac{H}{L} \right)^{-3} u_g^{-0.34} \qquad (4.21)$$

The final formula for E_L was obtained from Equation 4.18 by substituting u_{ls} and λ^2 from Equations 4.20 and 4.21, respectively:

$$E_L = 0.0115 \left(1 + \frac{H}{L} \right)^{-3} u_g^{-0.34} a_d \left[hu_g \left(\frac{h}{H} \right)^{1/2} \left(\frac{H}{W} \right)^{1/3} \right]^{m_d} (H + W) \qquad (4.22)$$

Table 4.8 Values of parameters a_d and m_d in Equations 4.20 and 4.22, where

$$\Phi = hu_g \left(\frac{h}{H}\right)^{1/2} \left(\frac{H}{W}\right)^{1/3}$$

Type of air diffuser	Φ(cm²/s)	m_d	a_d
Fine bubble types[1]	$\Phi \leq 20$	0.64	7.0
	$\Phi > 20$	0.46	12.0
Coarse bubble types[2]	$\Phi \leq 20$	0.78	3.5
	$\Phi \leq 20$	0.56	4.9

[1] Porous plates and tubes
[2] Perforated plates and tubes, single nozzles and others

Khudenko and Shpirt (1986). Various bench scale and full-scale tanks of a corridor type with diffused air systems, both fine and coarse porous plates, were studied. The width of the aerator band ranged from 12.5 to 100% of the tank width. The E_L coefficient was related to geometric and dynamic parameters through a general relationship:

$$E_L = f_1(H, W, L, w, u_g, u, v_1) \tag{4.23}$$

Using the Buckingham π-theorem, the following dimensionless equation, composed from the parameters listed in Equation 4.23, was derived:

$$\frac{E_L}{uL} = A_1 Re_g^{\alpha_1} Re_1^{\alpha_2} \left(\frac{L}{W}\right)^{\alpha_3} \left(\frac{H}{W}\right)^{\alpha_4} \left(\frac{w}{W}\right)^{\alpha_5} \tag{4.24}$$

where
w   - Width of aeration band, L
A_1 - Empirical constant,
Re_1 - Reynolds number associated with the fluid flow in reactor,
α1...α5   - Empirical constants,

Reynolds numbers associated with the aeration intensity (Re_g) and the fluid flow in reactor (Re_1) were defined as:

$$Re_g = \frac{u_g H}{v_1} \tag{4.25}$$

$$Re_1 = \frac{uH}{v_1} \tag{4.26}$$

Values of the E_L coefficient were found using Equation 4.10 and correlated with the hydrodynamic parameters of the reactor. The final form of Equation 4.24 was found to be:

$$\frac{E_L}{uL} = 4.2\,Re_g^{0.60}\,Re_l^{-0.75}\left(\frac{L}{W}\right)^{-0.9}\left(\frac{H}{W}\right)^{0.80}\left(\frac{w}{W}\right)^{0.28} \tag{4.27}$$

Equation 4.27 suggested that the longitudinal dispersion of flow in aeration tanks increased with an increase in the aeration intensity (or Re_g), the tank depth and the aeration band width. Longitudinal dispersion decreased with an increase in the velocity of flow along the reactor (or Re_l) and the tank length. The width of the tank produced only a slight effect on the mixing pattern.

Potier et al. (2005). Residence time distributions have been measured with tracer (lithium chloride) experiments in two aerated channel reactors (full-scale and bench-scale) for a wide range of gas and liquid flow rates and geometrical parameters. The total volume of the full-scale activated sludge bioreactor was 3300 m³, and its width and total unfolded length were 8 m and 100 m, respectively. The experiments were conducted for different liquid flow rates under a constant aeration rate and water depth. The bench-scale channel (width 0:18 m and length 3.60 m) had a variable water depth and maximum working volume of 0.08 m³. The experiments were conducted for different liquid and gas flow rates and water depths.

Based on the experimental data, a general empirical expression was obtained to calculate the E_L coefficient as a function of the gas and liquid velocities, and the geometrical parameters of the reactors:

$$E_L = (0.2032W - 0.008569)\left(\frac{Q_A}{L}\right)^{0.5}(100H)^{0.00473W^{-1.99}} \tag{4.28}$$

where E_L is expressed in m²/s, Q_A in m³/s and L, H, W in m.

In full-scale activated sludge reactors, with large values of H and W, the formula may be simplified to the following form:

$$E_L = 0.2032W\left(\frac{Q_A}{L}\right)^{0.5} \tag{4.29}$$

Lemoullec et al. (2008). The objective of that study was to develop a general correlation of axial dispersion in aerated channel reactors, covering a wide range of operating and geometrical parameters, based on available literature data (mostly described above) and extensive experimental data (175 data sets) gathered by the authors from two pilot plants. It was found that none of the previously developed formulae was able to represent properly the whole available set of axial dispersion coefficients (the relative error = 91–3300%), although all of them were acceptable for a certain range of operating conditions.

The general correlation was obtained by using the Buckingham π-theorem and selecting different parameters related to the studied systems and representing four categories: operating, physical properties of the fluids (gas and liquid), geometrical characteristics of the reactors, and physical parameters. Those parameters were grouped into six dimensionless terms:

$$\frac{E_L h'}{Q} = A_1 \left(\frac{\mu_L h'}{Q \rho_L}\right)^{\alpha_1} \left(\frac{h'}{v'}\right)^{\alpha_2} \left(\frac{g h'^5}{Q^2}\right)^{\alpha_3} \left(\frac{Q_A}{Q}\right)^{\alpha_4} \left(\frac{w}{W}\right)^{\alpha_5} \tag{4.30}$$

where

μ_L - Dynamic viscosity, $ML^{-1}T^{-1}$
h' - Hydraulic horizontal diameter, L
v' - Hydraulic vertical diameter, L
A_1 - Empirical constant,
$\alpha_1 \ldots \alpha_5$ - Empirical constants,
g - Gravity acceleration, LT^{-2}

The hydraulic horizontal and vertical diameters, h' and v', respectively, were defined as follows:

$$h' = 2\frac{LW}{L+W} \tag{4.31}$$

$$v' = 2\frac{HW}{H+W} \tag{4.32}$$

The regression process allowed to determine the E_L coefficient as a simple function of the geometrical parameters (H, L, W, w) and air flow rate (Q_A):

$$E_L = 0.33 v' \left(\frac{Q_A}{h'}\right)^{0.5} \left(\frac{w}{W}\right)^{0.5} \tag{4.33}$$

where E_L is expressed in m²/s.

Using Equation 4.33, the values of E_L coefficient were estimated with a relative error of 18% compared to the experimentally determined values for a large number of reactors ranging from bench scale to full-scale sizes (Figure 4.12). Moreover, the formula was valid for both swirling flow (low values of w/W) of the liquid phase induced by bubbles and for global horizontal liquid flow (w/W=1).

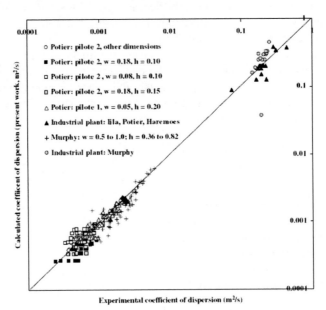

Figure 4.12 Comparison between E_L values experimentally determined and calculated using Equation 4.33 (Lemoullec *et al.* 2008).

A principal limitation of all these empirical formulae is that they are applicable only to the aerated tanks since the calculated E_L coefficient is related to the aeration intensity in the reactor. Most of the formulae were evaluated in terms of their capability to approximate mixing conditions based on the experimentally determined E_L values, reported in literature (Figure 4.13). For example, at the Rock Creek WWTP, Hillsboro (USA), the best accuracy in comparison to the results of three tracer studies (measured E_L=1040–1580 m²/h) was obtained for the formula of Fujie *et al.* (1983) (calculated E_L=990–1020 m²/h). When the calculated E_L coefficients were applied to the 1-D ADE, the average relative deviations (ARD) were higher by only less than 2.1% from the ARD corresponding to the optimum value of E_L (Makinia and Wells 2005). With the same formula, Zima *et al.* (2008) calculated E_L = 853 m²/h for a tracer test at the Wschod WWTP (Poland), but that value was approximately 40% lower compared to the numerically optimized coefficient (similar to one test at the Rock Creek WWTP).

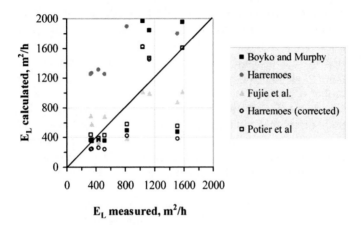

Figure 4.13 Comparison of the calculated values of E_L with the measurements reported in the studies of Iida (1988), Makinia and Wells (2005), and Zima *et al.* (2008).

4.3.2 Influent wastewater and biomass characterization

4.3.2.1 Integrated wastewater characterization

The Dutch Foundation of Applied Water Research (STOWA) formulated integrated guidelines based on a review of different techniques of wastewater characterization and their evaluation in terms of reproducibility and practical applicability (Roeleveld and Kruit 1998). Experiences with the guidelines including modifications to the original version were discussed by Roeleveld and van Loosdrecht (2002). The guidelines are based on a physical-chemical method to characterize the sum of the soluble COD fractions (S_I and S_S), combined with a BOD analysis (measuring the BOD as a function of time) for determining the biodegradable fraction of the influent COD ($S_S + X_S$) (Table 4.9). In such a case, the fraction of readily biodegradable COD (S_S) depends strictly on the choice of the filter pore size. At the early stages of wastewater characterization studies, it was assumed that the use of a 0.45 μm membrane filter would allow to obtain filtrate without the content of fine colloidal matter (e.g. Stensel 1992). Thus the STOWA recommended originally to use such a filter for the division between soluble and particulate COD fractions. Experiences with "*soluble*" COD analysis at six full-scale WWTPs revealed, however, that the concentration of S_S in pre-precipitated samples was reduced by up to 29% compared to the membrane filtrate. Further studies confirmed a similar reduction (up to 25%) in samples pre-treated by flocculation with $Zn(OH)_2$. Due to these discrepancies the revised guidelines advise to use a 0.1 μm filter or flocculation with $Zn(OH)_2$ (or another precipitant) instead of the originally proposed 0.45 μm membrane filter. It has been demonstrated at seven full-scale WWTPs that 0.1 μm filtration and flocculation generates comparable results with the difference of approximately 1%. Furthermore, it is emphasized in the updated guidelines that a variation in the X_S/X_I ratio is very sensitive to almost every modelled process, especially the sludge production. In the STOWA guidelines, the X_S/X_I ratio is determined indirectly by the measurement of biodegradable COD via the k_{BOD} coefficient. Due to the uncertainty of the BOD analysis the fractions X_S and X_I should be further calibrated. Meijer et al. (2001) demonstrated a proper way to evaluate the estimated X_S/X_I ratio using mass balances over the activated sludge system.

Wastewater characterization results, obtained with the STOWA guidelines for the COD fractions at 21 Dutch WWTPs, are summarized in Table 4.10. These results are related the wastewater after the screen, before any kind of pre-treatment and before mixing with internal flows. In the performance of the characterization, different methods were used to determine the soluble COD fraction: filtration (0.45 μm) of a pre-treated sample with $Zn(OH)_2$, only filtration (0.1 or 0.45 μm) or only flocculation (Roeleveld and van Loosdrecht 2002).

Carrette et al. (2001) used the original STOWA guidelines (Roeleveld and Kruit 1998) to perform the COD fractionation in the settled wastewater containing a high proportion (up to 41%) of COD originating from a textile industry. For comparison, the fractionation was performed with a combination of biological methods, i.e. the respirometric technique of Spanjers and Vanrolleghem (1995) for S_S and the comparison method of Orhon et al. (1992) for the inert fractions. When the industrial discharge was active, the contribution of the readily biodegradable fraction, determined using the STOWA guidelines, was extremely high accounting for 44% of total COD. Moreover, the resulted fractionation did not allow for a good calibration of the sludge mass balance. In contrast, the COD fractionation according to the biological methods allowed to fit both the sludge mass balance and observed measurement data after a minor adjustments of the X_S/X_I ratio, i.e. the contribution of X_I was decreased from 19% to 17% and the contribution of X_S was increased accordingly. During a temporary stop of the industrial discharge the influent composition was only characterized with the STOWA guidelines. In this case, the estimated contribution of S_S was considerably lower (12%) and the sludge mass balance fitted accurately.

Table 4.9 Equations for determination of the influent components according to the STOWA guidelines for wastewater characterization (cited in Brdjanovic *et al.* 2000)

Soluble components	Particulate components	Additional calculations
$S_A = COD_{VFA,in}$	$X_S = \alpha COD_{X,in}$	$\alpha = \dfrac{\dfrac{BOD_{u,in}}{1 - Y_{BOD}} - S_S}{COD_{X,in}}$
$S_I = 0.9 COD_{f,out}$	$X_I = (1-\alpha) COD_{X,in}$	$BOD_{u,in} = \dfrac{BOD_{5,in}}{1 - e^{-5k_{BOD}}}$
$S_S = COD_{f,in} - S_I$	$X_A = 0.1 - 1.0$	$COD_{X,in} = COD_{in} - COD_{f,in}$
$S_F = S_S - S_A$	$X_{PAO} = 0.1 - 1.0$	
$S_{NH} = TKN - \sum\limits_{i=1}(i_{N,Si}S_i + i_{N,Xi}X_i)$	$X_H = 0$	
$S_{PO4} = P_{tot.} - \sum\limits_{i=1}(i_{P,Si}S_i + i_{P,Xi}X_i)$	$X_{PHA} = 0$	
	$X_{PP} = 0$	

Table 4.10 Average results of the wastewater characterization at 21 WWTPs in the Netherlands (Roeleveld and van Loosdrecht 2002)

WWTP	Measurements						Fractions			
	COD_{in}	$COD_{f,in}$	$COD_{f,out}$	$BOD_{5,in}$	k_{BOD}	BCOD	S_I	S_S	X_S	X_I
Geestmerambacht[1]	593	134	32	189	0.35	269	29	105	164	295
Niedorpen[1]	393	150	39	199	0.42	267	35	115	152	91
Stolpen[1]	827	207	41	373	0.35	531	37	170	361	259
Ursem[1]	748	238	40	330	0.38	457	36	202	255	255
Wieringermeer[1]	592	140	40	227	0.42	304	36	104	200	252
Wieringen[1]	668	240	41	330	0.38	457	37	203	253	175
Deventer[2]	799	125	31	257	0.34	370	28	97	273	401
Nijmegen[2]	536	154	28	238	0.35	339	25	129	210	172
Franeker[2]	607	113	53	251	0.36	354	48	65	288	206
Gouda[2]	662	180	34	254	0.38	351	31	149	202	280
Venlo[3]	617	222	25	220	0.47	286	23	204	83	308
Apeldoorn[4]	468	195	33	173	0.70	210	30	165	45	228
Boxtel[4]	510	172	36	181	0.41	244	32	140	105	233
Nieuwgraaf[4]	613	297	43	277	0.59	344	39	258	86	230
Groote Lucht[4]	447	158	32	174	0.43	232	29	129	102	187
Nieuwe Waterweg[4]	430	113	30	148	0.46	194	27	86	108	209
Haarlem–WP[4]	328	118	23	120	0.39	165	21	97	67	143
Alphen KenZ[5]	512	145	30	176	0.33	256	27	118	138	229
Groesbeek[5]	428	180	30	180	0.34	259	27	153	106	142
Hardenberg[5]	555	219	38	217	0.40	295	34	185	110	226
Papendrecht[5]	241	127	28	119	0.60	147	25	102	46	68

Table 4.10 (cont.)

[1] Average values of 24–hour composite samples from five days; soluble analysis after filtration (0.45 µm) of pre-treated sample with $Zn(OH)_2$

[2] Average values of 24–hour composite samples from seven days; soluble analysis after filtration (0.45 µm) of pre-precipitated wastewater

[3] Grab sample; soluble analysis after filtration (0.1 µm)

[4] Average values of six grab samples; soluble analysis after flocculation

[5] Average values of 24–hour composite samples from four 2-day measuring campaigns; soluble analysis after filtration (0.45 µm)

4.3.2.2 Characterization of individual fractions

Most methods used for the determination of readily biodegradable COD rely on respirometric measurements conducted in continuous or batch reactors, under either aerobic or anoxic conditions (Orhon *et al.* 1997). It should be noted, however, that these methods are based on the approach proposed in ASM1 that the readily biodegradable COD is directly used for the growth of microorganisms without considering any storage phenomena. As mentioned earlier, in the presence of intracellular storage polymers the interpretation of OUR profiles becomes confusing (see: Section 3.3).

The fraction of soluble COD that is readily biodegradable can be measured through the OUR in the continuous reactors (Ekama *et al.* 1986; Sollfrank and Gujer 1991; Witteborg *et al.* 1996) as well as in the batch reactors (Kappeler and Gujer 1992; Kristensen *et al.* 1992; Orhon *et al.* 1994b; Xu and Hasselblad 1997). Similarly, a NUR in the batch reactors can be used to determine the concentration of readily biodegradable organic compounds (Ekama *et al.* 1986; Kristensen *et al.* 1992; Orhon *et al.* 1994b).

The value of RBCOD can also be estimated by measuring the change in OUR in a single completely mixed reactor operated at a very short SRT of 2 days under a daily cyclic square feeding pattern (12 h with feed, 12 h without feed) as earlier presented in Figure 3.5. A rapid drop in the OUR (ΔOUR) following feed termination, associated only with utilization of the RBCOD, can be used to find their concentration (Ekama *et al.* 1986):

$$S_S = \frac{V\,\Delta OUR}{Q_{in}(1-Y_H)} \tag{4.34}$$

where

S_S - Concentration of soluble, readily biodegradable organic compounds, $M(COD)L^{-3}$

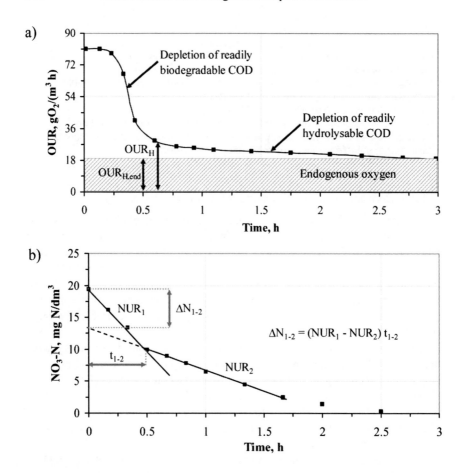

Figure 4.14 Interpretation of the batch tests results for the determination of RBCOD under aerobic conditions based on the method of Kappeler and Gujer (1992) (a) and under anoxic conditions based on the method of Orhon *et al.* (1994b) (b).

Another continuous flow method was proposed by Witteborg *et al.* (1996). The method is based on the measurement of OUR under different wastewater loading conditions (endogenous, intermediate, high) applied to a continuously operated respirometer. The concentration of S_S is calculated by solving a set of mass balance equations corresponding to the different loading conditions.

In the standard batch test (or "*bioassay*" as termed by Grady *et al.* (1999)) for determination of RBCOD, originally proposed by Ekama *et al.* (1986) and developed further by Kappeler and Gujer (1992), a sample of wastewater is mixed with endogenous sludge and the respiration rate is monitored until it reaches the endogenous level again (Figure 4.14a). When the test is carried out under aerobic conditions nitrification has to be suppressed by the addition of an inhibitor (e.g. allylthiourea, ATU). Then the selection of an appropriate initial S/X_V (substrate/ biomass) ratio provides a clear differentiation between the heterotrophic OUR, OUR_H, related to the utilization of RBCOD and SBCOD. In order to clearly separate these two phases the initial S/X_V ratio of 0.05–0.3 g COD/g VSS should be used (Carucci et al. 2001). Kappeler and Gujer (1992) recommended to use activated sludge biomass and wastewater in a volumetric ratio of approximately 1 to 2. The initial high OUR_H stays constant over a period, e.g. 1–3 h as suggested by Ekama et al. (1986), and after depleting S_S it rapidly drops to a lower level associated with hydrolysis of the remaining X_S and finally reaches the endogenous level, $OUR_{H,end}$. The concentration of readily biodegradable matter in wastewater, S_S, can be calculated from the following relationship (Ekama et al. 1986; Kappeler and Gujer 1992; Orhon et al. 1994b):

$$S_S = \frac{\int OUR_H \, dt - \int OUR_{H,end} \, dt}{1 - Y_H} \qquad (4.35)$$

The value of S_S can also be determined in a similar manner based on the NO_3-N profile in the anoxic batch reactor (Figure 4.14b). In this test, the initial nitrate utilization rate (NUR_1) is faster, due to oxidation of the readily biodegradable substrate. The amount of nitrate (ΔN_{1-2}) consumed during this period is used to calculate S_S from the following relationship (Orhon et al. 1994b):

$$S_S = \frac{2.86}{1 - Y_H} \Delta N_{1-2} \qquad (4.36)$$

where
ΔN_{1-2} - Amount of nitrate consumed due to oxidation of the readily biodegradable substrate, $M(N)L^{-3}$

Cokgor et al. (1998) evaluated parallel anoxic and aerobic respirometric measurements in batch reactors to quantify the RBCOD concentration in wastewater. Assuming equal values for aerobic and anoxic yield coefficients for heterotrophic biomass, it was found that the RBCOD concentrations in municipal wastewater obtained from the anoxic reactors were consistently higher by approximately 14% than the values obtained from the equivalent aerobic reactors. The authors concluded that this observation was associated with a lower anoxic

yield for heterotrophic biomass compared to the corresponding aerobic value
(see: Section 3.1.4).

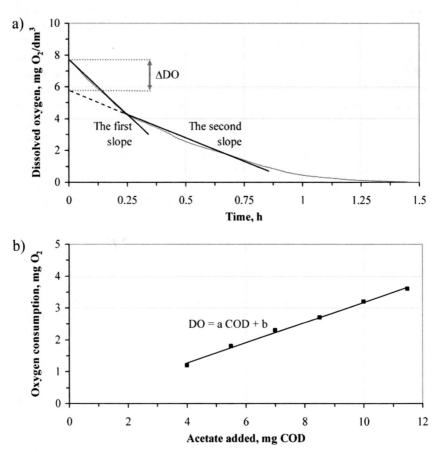

Figure 4.15 Estimation of the RBCOD according to the single-OUR method (Xu
and Hasselblad 1997).

   Xu and Hasselblad (1997) proposed a simple method, called the "*single-
OUR*" test, for determining the RBCOD concentration. The method is based on
the monitoring of a single depletion curve of DO concentration (Figure 4.15a).
Initially, the DO concentration decreases rapidly due to the utilization of RBCOD
(the first slope). Once most of the RBCOD is utilized a change in the OUR can
be observed (the second slope). The oxygen demand (ΔDO) that corresponds

to the amount of RBCOD can then be determined graphically from the oxygen concentration profile. Finally, the ΔDO value is compared with a calibration curve to obtain the concentration of readily biodegradable substrate (Figure 4.15b). This curve is established earlier by performing a number of "*single-OUR*" tests with a substrate (e.g. acetate) having a known COD and oxygen demand for its removal. The advantage of this method is a relatively short time (approximately 30 min) compared to the traditional "*multi-OUR*" methods of Ekama *et al.* (1986) and Kappeler and Gujer (1992) which require 2–3 hours to determine the concentration of readily biodegradable substrate. Ziglio *et al.* (2001) validated the "*single-OUR*" test by comparing it with the traditional batch method of Ekama *et al.* (1986) at a full-scale WWTP (100,000 PE). The comparison between these two methods revealed an average difference of only 2% (±1.5 mg COD/dm^3) for 4 measurements.

The value of S_S can also be determined by a physical-chemical ("*flocculation*") method developed by Mamais *et al.* (1993). With this method, a sample of wastewater is flocculated by adding 1 cm^3 of a solution of $ZnSO_4$ (100 mg/dm^3) to 100 cm^3, mixing vigorously for approximately 1 min, and adjusting the pH with NaOH (6M solution). The sample is then allowed to settle quiescently before clear supernatant is withdrawn and filtered through a 0.45 µm membrane filter. The concentration of the readily biodegradable COD, S_S, is obtained by subtracting the inert soluble COD, S_I, from the COD of the filtrate. The rationale for this method is that readily biodegradable organic matter consists of simple molecules such as VFAs and low molecular weight carbohydrates that pass through the cell membrane and are metabolized immediately (Mamais *et al.* 1993). Hu *et al.* (2002a) indeed confirmed that the COD quantified with this method corresponded closely with the low (<1000 Da) molecular weight fraction. For domestic wastewater, the flocculation method gives results that correlate well with the traditional batch test of Ekama *et al.* (1986) (Grady *et al.* 1999). This statement remains in contradiction to the observations from other studies. Even though colloidal particles, which normally pass the filter, are removed before filtration, the filtered COD is equal to the truly soluble COD rather than only to the readily biodegradable fraction (Xu and Hultman 1996). The truly soluble COD (COD_{sol}) is a sum of three fractions:

$$COD_{sol} = S_S + S_H + S_I \qquad (4.37)$$

where
S_H - Concentration of soluble, rapidly hydrolysable organic compounds, $M(COD)L^{-3}$

Due to the presence of rapidly hydrolysable COD in the membrane filtrate different respiration rates may occur for the same concentration of soluble COD

(Sollfrank and Gujer 1991). Ginestet *et al.* (2002) characterized samples of raw wastewater originating from seven French WWTP. Respirometric measurements were carried out with the samples of the raw, settled and "*coagulated*" (i.e. settled and precipitated with $FeCl_3$) wastewater. The latter group predominantly consisted of the readily hydrolysable fraction (37–90%), whereas the readily biodegradable and inert fractions accounted for 2-27% and 2-47% of soluble COD, respectively. Naidoo *et al.* (1998) examined the raw wastewater entering eight WWTPs in different European countries (Table 4.11). The following four parameters were determined: total COD (COD_{tot}), COD after centrifugation (COD_{cent}), COD after filtration through a 0.45 μm (COD_f) and COD after coagulation and centrifugation ($COD_{coag+cent}$). In addition, the concentration of readily biodegradable substrate, $RBCOD_{NUR}$, was determined from Equation 4.36 based on the specific NUR kinetics. The estimated contributions of $RBCOD_{NUR}$ ranged between 7 and 19% of total COD and these values were significantly lower than the soluble COD determined by the other methods, i.e. physical (filtration) or physical-chemical (coagulation-centrifugation). Discrepancies between the physical-chemical methods and respirometric measurements may especially be observed for samples with a high content of industrial wastewater (Carrette *et al.* 2001).

Table 4.11 Results of raw wastewater characterization at eight European WWTPs (Naidoo *et al.* 1998)

WWTP	COD_{tot} g COD/m³	COD_{cent} % COD_{tot}	COD_f % COD_{tot}	$COD_{coag+cent}$ % COD_{tot}	$RBCOD_{NUR}$ % COD_{tot}
Crespieres (France)	549	57	39	31	7
Morainvilliers (France)	344	49	48	41	15
Boran (France)	707	65	57	50	10
Plaisir (France)	691	32	30	25	11
Rostock (Germany)	953	30	27	24	19
Berwick (England)	913	n.d.	51	41	n.o.
Orense (Spain)	407	n.d.	32	17	14
Brno (Czech Republic)	250	n.d.	40	32	10

(Note: n.d. – not determined, n.o. – not observed).

Spanjers and Vanrolleghem (1995) developed a rapid procedure to estimate wastewater characteristics (readily biodegradable and rapidly hydrolysable fractions) in a combination with several kinetic and stoichiometric parameters in a model similar to ASM1. The procedure was based on two respirometric measurements at low S/X_V ratios, i.e. 1:20 and 1:200 on the COD basis, referred further to as $S/X_{V,20}$ and $S/X_{V,200}$. The addition of nitrification inhibitor was avoided in the $S/X_{V,200}$ by applying a calibrated model of nitrification to the measured respiration rate data resulting from both heterotrophic and autotrophic activities. In such a way, the sludge can be reused in a series of consecutive experiments and there is no risk for inhibition of heterotrophic activity by the nitrification inhibitor. The information obtained from the $S/X_{V,200}$ experiment is insufficient to determine the contribution from hydrolysis. Therefore, the authors proposed an additional $S/X_{V,20}$ measurement. Due to an increased model complexity the nitrification suppression would be required to distinguish between the respiration caused by nitrogen hydrolysis (or ammonification) and the respiration of organic material from hydrolysis.

The concentration of soluble inert organics (S_I) can be determined by measuring soluble (filtered) COD in the effluent from a low-loaded activated sludge plant (Ekama et al. 1986, Sollfrank et al. 1992; Stensel 1992). Actually, however, the inert organic fraction constitutes 90–95% of the effluent soluble COD (Henze et al. 1994). A more precise method of the S_I estimation was used by Henze (1992) and Petersen et al. (2002):

$$S_I = COD_{sol,out} - CBOD_{sol,out} = COD_{sol,out} - 1.5 CBOD_{5,sol,out} \qquad (4.38)$$

where
$CBOD_{5,sol,out}$ - Soluble carbonaceous BOD_5 in secondary effluent, $M(BOD)L^3$
$CBOD_{sol,out}$ - Soluble carbonaceous BOD in secondary effluent, $M(BOD)L^3$
$COD_{sol,out}$ - Soluble COD in secondary effluent, $M(COD)L^3$

Henze et al. (1987) proposed another method that included taking a sample of aliquot from a continuously fed completely mixed reactor operated at an SRT in excess of 10 days and aerating the sample in a batch reactor until the concentration of soluble COD remained constant. The soluble inert organics can also be determined from a batch test after continuous aeration of a filtered sample of the influent for several days (Lesouef et al. 1992). In both cases, the final residual soluble (filtered) COD is assumed to be equal to the concentration of inert soluble organics in the influent wastewater.

None of the above mentioned methods differentiates between soluble inerts in the influent and soluble residuals of microbial products. Even though this approach is acceptable for domestic wastewater, it may cause serious problems in the characterization of strong industrial wastewater (Orhon and Cokgor, 1997). In such a case, the effluent concentration of soluble inert organics is assumed to consist of a sum of the influent $(S_{I,in})$ and produced $(S_{I,prod})$ fractions:

$$S_{I,out} = S_{I,in} + S_{I,prod} \qquad (4.39)$$

where
$S_{I,in}$ - Influent concentration of soluble inert organic compounds, $M(COD)L^{-3}$
$S_{I,out}$ - Effluent concentration of soluble inert organic compounds, $M(COD)L^{-3}$
$S_{I,prod}$ - Concentration of soluble inert organic produced in the activated sludge system, $M(COD)L^{-3}$

The nature of the soluble products is not very well recognized, i.e. if they are indeed residuals or they undergo biodegradation at a much lower rate than the biodegradable compounds in the influent wastewater (Orhon and Cokgor, 1997).

Henze et al. (1987) proposed that the concentration of inert suspended organic fraction in wastewater, X_I, should be estimated by comparing the observed and predicted sludge production in a real WWTP as a function of time. This fitting acts to tune the model to the particular wastewater under study and compensates for any error made in the estimation of Y_H or b_H.

Another possibility is to carry out a batch test just with wastewater. The initial concentrations S_S, X_S, X_H can be determined by curve fitting, whereas the soluble inert COD, S_I, and the total COD can be measured directly. Finally, the X_I value can be calculated from the following relationship (Kappeler and Gujer, 1992):

$$X_I = \text{Total COD} - S_S - S_I - X_S - X_H \qquad (4.40)$$

where
X_I - Concentration of inert particulate organic compounds, $M(COD)L^{-3}$

Lesouef et al. (1992) determined the X_I fraction based on long-term (e.g. 14-day) aeration of filtered and non-filtered wastewater in two parallel batch reactors. The final total COD of the non-filtered sample consists of both soluble and particulate inert fractions, and the produced biomass:

$$\text{Final COD} = S_I + X_I + Y'(\text{Initial total COD} - S_I - X_I) \qquad (4.41)$$

where
Y' - Apparent yield coefficient for degrading the matter remaining in the filtrate, $M(COD)[M(COD)]^{-1}$

Rearranging Equation 4.41 the particulate inert fraction can be calculated as follows:

$$X_I = \frac{\text{Final COD} - S_I - Y'(\text{Initial total COD} - S_I)}{1 - Y'} \quad (4.42)$$

The Y' coefficient is calculated as:

$$
\begin{aligned}
Y' &= \frac{\text{Particulate COD formed}}{\text{COD degraded}} \\
&= \frac{\text{Final COD (filt. sample)} - \text{Final SCOD (filt. sample)}}{\text{Initial SCOD (non-filt. sample)} - S_I}
\end{aligned}
\quad (4.43)
$$

Orhon *et al.* (1994a) developed a very similar procedure which also accounted for soluble residual products. For this purpose, the procedure involved three aerated batch reactors, the first one fed with the raw wastewater, the second with soluble/filtered wastewater and the last with glucose. The fraction of soluble inert COD generated in biomass decay was calculated with the results from the soluble wastewater and glucose reactors.

Xu and Hultman (1996) determined the concentration of heterotrophic biomass, X_H, in the influent wastewater by comparing the measured OUR values with a conversion factor of 150 mg O_2/(g VSS•h) assumed to correspond to the maximum OUR for active heterotrophic biomass (Henze, 1986). Prior the measurement, the wastewater is aerated to reach the DO concentration approximately 6–8 g mgO_2/dm^3 and nitrification is suppressed by the addition of ATU. The X_H concentration is then given by the following equation:

$$X_H = \frac{OUR_w}{150} \cdot 1000 \cdot i_{CV} \quad (4.44)$$

where
OUR_w - Oxygen uptake rate obtained by performing a respirometric measurement with only wastewater, $M(O_2)L^{-3}T^{-1}$
i_{CV} - COD/VSS ratio, $M(COD)M^{-1}$

Alternatively, the X_H concentration can be estimated, along with the concentrations of S_S and X_S, by curve fitting against the experimental data from a batch test performed just with wastewater (Kappeler and Gujer, 1992; Wentzel *et al.* 1995). Von Munch and Pollard (1997) measured bacterial biomass COD based on counting the number of bacteria with a classic microbiological method, acridine orange stain direct counting (AODC). The average cell volume of wastewater bacteria was used to determine a conversion factor of $20 \cdot 10^{-11}$ mg-COD/cell for calculating bacterial biomass COD.

4.3.3 Estimation of kinetic and stoichiometric parameters in the biokinetic models

The information needed for the determination of kinetic and stoichiometric parameters in the biokinetic models can be obtained from three sources (Petersen *et al.* 2002):

- Default parameter values from literature;
- Full-scale plant (field) data, such as average or dynamic data from grab or time/flow proportional samples, conventional mass balances of the full-scale data, on-line data, measurements inside reactors to characterize process dynamics (primarily relevant for SBRs and other alternating systems);
- Data obtained from different kinds of laboratory scale experiments with wastewater and biomass from the studied WWTP.

The required quality and quantity of the information depends on the purpose for which the model is intended to be used (Gernaey *et al.* 2004). The default parameters are sufficient for educational purposes aiming to increase the process understanding as well as for qualitative comparisons, such as a comparison of design alternatives for non-existing plants. In the case, when a more accurate description of the process is required (e.g. process performance evaluation and optimization) the data should be obtained from field measurements and/ or laboratory scale experiments (Petersen *et al.* 2002). Numerous applications of the biokinetic models to specific WWTPs have demonstrated, that the parameters of these models are not universal, i.e. the same parameter set is not applicable to every system (Brun *et al.* 2002). In most cases, some of the model parameters need to be adjusted in the process called system identification (see: Section 4.1).·

The complex biokinetic models with many parameters are considered to be "*over-parameterized*" (Insel *et al.* 2002), which results in the problem known in control theory as the lack of identifiability of the model parameters (Lessard and Beck 1991; Brun *et al.* 2002; Gernaey *et al.* 2004; Gujer 2006). This means that different combinations of influent components and model parameters can give a comparable prediction of the effluent concentrations, in-process concentrations and sludge production. The lack of identifiability of the model parameters is a major problem encountered in calibration of the biokinetic models (Gernaey *et al.* 2004). In order to overcome this constraint two approaches are possible including model simplification and selection of small subsets of parameters to be adjusted.

The biokinetic models can be simplified by order reduction, elimination/combination of some processes or variables, separation of equations into subsets and linearization of kinetic expressions to substitute the complicated Monod terms (Kim *et al.* 2001). Simplified models still retain process fundamentals, but the model parameters can be easier estimated with a lesser amount of experimental data. A simplified model, however, may not predict the performance of a studied system as accurate as the full model.

In most applications, no attempts are made to estimate all parameters simultaneously from the data collected from a single system. Most parameters remain fixed at the default values and only small subsets of parameters are adjusted, either by "*ad hoc*" tuning or a parameter estimation algorithm (Brun *et al.* 2002). It has indeed been demonstrated that even with complex biokinetic models it is possible to select only a few parameters for calibration (Cinar *et al.* 1998; Satoh *et al.* 2000; van Veldhuizen *et al.* 1999; Brdjanovic *et al.* 2000; Meijer *et al.* 2001; Meijer *et al.* 2002b; Petersen *et al.* 2002; Wichern *et al.* 2001, Wichern *et al.* 2003; Meijer 2004). Such an approach overcomes the problem of poor parameter identifiability (Weijers and Vanrolleghem 1997), but it has two important disadvantages (Brun *et al.* 2002):

- Values of the adjusted parameters values may be biased by the values of the default parameters and their uncertainty is usually underestimated,
- It is often unclear which parameter subset is best to be estimated from the available measured data.

The uncertainty of parameter estimation can be substantially reduced by simultaneously evaluating the multitude of experiments with different carbon sources (single substrate and mixed substrate), different electron acceptors (oxygen and nitrate), different time constants (long-term and short-term) and temperatures (Koch *et al.* 2000b).

Van Veldhuizen *et al.* (1999) listed two possible approaches for selecting the parameters to be adjusted during model calibration: a system engineering approach (purely mathematical optimization algorithm based on a sensitivity analysis) or a process engineering approach (based on detailed knowledge of the process and experience of the modeller). The first approach cannot differentiate between better or less defined parameter values, even though some parameters (e.g. yield coefficients) are known to have well established values. The authors concluded that not only the choice of parameters to be adjusted would be different, but also the number of these parameters would be greater, when following the sensitivity based optimization algorithm compared to the process engineering approach.

The system engineering approach is recognized as scientifically sound, but its application has been limited to few academic studies until now. It may not be feasible for practical applications, primarily due to huge computational efforts needed (Ruano *et al*. 2007). Furthermore, a purely mathematical optimization may often end up with relatively small adjustments made to a considerable number of parameters (Gernaey *et al*. 2004).

Two complex mathematical techniques can be used to guide parameter subset selection for parameter estimation. Both techniques are based on a local sensitivity analysis (see: Section 4.5.3). The first one facilitates the identification of the most important model parameters and the analysis of parameter interdependencies based on three diagnostic measures: sensitivity measure, collinearity index and determinant measure (Brun *et al*. 2002). Another technique, called the optimal experimental design (OED) procedure, based on the Fisher Information Matrix (FIM), is used to improve the confidence interval for parameter estimation (Dochain and Vanrolleghem 2001). The FIM is calculated from the following equation:

$$\text{FIM} = \sum_{i=0}^{N} \left(\frac{\partial y}{\partial x} \right)_i^{T} Q_i^{-1} \left(\frac{\partial y}{\partial x} \right)_i \qquad (4.45)$$

where
y - Model predictions
x - Model parameters to be adjusted
Q - Weighting matrix, typically the measurement error matrix
N - Number of measurement points over time

The matrix represents the information content of a hypothetical experiment under different combinations of the available measurements and manipulations. There are several scalar functions that can be derived from the FIM for evaluation, e.g. the determinant of the FIM (so called "*D-criterion*") or the condition number of the FIM (so called "*Mod-E criterion*"). Once the optimal hypothetical experiment is found, it can be "*repeated*" in reality to collect data. Using these data, the model can be recalibrated to find more accurately estimated parameters. Practical applications of the FIM can be found in several simulation studies (e.g. Weijers and Vanrolleghem 1997, Chandran and Smets 2000a; Insel *et al*. 2003; Chandran and Smets 2005; De Pauw 2005; Sin *et al*. 2005b; Machado *et al*. 2009).

A reasonable calibration of the complex mechanistic activated sludge models will only be possible with a detailed understanding of the principles of the models (Henze *et al*. 1995a). Van Veldhuizen *et al*. (1999) emphasized that a calibration based on process knowledge is more sensible than the mathematical optimization but requires much more experience of the modeller (so-called "*heuristic*" calibration approach). In such a case, a model may be calibrated using the "*human expert method*" described by Cinar *et al.* (1998). The principle of this method is that unknown parameter values are determined by a sequential "*trial and error*" adjustments, which are repeated until an acceptable fit (most often judged by the modeller) is obtained between the model predictions and measured data. However, it is often unclear which parameter subset is best to be estimated from the available data and values of the adjusted parameters may be biased by the values of the fixed parameters (Brun *et al.* 2002). Sin *et al.* (2008a) attempted to automate the manual "*trial and error*" calibration procedure by incorporating Monte Carlo simulations. This method is difficult to implement in practice (at this moment) as a specialised software tool and a huge computational power are required to perform the simulations. For example, the authors noted that 500 Monte Carlo runs comprising a 5-month simulated period would require approximately two weeks of computation time.

In terms of the amount of available data and the planned use of simulation results, calibration of the biokinetic models can be accomplished at different levels as presented by Henze *et al.* (1995a). The authors recommended to use a logical stepwise approach that assumes changing values of a few parameters at each calibration level. The growth rate constants and saturation coefficients for autotrophs and "*ordinary*" heterotrophs can be adjusted using non-dynamic data, either from steady-state experiments or from 24-hour composite samples from dynamic experiments (level 1). Moreover, the average sludge production or MLVSS concentrations could be calibrated based on adjustments of some stoichiometric coefficients. The authors did not recommend to calibrate the PAO kinetics based on steady-state experiments. This should be rather done at level 2 (dynamic experiments) along with the calibration of kinetics of the other biomass components.

A combination of sensitivity analysis and process knowledge appears to be an appropriate approach for selection of parameters to be adjusted during the calibration phase (van Veldhuizen *et al.* 1999). This approach was indeed applied in several studies (van Veldhuizen *et al.* 1999; Meijer *et al.* 2001; Petersen *et al.* 2002; Meijer 2004; Makinia *et al.* 2005a) to verify whether the model is indeed sensitive to changes in the values of adjusted parameters. For example, Figure 4.16 illustrates the effect of $\eta_{NO3,PAO}$ on the anoxic P uptake and nitrate utilization during a P release/uptake batch test. The sensitivity analysis revealed that only in the first case the effect turned out to be influential (see: Section 4.5.3). These results also confirmed that the measurements of maximum anoxic P uptake rate are suitable for estimating the value of $\eta_{NO3,PAO}$.

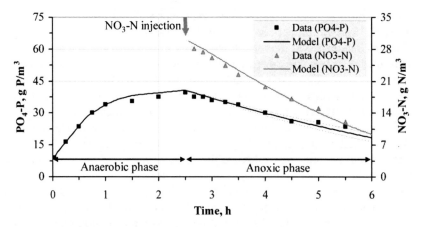

Figure 4.16 Measured vs. predicted PO$_4$-P concentrations during a P release/ uptake batch test carried out at the Wschod WWTP (dash line during the anoxic P uptake – results of the sensitivity analysis with $\eta_{NO3,PAO}$ = 0.32 ± 10%) (Makinia *et al.* 2006c).

The starting point for calibration of the biokinetic models is usually the default parameter and the models should be calibrated iteratively in the following order: sludge production and sludge balance, nitrification, denitrification and finally the EBPR process (Gernaey *et al.* 2004; Langergraber *et al.* 2004). Practical applications of such iterative stepwise approaches were presented in literature. Meijer *et al.* (2001) used a stepwise procedure to calibrate the TUDP model of a full-scale WWTP with the average data of a 2-day measuring campaign. The procedure involved sequential fitting of the following elements: the solids and COD balances, nitrification, denitrification (including this process in the secondary clarifier), phosphate uptake and the total oxygen consumption. Calibration parameters were selected based on process knowledge rather than sensitivity analysis. The proposed stepwise procedure provided an orderly and faster calibration, reducing the number of iteration loops. The same approach was used by Meijer *et al.* (2002b) to calibrate the TUDP model for the start-up conditions at a full-scale BNR plant. Petersen *et al.* (2002) conducted respirometric lab-scale experiments with wastewater and the process biomass during the measurement campaign at the full-scale reactor. The experiments aimed to determine two kinetic parameters including $\mu_{H,max}$ and $\mu_{A,max}$, which are the most important parameters in predicting dynamic situations. In addition, the b_H coefficient was also estimated under lab-scale conditions to support the calibration of sludge production in the full-scale bioreactor. In the Swiss studies, Koch *et al.* (2000b) and Rieger *et al.* (2001) used an iterative calibration procedure with the results of multiple full-scale measurements and batch experiments. First, a steady state run of the studied plant was simulated to get the initial sludge characteristics. Next, a series of simulations was run in a closed loop until the plant performance and batch tests could be modelled with the same parameter set. The most important parameters of ASM3 were estimated using the results of various aerobic and anoxic batch experiments with activated sludge originating from several WWTPs (Koch *et al.* 2000b). The following data were selected to calibrate the biokinetic model including the EBPR process (Rieger *et al.* 2001): results of single batch experiments (anaerobic phosphate release and aerobic/anoxic phosphate uptake), results of series of batch experiments (the wash-out and grow-in of PAOs), diurnal variations based on garb samples as well as long-term variations (one week or more) based on on-line measurements.

Table 4.12 Overview of papers presenting detailed practical experiences with full-scale WWTP model calibration (Gernaey *et al.* 2004)

		Data			(Phase)		
Reference	Model	Static	Dynamic	Batch exp.	Calibration	Validation	Remarks
Coen *et al.* (1996)	ASM1	+	+	+	D	S	Dynamic influent profile was designed based on the measurement campaign
Cinar *et al.* (1998)	ASM2	+			S	S	
van Veldhuizen *et al.* (1999)	TUDP	+	+		S	D	
Brdjanovic *et al.* (2000)	TUDP	+	+	+	S	S	Use of batch experiments mainly during model unfalsification phase
Koch *et al.* (2000b)	ASM3	+	+	+	D	D	Use of batch experiments before the full-scale model calibration phase
Meijer *et al.* (2001)	TUDP	+	+		S		
Rieger *et al.* (2001)	ASM3P	+	+	+	S/D	S/D	Use of batch experiments during calibration; iterative calibration procedure
Hulsbeek *et al.* (2002)	ASM1/ general	+	+		S/D	S/D	Type of data depending on model purpose
Petersen *et al.* (2002)	ASM1/ general	+	+	+	S/D	S	Type of data depending on model purpose; Use of batch experiments for calibration and unfalsification

D – dynamic, S – steady state

An overview of practical experiences with full-scale WWTP model calibration is presented in Table 4.12. Gernaey *et al*. (2004) emphasized the importance of obtaining informative data that allow constraining the model parameters within realistic boundaries. According to the protocol of Hulsbeek *et al*. (2002), a calibration phase of ASM1 can be accomplished using only the data concerning the sludge composition and production, and the effluent quality measurements (NH_4–N and NO_3–N). In practice, however, the informative data on the reaction kinetics are often difficult to obtain in full-scale facilities due to little effluent dynamics, which is especially observed in underloaded WWTPs (Gernaey *et al*. 2004). In such cases, internal in-process measurements inside the bioreactor can provide much more informative data for the calibration of biokinetic models than exclusively the effluent concentrations. Van Veldhuizen *et al*. (1999) demonstrated that the calibration should not only be based on the effluent concentrations but also on internal in-process concentrations. The adjustment of three parameters allowed to fit the predicted and measured concentration profiles along the process units, even though these parameters did not affect considerably the effluent concentrations. Also Langergraber *et al*. (2004) proposed to check concentrations in particular in-process points, e.g. PO_4-P at the end of the anaerobic zone for the EBPR process or NO_3-N in returned sludge for denitrification in the secondary clarifier, in addition to the calibration of biokinetic models based on the sludge production, effluent concentrations of NH_4-N, NO_3-N and, if implemented, PO_4-P.

Vanrolleghem and Coen (1995) and Vanrolleghem *et al*. (1999) stated that full-scale measurements are not practicable or not informative enough for characterization of the biokinetic models. The IAWQ (former name of IWA) Task Group on Mathematical Modelling for Design and Operation of Biological Wastewater Treatment Processes (Henze *et al*. 1995a) recommended both continuous and batch experiments with the biomass from the process as a very reliable technique for calibration of the kinetic and stoichiometric parameters of the biokinetic models. Also several studies have emphasized the importance of laboratory scale experiments for determining model parameters (Gernaey *et al*. 2004). Continuous procedures are primarily suitable for determining stoichiometric coefficients, but they require a long measuring period since steady state must be attained (Spanjers and Vanrolleghem 1995) Moreover, they are also quite resources demanding compared to batch tests (Kristensen *et al*. 1998). Batch experiments are considered a valuable tool for determining kinetic parameters and wastewater biodegradability (Gernaey *et al*. 2004). During batch tests the substrate and biomass are placed together into a batch reactor and three types of data may be collected including substrate disappearance, biomass growth and oxygen consumption (Grady *et al*. 1999). This follows directly from proportionality of the three rates during balanced growth. In unbalanced growth, generation and disappearance of storage polymers should also be taken into account separately from the biomass growth (see: Section 3.1). Many different procedures of laboratory scale experiments have been proposed for determining both the kinetic and stoichiometric parameters (Kappeler and Gujer 1992; Kristensen *et al.* 1992; Spanjers and Vanrolleghem 1995; Kristensen *et al.* 1998; Spanjers *et al.* 1999; Vanrolleghem *et al.* 1999, Brdjanovic *et al.* 2000; Koch *et al.* 2000b; Rieger *et al.* 2001; Petersen *et al.* 2002). In general, two approaches for the determination of model parameters are possible (Vanrolleghem *et al.* 1999):

- Direct methods, which focus on specific parameters that can directly be evaluated from the measured data;
- Optimization methods, which involve a procedure of fitting model predictions to measured data. These methods require numerical techniques for adjusting parameter values in order to achieve the lowest deviation between the predictions and measured data.

Some authors (Novak *et al.* 1994; Vanrolleghem *et al.* 1999; Gernaey *et al.* 2004; Sin *et al.* 2005a) have claimed that care should be taken in the transfer of results derived from laboratory scale experiments to a model of the full-scale plant. The laboratory scale behaviour may deviate from the full-scale behaviour due to differences in feeding patterns, environmental conditions (e.g. pH, temperature), mixing conditions and sludge history (Vanrolleghem *et al.* 1999). On the other hand, Funamizu *et al.* (1997) managed to calibrate a biokinetic model (ASM2) based on the results of laboratory scale batch experiments with multiple anaerobic/ anoxic and aerobic phases. It was demonstrated that no modification of parameter values was required to evaluate the performance of the full-scale train (51,600 m^3/d) and internal composition of the activated sludge.

The ratio between initial substrate concentration (S_0) and initial biomass concentration ($X_{V,0}$) is considered to be one of the important factors in designing laboratory scale batch experiments in order to get a system response sufficient for interpretation (Vanrolleghem *et al.* 1995). The $S_0/X_{V,0}$ ratio directly affects the actual system behaviour and determines whether the catabolic or anabolic processes prevail (Chudoba *et al.* 1992). Grady *et al.* (1999) discussed two extreme conditions representing the limits of the physiological state that microorganisms may attain during a batch experiment. At one extreme, the test conditions need to be such that few changes occur in the physiological state or composition of the biomass during the test. This can be achieved by keeping the value of $S_0/X_{V,0}$ small during batch tests, with 0.02 being a typical value. Because kinetic parameters tend to reflect the conditions of the biomass in the bioreactor from which they are obtained, those parameters are called "*extant*". At the other extreme, if the physiological state of the biomass is allowed to change during the test to the point that microorganisms grow at the fastest rate possible at the given environmental conditions, the resulting kinetic parameters are said to be "*intrinsic*". The term "*intrinsic*" follows from the fact that the parameter values are dependent only on the nature of the substrate and the types of microorganisms in the biomass. The key to the determination of intrinsic kinetic values during a batch test is to provide sufficient substrate to allow the bacteria performing the biodegradation to fully develop their metabolism. This can usually be accomplished when the initial substrate to biomass ratio ($S_0/X_{V,0}$) is at least 20 when both concentrations are expressed as COD.

It is unclear which type of kinetic parameter set, intrinsic or extant, is more useful for modelling a full-scale process behaviour. It appears that intrinsic parameters (determined at high S/X_V) are most useful for comparing the biodegradability of organic compounds, whereas extant parameters (determined at low S/X_V) are most useful for predicting the reaction kinetics with respect to the removal of a specific organic compound (Grady *et al*. 1999). However, attention has to be paid to the fact that batch cultivations at high S/X_V ratios (more than 4:1) can change the proportion between groups of microorganisms in the microbial population by favouring fast growing groups. This means that the observed kinetic characteristics may be representative for the microorganisms that become dominant during the experiment rather than for the entire microbial population (Vanrolleghem *et al*. 1999). Novak *et al*. (1994) gave practical evidence for this hypothesis by comparing the $\mu_{H,max}$ values in a continuously operated system and in a batch reactor (at a high S/X_V ratio). The determined value of $\mu_{H,max}$ for the batch cultivation (10 d^{-1}) was 2.5 higher than the value obtained for the continuous system (4 d^{-1}). On the other hand, the disadvantage of determining kinetic parameters at a low S/X_V ratio is that the substrate degradation or respirometric response may be too short for a reliable measurement (Spanjers and Vanrolleghem 1995; Vanrolleghem *et al*. 1999).

A popular technique that provides the experimental information required for calibration of biokinetic models is respirometry (Spanjers and Vanrolleghem 1995; Vanrolleghem and Coen 1995; Brouwer *et al*. 1998; Coen *et al*. 1998; Spanjers *et al*. 1998, Vanrolleghem *et al*. 1999; Koch *et al*. 2000b; Insel *et al*. 2002; Insel *et al*. 2003). A comprehensive overview of respirometric methods for determining some of the ASM1 parameters and component concentrations was presented by Vanrolleghem *et al*. (1999). Traditionally, respirometry meant the measurement and interpretation of the biological oxygen consumption (also called respiration, uptake or utilization) rate under well defined experimental conditions (Spanjers *et al*. 1998), even though respirometric techniques actually involve observing utilization profiles (respirograms) both electron acceptors, i.e. oxygen or nitrate

(Insel *et al.* 2002). The respiration rate is expressed as the amount of electron acceptor consumed by the microorganisms measured per unit volume and unit time. The OUR is measured in an aerated reactor for the characterization of aerobic processes, whereas the NUR is primarily used for the determination of reduction factors under anoxic conditions (Insel *et al.* 2002). Oxygen consumption is a key activity in both carbon oxidation and nitrification reflecting two of the most important biochemical processes occurring in a bioreactor, i.e. biomass growth and substrate consumption (Olsson and Newell 1999). Various instruments, called respirometers, based on DO probes and measurement techniques have been developed to monitor the respiration rate under aerobic conditions. Respirometric measurements can be conducted in continuous or batch devices (Spanjers *et al.* 1993). In continuous respirometers, a wastewater sample is continuously fed into a flow-through reactor having a volume of several dm³. In batch methods, the OUR is calculated from a respirogram obtained after the addition of a sample of wastewater or artificial substrate to a respirometer. Most respirometric techniques for short-term batch experiments use cyclic operation: aeration is stopped, the respiration rate is calculated and the sludge is reaerated for a new cycle. This procedure may not, however, enable a sufficient frequency of the measurements (Spanjers and Vanrolleghem 1995). Respirometers can also be classifies as closed and open aerated (Spanjers *et al.* 1993). The closed respirometers, which are not aerated, use a DO probe or a manometer to record the respirogram. These devices are laborious to operate and unsuited for continuous measurement. The open aerated respirometers need either a calibration, additional DO measurements at a separated closed respiration chamber or knowledge of the aeration constants.

When running the respirometric tests, a S/X_V ratio needs to be selected in the proper range (Ekama *et al.* 1986). If that ratio is too high, the difference between the rates during the two phases (associated with the utilization of readily biodegradable and slowly biodegradable substrate) will be too low to be clearly distinguishable. On the other hand, at too low S/X_V ratios the readily biodegradable substrate is removed too fast to allow an accurate measurement of the OUR.

4.3.4 Estimation of settling parameters

In order to determine the settling parameters for a specific activated sludge sample (see: Section 2.5.1), a series of column (or batch) settling tests (usually three to eight) over a range of sludge concentrations is needed (Bye and Dold, 1999). Different sludge concentrations are usually obtained by diluting the RAS with secondary effluent up to some dilution of the bioreactor MLSS concentration (Giokas *et al.* 2003). After placing the sludge in the column and initial vigorous mixing, the displacement of the liquid-sludge interface (sludge blanket) is monitored for

30-60 minutes depending on the settling column height (Figure 4.17). A slow stirring (1-2 rpm) is maintained during the whole test to minimise wall effects. A short time (a few minutes) after placing the sludge in the column, a sharp interface is formed (at time t_1) and all sludge particles below the interface start to settle with the same velocity, so that the interface is also displaced at a constant rate. The gradient of the linear (or linearised) part of the curve of the interface displacement is defined as the zone settling velocity (ZSV). Simultaneously, sludge with a higher concentration accumulates at the bottom of the settling column. After some time (t_2), the liquid-sludge interface approaches the region of thickened sludge and its rate of displacement starts to decrease gradually.

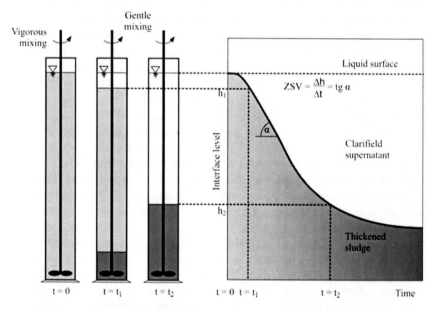

Figure 4.17 Schematic representation of the batch settling test (left) and progression of displacement of the sludge-liquid interface (right).

The use of different sludge concentrations allows the experimental determination of the relationship between the sludge concentration and the ZSV (Figure 4.18). Vanderhasselt and Vanrolleghem (2000) noted that estimation of the settling parameters based on batch settling curves appears to have the potential of being a good source of information on sludge settling characteristics for use in the solid flux theory. However, since this procedure is time consuming, more rapidly measurable parameters, such as SVI-type measurements are preferred in practice. These measurements can be further correlated with the settling parameters as presented in Section 2.5.1.

The compression rate parameters are obtained following the same procedure used for the determination of the zone settling parameters but using a RAS or WAS sample. This procedure is not always successful since denitrification and rising of the sample may occur in the settling column. A conservative approach assumes that the compression rate parameters equal to the zone settling parameters (Griborio *et al.* 2007). However, dividing the complete settling process of activated sludge into the zone settling and compression settling stages, and describing them with different parameter sets appears more reasonable for characterizing the complete settling process of activated sludge (Zhang *et al.* 2006).

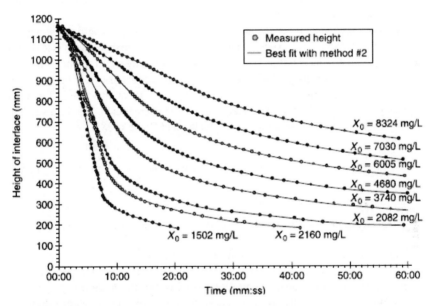

Figure 4.18 Example of the batch settling test results at different sludge concentrations (Stricker *et al.* 2007).

4.3.5 Estimation of the K_La coefficient

4.3.5.1 Estimation of the K_La coefficient in the field studies

Steady state method. The oxygen transfer coefficient may be determined by conducting a mass balance on a completely mixed volume of the aeration basin at steady state. For this purpose, a mass balance equation over a CSTR may be rearranged to calculate K_La as follows (Mueller and Boyle 1988):

$$K_L a = \frac{OUR_T - \dfrac{Q}{V}(S_O - S_{O,R})}{S_{O,sat} - S_{O,R}} \qquad (4.46)$$

where

$S_{O,R}$ - DO concentration that most accurately represents the driving force in the transfer zone of reactor, $M(O_2)L^{-3}$

Although simple to apply, this method can be inaccurate due to a number factors including spatial variations in the DO concentrations, variable influent characteristics affecting the α factor, and the requirement of maintaining an excess DO concentration in the aeration basin to obtain $S_{O,R}$ between 2 mg O_2/dm^3 and 0.75 $S_{O,sat}$ (Mueller and Boyle 1988).

Dynamic methods. Under dynamic process conditions two approaches may be used to estimate the overall oxygen transfer coefficient. The first approach is to obtain a changing oxygen concentration with time until a new steady-state DO concentration, $S_{O,SS}$, is achieved. This may be done by increasing power levels or adding hydrogen peroxide to the aeration basin. In such cases, the DO saturation value in Equation 4.46 may be replaced by the new steady-state DO concentration. The $K_L a$ value is determined from the following equation (Mueller and Boyle 1988):

$$S_{O,SS} - S_O = (S_{O,SS} - S_{O,in})\exp\left[-\left(K_L a + \frac{Q}{V}\right)t\right] \qquad (4.47)$$

where

$S_{O,SS}$ - Steady-state dissolved oxygen concentration in reactor, $M(O_2)L^{-3}$

Equation 4.47 requires constant flow and oxygen uptake conditions, but it does not require the measurement of the oxygen uptake rate nor the accurate determination of saturation concentration (Mueller and Boyle 1988).

The opposite conditions are required in the alternative approach proposed by Reinius and Hultgren (1988). The $K_L a$ value may be determined from a numerical approximation of the mass balance equation:

$$K_L a^{n-1}(S_{O,sat} - S_O^{n-1}) = OUR_T^{n-1} + \frac{(S_O^n - S_O^{n-1})}{\Delta t} + Q^{n-1}\frac{(S_O^{n-1} - S_{O,in}^{n-1})}{V} \qquad (4.48)$$

Off-gas technique. The off-gas technique is the process water aeration efficiency measurement with the highest accuracy and precision (ASCE 1997). This technique, limited only to diffused air systems, involves collection of gas exiting an aeration tank and measurement of the oxygen content of the gas entering and exiting the treatment units. The principles of the off-gas measurements have been presented by Redmon *et al.* (1983), Boyle *et al.* (1989), and Daigger and Buttz (1992).

A gas phase mass balance over the liquid volume may be written as follows (Redmon *et al.* 1983):

$$V\rho_a \frac{d\bar{Y}}{dt} = \rho_a Q_{A,i} \bar{Y}_{in} - \rho_a Q_{A,out} \bar{Y}_{out} - K_L a(S_{O,sat} - S_O)V \qquad (4.49)$$

where
$Q_{A,out}$ - Total air flow rate of outlet gases, $L^3 T^{-1}$
$Q_{A,in}$ - Total air flow rate of inlet gases, $L^3 T^{-1}$
$\bar{Y}$ - Mole fraction of oxygen gas,
$\bar{Y}_{out}$ - Mole fraction of oxygen gas in outlet,
$\bar{Y}_{in}$ - Mole fraction of oxygen gas in inlet,
ρ_a - Air density, ML^{-3}

Under steady-state conditions and assuming that nitrogen and other inerts are conservative, the $K_L a$ value may be determined by rearranging Equation 4.49 (Redmon *et al.* 1983):

$$K_L a = \frac{\rho_a \left(Q_{A,in} \bar{Y}_{in} - Q_{A,out} \bar{Y}_{out} \right)}{(S_{O,sat} - S_O)V} \qquad (4.50)$$

Gas flow measurements ($Q_{A,in}$, $Q_{A,out}$) may be omitted by using molar ratios of inlet and outlet oxygen to the inlet gas fractions. In such a case, oxygen transfer is expressed as a function of oxygen transfer efficiency (OTE), defined as (Redmon *et al.* 1983):

$$OTE = \frac{MR_{in} - MR_{out}}{MR_{in}} = \frac{\text{Mass Oxygen Transferred}}{\text{Mass Oxygen Supplied}} \qquad (4.51)$$

where
OTE   - Oxygen transfer efficiency
MR_{in} - Mole ratio of oxygen to inerts in inlet
MR_{out} - Mole ratio of oxygen to inerts in outlet

The OTE may be also written as (Boyle *et al.* 1989):

$$OTE = \frac{K_L a(S_{O,sat} - S_O)V}{Q_A \rho_a 0.23} \qquad (4.52)$$

where $Q_A = Q_{A,i} = Q_{A,e}$.

Figure 4.19 illustrates a comparison of the K_LA values derived from Equation 4.52 and those obtained from Reinius and Hultgren's (1988) method. A good agreement was observed in the range of medium air flowrates (between 700 and 2500 m^3/h). At low air flowrates, however, Reinius and Hultgren's method produced higher values compared to the off-gas technique. This may be attributed to small accuracy of the OUR measurements at low DO concentrations maintained in some compartments of the bioreactor.

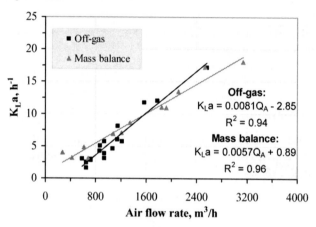

Figure 4.19 Comparison of the K_La values using the off-gas and mass balance methods applied at the Rock Creek WWTP (Makinia 1998).

Tracer methods. Inert gas tracers are characterized by no net absorption or desorption (Boyle *et al.* 1989). The appropriate gas transfer equation for them does not contain a reaction term. Various gaseous tracers have been proposed to evaluate gas transfer in natural waters and wastewater treatment systems. The tracers used include the radioactive krypton-85, low molecular weight hydrocarbon gases such as ethylene and propane, and the noble gases such as helium (Mueller and Boyle 1988). In a continuously flowing systems, the value of gas transfer coefficient may be corrected for dispersion by measuring a second conservative, non-volatile dissolved tracer (e.g., lithium chloride) introduced at the same time (Mueller and Boyle 1988).

Tsivoglou *et al.* (1968) established the following relationship for krypton-85 and oxygen in the surface transport under the same mixing conditions:

$$\frac{K_{Kr}}{K_La} = 0.83 \pm 0.04 \tag{4.53}$$

In diffused aeration systems this ratio must be corrected downward because of gas side krypton stripping (USEPA 1983).

4.3.5.2 Estimation of the $K_L a$ coefficient from dimensionless analysis

Assuming that the oxygen transfer coefficient depends on the major geometric and dynamic parameters and their relation to each other, a criterial equation consisting of dimensionless groups of these parameters can be developed. These dimensionless groups can reflect the geometric, kinematic, dynamic, and physicochemical similarity of aeration systems (Khudenko and Shpirt 1986). The dimensionless group as a whole represents contributions of several parameters to a particular effect (Winkler 1981).

Based on the literature data, Eckenfelder (1989) developed a relationship between three dimensionless groups of parameters: the Sherwood number $(Sh = K_L d_b / D)$, which is the ratio of turbulent mass transfer and molecular diffusion, the Reynolds number $(Re = u_b d_b \rho_l / \mu_l)$, which is the ratio of inertial forces initiated by air bubbles of specific size db to friction forces determined by the viscosity of liquid, and the Schmidt number $(Sc = \mu_l / \rho_l D)$, which is the ratio of the dynamic viscosity to the diffusivity. The relationship may be rearranged to solve the $K_L a$ coefficient as follows (Eckenfelder 1989):

$$K_L a = \frac{6\beta' H^{2/3} Q_A \rho_l^{1/2} D^{1/2}}{d_b V \mu_l^{1/2}} = C' \frac{H^{2/3} Q_A}{V} \tag{4.54}$$

where

$$C' = \frac{6\beta'}{d_b} \left(\frac{\rho_l D}{\mu_l} \right)^{1/2} \tag{4.55}$$

ρ_l - Liquid density in reactor, ML^{-3}
D - Molecular diffusion coefficient, $L^2 T^{-1}$
d_b - Air bubble diameter, L
μ_l - Dynamic viscosity of liquid, $ML^{-1}T^{-1}$

The β' factor is a part of the dimensionless proportionality factor β, but exact values under various conditions are not available. Schroeder (1977) estimated the β' value equal to approximately 0.06.

Since d_b is proportional to air flows normally encountered in operation practice, Equation 4.54 may be further modified (Eckenfelder 1989):

$$K_L a = C'' \cdot \frac{H^{2/3} Q_A^{(1-n_2)}}{V} \tag{4.56}$$

The value of $(1-n_2)$ generally ranges from 0.8 to 1.0 (Eckenfelder 1989). However, with nozzle-type spargers increased air flow rates and increased turbulence tend to break up larger bubbles to form smaller ones. This results in negative n_2 coefficient, and an increasing oxygen transfer with increasing gas rate (Weber 1972).

Khudenko and Shpirt (1986) added one more dimensional complex (the Froude number) in their analysis. The Froude number ($Fr = u_g W / w(gh)^{0.5}$) relates the specific kinetic energy of air bubbles in the aeration basin to the specific potential energy of the liquid column of the same depth as the depth of the aeration basin.

Based on experimental data Khudenko and Shpirt (1986) developed the following relationship:

$$K_L a = 0.041 \alpha \theta \left(\frac{h}{d_b} \right)^{0.67} \left(\frac{w}{W} \right)^{0.18} \left(\frac{u_g}{H} \right) \tag{4.57}$$

The correlation coefficient obtained from the experimental values and values computed from Equation 4.57 was 0.95. The oxygen transfer coefficient varied from 1 to 100 1/hr which corresponds to the oxygen requirements from 0.07 to 8 kg $O_2/(m^3 d)$ under operational conditions (Khudenko and Shpirt 1986).

4.4 GOODNESS-OF-FIT MEASURES

The goodness of fit of a model describes degree to which model predictions fit the observed data. In most studies, only verbal statements about the goodness of model fit are made, e.g. *"the model fits well; reached an acceptable level of accuracy; modelling results correspond to the measured values. . ."*. This often results in a different interpretation of model accuracy and emphasizes the need for a standardized procedure for the evaluation of simulation results (Ahnert *et al.* 2007).

Schunn and Wallach (2005) discussed advantages and disadvantages of various kinds of visual display techniques and numerical measures of goodness-of-fit. Both approaches provide important, non-overlapping information. Visual displays in the form of graphs (overlay scatter, overlay line, interleaved bar, side-by-side, distant) are useful for estimating roughly the degree of fit, indicating situations when the fits are most problematic and diagnosing a variety of types of problems (e.g. systematic biases in model predictions).

Numerical measures of goodness-of-fit can be divided into two groups in terms of providing information about fit to relative trend magnitudes or deviation from exact location. In a special case of the latter group, some measures (mean error or linear repression coefficient) can provide information whether model predictions differ from the exact location of the data points in a systematic fashion rather than simply mismatching those points. The authors recommended a combination of r^2 for trend relative magnitude and RMSE (root mean squared error) or RMSSE (root mean squared scaled error) for deviation from exact data location (Table 4.13). These measures, in combination, can assure the most complete and accurate evaluation of model fits (in typically situations) because they are:

- Scale invariant;
- Rewarding good data collection practice (large N, low noise);
- Reducing overfitting problems;
- Functioning when the performance of the model and the data are measured in different units;
- Appropriate for non-interval scale or arbitrary model dependent measures;
- Making use of data uncertainty information.

Table 4.13 Some numerical measures for evaluating goodness-of-fit of a model (Schunn and Wallach 2005; Sin *et al.* 2005b; Makinia 2006; Ahnert *et al.* 2007)

Criterion	Formula
Mean relative error (MRE)	$\dfrac{1}{n}\displaystyle\sum_{i=1}^{n}\dfrac{\left\|y_{i,obs}-y_i\right\|}{y_{i,obs}}$
Mean absolute error (MAE)	$\dfrac{1}{n}\displaystyle\sum_{i=1}^{n}\left\|y_{i,obs}-y_i\right\|$
Root mean squared error (RMSE)	$\sqrt{\dfrac{1}{n}\displaystyle\sum_{i=1}^{n}\left(y_{i,obs}-y_i\right)^2}$
Root mean squared scaled error (RMSSE)	$\sqrt{\dfrac{1}{n}\displaystyle\sum_{i=1}^{n}\dfrac{m_i\left(\bar{y}_{i,obs}-y_i\right)^2}{(std)_i^2}}$
Coefficient of efficiency (E$_j$), with j = 1 (original coefficient of efficiency with absolute differences) or j = 2 (modified coefficient of efficiency with squared differences)	$1-\dfrac{\displaystyle\sum_{i=1}^{n}\left\|y_{i,obs}-y_i\right\|^j}{\displaystyle\sum_{i=1}^{n}\left\|y_{i,obs}-\bar{y}_{obs}\right\|^j}$
Janus coefficient (J^2)	$\dfrac{\dfrac{1}{n\_val}\displaystyle\sum_{i=1}^{n\_val}\left(y_{i,obs}-y_i\right)^2}{\dfrac{1}{n\_cal}\displaystyle\sum_{i=1}^{n\_cal}\left(y_{i,obs}-y_i\right)^2}$

n	- Total number of experimental data points for the output variable y
$y_{i,obs}$	- Observed (measured) data in point i
y_j	- Model prediction in point i
$\bar{y}_{i,obs}$	- Observed (measured) data mean in point i,
$\bar{y}_{obs}$	- Overall mean of the observed (measured) data
m_i	- Number of data values contributing to $\bar{y}_{i,obs}$
std	- Standard deviation for the observed (measured) mean in point i,
n_cal and n_val	- Total number of measurements used in the calibration and validation periods, respectively.

Ahnert *et al.* (2007) illustrated different approaches for a visual and numerical interpretation of three different simulation results. Selected goodness-of-fit measures are compared for different simulations of an effluent concentration from a bioreactor. Figure 4.20a shows measured concentration data and the results of three different simulation runs estimating some effluent concentration from a bioreactor. Based on a first visual check of time series, V1 predictions are explicitly best fits of the measured data and this was confirmed by the plot of residuals (Figure 4.20b).

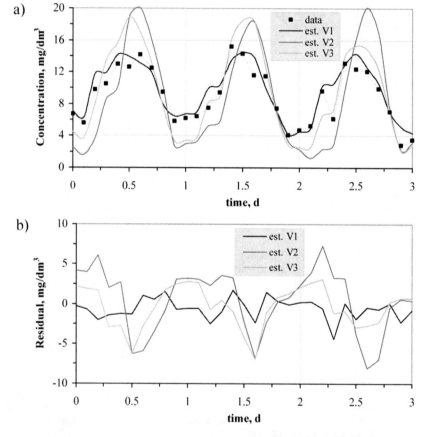

Figure 4.20 Sample data from three different simulations for analysis of the goodness-of-fit (a) and plot of residuals (b) (Ahnert *et al.* 2007).

Table 4.14 shows a quantitative assessment by means of different numerical goodness-of-fit measures with the best predictions listed in the last column. Considering the mean value, both V2 and V3 predictions are closer to the measured mean than V1 which suggests that this conventional measure does not adequately represent the quality of the model fit. As a minimum for evaluation of the quality of a model fit, the authors recommended a visual verification (time series, scatterplot, plot of residuals, etc.) followed by the quantitative assessment using the same measures, RMSE and r^2, as proposed by Schunn and Wallach (2005). However, the r^2 coefficient can be used provided that that the calculations of slope and intercept of the linear regression are included. In addition, the modified coefficient of efficiency (E_1) was also recommended as a numerical measure. For a more comprehensive evaluation procedure, the Janus coefficient (Table 4.13) can be considered, as proposed by Sin et al. (2005b), which is the criterion related to model validity and should be close to 1.

Table 4.14 Goodness-of-fit measures for example simulations (grey marked are hints for III model fits while bold values show a good evaluation) (Ahnert et al. 2007)

Goodness-of-fit measure	Measured	Est. V1	Est. V2	Est. V3	Best est.
Mean	9.04	9.64	8.63	9.42	V3,V2
Median	9.49	10.20	7.18	8.10	V1
Standard deviation	3.53	**3.43**	6.45	5.66	V1
Coeff. of determination, r^2		0.86	0.67	**0.88**	V3,V1
Linear regression, a (y=ax+b)		**0.90**	1.50	1.50	V1
Linear regression, b (y=ax+b)		**1.47**	−4.90	−4.18	V1
Root Mean Squared Error		**1.42**	4.06	2.64	V1
Coeff. of efficiency, E_1		**0.65**	−0.13	0.29	V1
Coeff. of efficiency, E_2		**0.83**	−0.37	0.42	V1
Index of agreement, d_1		**0.83**	0.61	0.73	V1
Index of agreement, d_2		**0.96**	0.82	**0.91**	V1

4.5 UNCERTAINTY AND SENSITIVITY ANALYSIS

4.5.1 Background

It is not realistic to expect that a model works perfectly, and therefore, uncertainty in model predictions becomes an inherent property of modelling. The uncertainty analysis is necessary to determine the confidence (reliability) with which the model can be used. This is essential for decision making and allows to recognize the model constrains and avoid incorrect interpretations of the model predictions.

Vanrollenghem and Keesman (1996) noted that uncertainty is "*ubiquitous*" in wastewater treatment systems, but received little attention (unlike sewer system and water quality modelling). This statement has not changed too much since then (Griborio *et al*. 2007).

The overall uncertainty is represented by a combination of the errors resulting from an incorrect model structure, incomplete or inadequate input data, spatial and temporal variability of model parameters, and a poor definition of initial and boundary conditions. (Izquierdo *et al*. 2004). Sin *et al*. (2009b) classified uncertainty associated with the model predictions under three general categories:

- **Structural** uncertainty that relates to the mathematical formulation of the model (model structures are an imperfect form of approximation of a real system);
- **Stochastic** uncertainty that arises from stochastic components of the model itself (e.g., random failure events of the system elements, such as pumps, air supply blowers, etc.);
- **Input** (or subjective) uncertainty that represents incomplete knowledge about the fixed values used as model inputs (e.g., kinetic and stoichiometric coefficients, physical-chemical constants, operational parameters, etc.).

4.5.2 Uncertainty analysis

The effects of parameter uncertainty on the confidence of model predictions can be determined using two well-known approaches, such as uncertainty and sensitivity analysis (Abrishamchi *et al*. 2005). With the first approach, propagation of the various sources of uncertainty to the model output is evaluated which, however, requires substantial computational efforts. The uncertainty analysis provides probability distributions of model outputs, which are subsequently used to derive the mean, variance and quantiles of model predictions (Sin *et al*. 2009a). In general, the methods available for this kind of analysis can be classified into three main groups (Abrishamchi *et al*. 2005):

- Bayesian methods, typically used when parameter values can only be specified by expert judgments, require many assumptions concerning the applications. Sin *et al*. (2009b) noted that these methods in combination with evolutionary optimization algorithms are appropriate for performing uncertainty analysis in complex numerical models. An example of applying a Bayesian approach in environmental projects was presented by Abbaspour *et al*. (1996).
- First-order reliability analysis (FORA) based on a Taylor series expansion and truncated after the first-order term. This requires knowledge of the sensitivity coefficient and covariance structure for each parameter.

- Statistical sampling (or simulation-based) methods, which evaluate the range of likely output estimates by defining a representative set of values for the uncertain parameter as inputs for the model. The set of values is determined using random sampling methods, taking into account the probability distribution functions and correlation between the parameters.

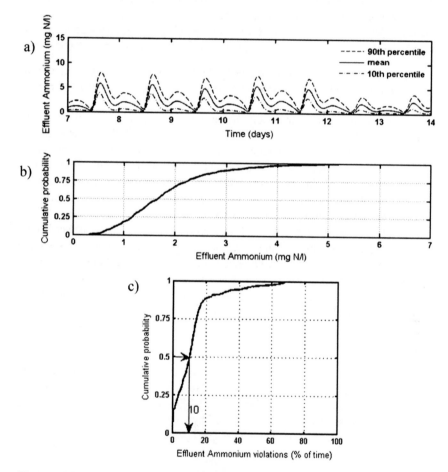

Figure 4.21 Representation of uncertainty obtained from the Monte Carlo simulations: (a) mean, 10th and 90th percentiles, (b) cumulative distribution function (CDF), (c) performance criterion expressed as the effluent violation statistics (Sin *et al.* 2009a).

A well-known method belonging to the latter approach is Monte-Carlo simulation, which is especially appropriate for complex nonlinear models or models that involve many uncertain parameters. The basic idea of this method is to evaluate iteratively a deterministic model using sets of inputs that are randomly generated from probability distributions to simulate the process of sampling from an actual population. The ultimate goal is to find a distribution for the inputs that most closely fits to measured data or best represents the current state of knowledge. Following the selection of an appropriate mathematical model structure, uncertainty analysis with Monte Carlo simulations involves the following steps (Sin *et al.* 2009a):

- Specifying input uncertainty based on an expert review process (consulting the opinion of process experts and/or finding the relevant literature resources);
- Sampling input uncertainty using Latin Hypercube Sampling (LHS) method which provides an effective coverage of the full parameter space;
- Propagating input uncertainty through the model structure to obtain prediction uncertainty for output variables (Monte Carlo simulations). (The goal of this step is to determine how random variation, lack of knowledge, or error affects the sensitivity, performance or reliability of the system that is being modelled).
- Presenting and interpreting of simulation results, which can be performed in different ways as presented in Figure 4.21. This includes simple statistics (e.g. mean and 90th and 10th percentiles of outputs), cumulative distribution function of an aggregated measure of the data (e.g. average effluent concentrations) and violation statistics (e.g. fraction of time that a given threshold effluent concentration is exceeded).

4.5.3 Sensitivity analysis

The second approach (sensitivity analysis) essentially comprises perturbing input parameters in a defined range of values and observing the effect of the perturbation on model predictions (Abrishamchi *et al.* 2005). The sensitivity of an output variable y_j to an input parameter x_i can be measured by a general sensitivity function (the first-order derivative of the output with respect to the input parameter):

$$\delta_{i,j}(t) = \frac{\partial y_j(t)}{\partial x_i} \qquad (4.58)$$

where
$\sigma_{i,j}$ - General sensitivity function
x_i - Value of the manipulated input variable i
y_j - Predicted value of the output variable j

Alternatively, the sensitivity can also be expressed by a sensitivity coefficient, $S_{i,j}$, representing as a relative change in the output variable y_j for a change (perturbation) in the input parameter x_i (USEPA 1987b; van Veldhuizen *et al.* 1999; Makinia 2006; Takacs 2008):

$$S_{i,j}(t) = \frac{x_i}{y_j(t)} \frac{\partial y_j(t)}{\partial x_i} \qquad (4.59)$$

A parameter with high sensitivity (or highly sensitive parameter) is one for which a small variation in its value results in a large variation in the response predicted by the model. A parameter with low sensitivity (also termed low sensitive parameter) is one that may be varied over a relatively wide range while producing only a relatively small variation in the predicted response (USEPA 1993). Sensitivity analysis is usually associated with a system engineering (purely mathematical) approach to model calibration without differentiating between better or less defined parameter values (van Veldhuizen *et al.* 1999). Furthermore, the analysis can be used to investigate if the parameters that were modified during the model calibration are indeed influencing the model outputs significantly (Petersen *et al.* 2002; Meijer *et al.* 2001 and 2002b; Makinia *et al.* 2005a). For newly developed models, sensitivity analysis also provides useful information about the model response to changes in the values of specific parameters. For example, Koch *et al.* (2000b) applied this technique to evaluate which parameters in ASM3 could be determined using the available batch experiments.

Even though sensitivity analysis reveals which of the input parameters contribute most significantly to the variance in model predictions, the obtained information is not appropriate for determining the sources of uncertainty that affect model outputs to the greatest extent. This can be explained by the fact that sensitivity analysis does not account for the likelihood that the input parameters deviate from their "*correct*" values. Consequently, a highly sensitive parameter that is known with low uncertainty may have much less effect on the uncertainty of model output than a much less sensitive parameter that is highly uncertain (Abrishamchi *et al.* 2005).

Takacs (2008) identified three main practical benefits of using sensitivity analysis in the following areas:

- Selection of model parameters that can be estimated with the most accuracy in terms of available measurement data;
- Development of experimental work to gather additional data for targeting the most sensitive elements of the model in terms of the particular process objective;
- Identification of the model parameters that have negligible effect on state variables (those parameters usually can be unchanged during simulations).

Conversely, the state variables insensitive to any model parameters can be identified (those variables do not warrant consideration while planning the experimental work).

Sensitivity analysis can be divided into two general categories: global and local. Global sensitivity techniques evaluate the partial derivative (Equation 4.58) in various points of the parametric domain, which is associated with the combined effect of simultaneous, wide range (possibly orders-of-magnitude) parameter changes. Consequently, these techniques require an enormous number of simulations before providing the parameter estimates. Local sensitivity analysis techniques refer to small changes of input parameters and evaluate the partial derivative (Equation 4.58) around one specific (nominal) set of values for input parameters. Various techniques for the local sensitivity analysis are available including the finite difference method, direct differential method, Green's function method, polynomial approximation method, automatic differentiation and complex-step derivative approximation method. Apart from the finite difference approximation, all the other methods require (complicated) manipulations of the model equations. In many cases, this is not practically feasible because the models are too complicated or the equations cannot be accessed directly (e.g. when compiled in executable commercial codes) (De Pauw 2005).

The finite difference approximation is the simplest and most often used in practice method of calculating local sensitivities (De Pauw 2005), which can implemented using a forward difference numerical scheme for Equations 4.58 and 4.59:

$$\delta_{i,j}(t) \approx \frac{y_j(t, x_i + \Delta x_i) - y(t, x_i)}{\Delta x_i} = \frac{\Delta y_j(t)}{\Delta x_i} \tag{4.60}$$

$$S_{i,j}(t) \approx \frac{x_i}{y_j(t, x_i)} \frac{y_j(t, x_i + \Delta x_i) - y(t, x_i)}{\Delta x_i} = \frac{\Delta y_j(t)/y_j(t)}{\Delta x_i/x_i} \tag{4.61}$$

where

Δx_i - Change in the value of manipulated input variable i
Δy_j - Calculated change in the value of predicted output variable j

Input parameters are perturbed by a perturbation value Δx_i using a "*one-variable-at-a-time*" approach (USEPA 1987b). The influence of a selected input parameter on the output variable can be interpreted quantitatively in terms of the calculated $S_{i,j}$ value as proposed by Petersen *et al.* (2002). For $S_{i,j} < 0.25$, a parameter is considered to have no significant influence on a certain model output; if $0.25 \leq S_{i,j} < 1$, the parameter is considered to be influential; if $1 \leq S_{i,j} < 2$, the parameter is considered to be very influential; and if $S_{i,j} \geq 2$ the parameter is considered to be extremely influential. Also Takacs (2008) noted that parameters for which $S_{i,j} > 1$ are generally considered very sensitive. Sensitivity analysis-based parameter ranking is a very attractive and highly relevant approach for process characterization and model validation (Sin *et al.* 2009b).

In addition to the sensitivity coefficient $S_{i,j}$, Brun *et al.* (2002) introduced the sensitivity measure, δ_i^{msqr}, which is defined as:

$$\delta_i^{msqr} = \sqrt{\frac{1}{n} \sum_{j=1}^{m} S_{i,j}^2} \qquad (4.62)$$

This parameter measures the mean sensitivity of the model outputs to a change in the input variable x_i (in the mean square sense). A high value of δ_i^{msqr} indicates that x_i has an important influence on the simulation result, whereas a value of zero means that the simulation result is independent from x_i (Brun *et al.* 2002).

Table 4.15 Various input parameters used in sensitivity analysis of activated sludge models (modified from Takacs (2008))

General category	Types of parameters	Used in the analysis
Physical parameters	Number of tanks-in-series within the given volume	Sometimes (in tracer studies)
Operational parameters	Process temperature, internal flow rates, air flow rate, DO setpoints, chemical dosages	Sometimes
Boundary conditions	Influent flow rates and concentrations (COD, TN, TP), distribution of the influent fractions	Frequently
Initial conditions	Initial concentrations of different groups of bacteria (heterotrophs, autotrophs, PAO)	Sometimes (in short-term batch experiments)
Model parameters	Stoichiometric coefficients (nutrient contents, yields), kinetic coefficients (max. growth rate constants, half-saturation constants), settling coefficients	Frequently

In activated sludge simulation studies, the sensitivity can be analyzed with respect to perturbations in the input parameters grouped in five general categories: physical parameters, operational parameters, boundary conditions, initial conditions and model parameters (Table 4.15). Different perturbation values, Δx_i, were reported in the literature ranging from 1 to 100%, e.g. 1% (Petersen *et al.* 2002; Takacs 2008), 10% (van Veldhuizen *et al.* 1999; Makinia and Wells 2000; Makinia 2006), 20% (Meijer *et al.* 2002b), 100% (Takacs 2008). It should be emphasized, however, that assuming Δx_i too small will result in numerical inaccuracies. On the other hand, Δx_i should not be too large because then the nonlinearity of the model will affect the sensitivity calculations (De Pauw 2005). For example, USEPA (1987b) defined a normalized sensitivity coefficient which represents the percentage change in the output variable, y_j, resulting from a 10% change in the input variable, x_i.

There are four major categories of output variables in activated sludge models that could be considered while performing the sensitivity analysis (Takacs 2008):

- State variables in reactors (concentration of DO, NH_4-N, NO_3-N, heterotrophs, nitrifiers, PAO, etc.);
- Composite variables in reactors (concentration of MLSS, MLVSS, etc.);
- Composite variables in settlers (suspended solids concentration in the effluent and RAS, sludge blanket level, etc.);
- Internal model expressions, such as process rates, saturation and inhibition functions (OUR, NUR, AUR, PRR, PUR, etc.).

The sensitivity analysis can be performed under both steady-state and dynamic conditions (Makinia 2006; Takacs 2008). The simplest practical approach is to develop the sensitivity functions at steady state, around a certain (long-term average) operating point described by a "*nominal*" set of parameters. The result of a steady-state sensitivity analysis becomes a two-dimensional matrix (n x m), with one dimension containing **n** input parameters, and the other containing **m** model outputs. Processing of such a matrix can provide valuable insight into the model structure and importance of parameters (Takacs 2008).

It is sometimes important to observe the sensitivity functions under dynamic conditions. This refers to processes that are never in steady state (e.g. SBRs or alternating aeration systems - BioDenitro) or situations when the effects of input parameters on model outputs are evaluated in terms of short-term system disturbances, such as diurnal influent variations, stormwater events and process upsets. The dynamic sensitivity analysis provides more detailed information, but requires more computational efforts, although the number (**n**) of dynamic simulation runs does not change compared to steady state. The resulting sensitivity matrix is extended by incorporating one more dimension (time) and such a multidimensional matrix loses an illustrative value. Therefore, Takacs (2008) proposed a practical approach including a preliminary steady-state sensitivity analysis (to identify most important input parameters), which is followed by a dynamic sensitivity analysis for only the selected subset of input parameters.

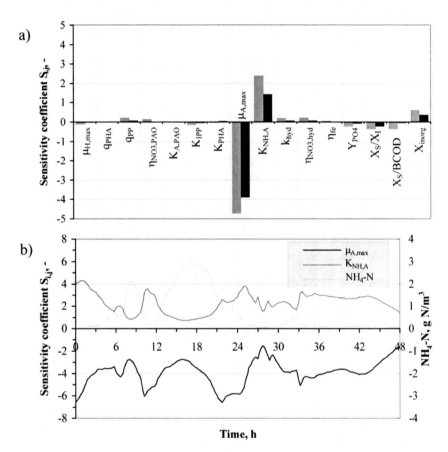

Figure 4.22 Results of the sensitivity analysis performed with the calibrated ASM2d: (a) steady-state data (gray columns) vs. averaged dynamic data (black columns), (b) temporal variations of $S_{i,j}$ for $\mu_{A,max}$ and $K_{NH,A}$ (partially adopted from Makinia (2006)).

Alternatively, it is possible to generate a two-dimensional matrix of the averaged dynamic sensitivity coefficients. It should be emphasized, however, that their values do not correspond to the values determined under steady state, although a strong correlation between the coefficients could be expected under normal operating conditions. For example, Figure 4.22 illustrates the sensitivity of effluent NH_4-N concentration to changes in the values of the maximum growth rate constant of autotrophs, $\mu_{A,max}$, and ammonia half-saturation constant for growth of autotrophs, $K_{NH,A}$ (Makinia 2006). Very similar observations were also reported by Takacs (2008) and the following conclusions could be derived based on the results of both studies:

- The effluent NH_4-N concentrations are considerably more sensitive to changes in $\mu_{A,max}$ and $K_{NH,A}$ compared to other considered input parameters (Figure 4.22a);
- $\mu_{A,max}$ and $K_{NH,A}$ have opposite effects on the effluent NH_4-N concentrations;
- Sensitivities under dynamic and steady-state conditions are not equal (e.g., differences for $\mu_{A,max}$ and $K_{NH,A}$ were 20% and 35%, respectively);
- The variations in the values of $S_{i,j}$ for $\mu_{A,max}$ were larger compared to $S_{i,j}$ for $K_{NH,A}$ (Figure 4.22b);
- The effluent NH_4-N concentration was reaching the highest sensitivity to changes in $\mu_{A,max}$ and $K_{NH,A}$ during recovery from high loaded periods, when the negative concentration gradient was greatest. In contrast, the effluent NH_4-N concentration was least sensitive during low loaded periods.

The approximation using Equation 4.60 has two principal drawbacks. The resulting sensitivity function relates to the $(x_i+\Delta x_i/2)$ parameter set and it does not provide any information on the quality of the sensitivity function (Equation 4.58). For sensitivities around the nominal values of the input parameters, the central difference formula is required:

$$\frac{\partial y_j}{\partial x_i} \approx \frac{y_j(t, x_i + \Delta x_i) - y(t, x_i - \Delta x_i)}{2\,\Delta x_i} \qquad (4.63)$$

Instead of making two evaluations of $y(t,\ x_i)$ and applying Equation 4.63, De Pauw (2005) proposed to derive the central difference from two sensitivity functions. The first sensitivity function is calculated by increasing the nominal input parameter by Δx_i, whereas the second sensitivity function is calculated by decreasing the nominal input parameter by the same value:

$$\delta_P^+ = \frac{y(x_i + \Delta x_i) - y(x_i)}{\Delta x_i} \qquad (4.64)$$

$$\delta_P^- = \frac{y(x_i) - y(x_i - \Delta x_i)}{\Delta x_i} \qquad (4.65)$$

Then the centralised sensitivity function is calculated by averaging both sensitivity functions:

$$\delta_P = \frac{\delta_P^+ + \delta_P^-}{2} \tag{4.66}$$

Although this method requires more calculation efforts, it also provides additional information concerning the choice of the perturbation factor and quality of the sensitivity function. In order to minimize effects of the numerical error and the error introduced by the nonlinearity, the difference between the two sensitivity functions (Equations 4.64 and 4.65) should be as small as possible. This difference can be used to select the proper perturbation factor based on one of several criteria presented in Table 4.16.

Table 4.16 Criteria used to quantify the quality of sensitivity calculations (De Pauw, 2005)

Criterion	Definition	Formula
Sum of squared errors (SSE)	SSE is calculated and summed over all times where the sensitivity is desired (N)	$SSE = \dfrac{\sum\limits_{i=1}^{N}\left(\delta_P^+\left(i\right) - \delta_P^-\left(i\right)\right)^2}{N}$
Sum of absolute errors (SAE)	SAE is calculated between both sensitivity functions and summed over all times where the sensitivity is desired (N)	$SAE = \dfrac{\sum\limits_{i=1}^{N}\left|\delta_P^+\left(i\right) - \delta_P^-\left(i\right)\right|}{N}$
Maximum relative error (MRE)	MRE returns the maximum value of the relative difference between both sensitivity functions. One should be careful with this criterion because $\partial y = \partial \theta_+$ may become 0. In these special cases the criterion returns 0.	$MRE = \left|\dfrac{\delta_P^+\left(i\right) - \delta_P^-\left(i\right)}{\delta_P^+\left(i\right)}\right|_{MAX}$
Sum of relative errors (SRE)	SRE is also based on the ratio of the sensitivity functions. The ideal case is when this ratio equals 1, because then both sensitivity functions are equal. The criterion returns the sum of deviations from this ideal situation over all times where the sensitivity is desired (N). Like the MRE criterion one should be careful if $\partial y = \partial \theta_+$ becomes 0. In this special case no contribution is made to the total sum.	$SRE = \dfrac{\sum\limits_{i=1}^{N}\left|1 - \dfrac{\delta_P^-\left(i\right)}{\delta_P^+\left(i\right)}\right|}{N}$

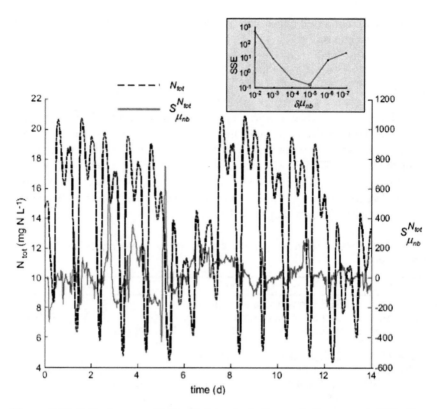

Figure 4.23 Trajectory sensitivity of total output nitrogen to the second nitrification rate μ_{NOB}, showing a good identifiability and a counter-phase behaviour of the sensitivity with respect to nitrogen concentration. The inset shows that the optimal perturbation factor yields the minimum of the sum of squared errors (SSE) (Iacopozzi *et al.* 2007).

Using the above approach, Iacopozzi *et al.* (2007) performed a dynamic sensitivity analysis to evaluate the effect of the second-step nitrification rate constant, μ_{NOB}, on the efficiency of nitrogen removal using a benchmark configuration of the pre-denitrification process. The dynamic (or trajectory) incremental sensitivities (δ_p^+, δ_p^-) with respect to the selected input parameter x_i (μ_{NOB}) were computed with different perturbation factors $\pm\Delta x_i$ on both sides of the nominal trajectory $y(t,x_i)$ of the effluent total nitrogen (TN) concentration. Based on a series of calculations, the optimal value of Δx_i was determined $(\delta\mu_{NOB} = 10^{-5})$ by minimizing the SSE (Table 4.16) of the two trajectories (δ_p^+, δ_p^-). The optimal

Δx_i was eventually used to compute the trajectory sensitivity δ_p of the effluent TN concentration with respect to the changes of μ_{NOB} (Figure 4.23), exhibiting a counter-phase behaviour, i.e. the lower was the TN concentration the higher was its sensitivity to μ_{NOB}.

Chapter 5

Practical model applications

Practical applications of the complex biokinetic models, which have been reported in literature, can generally be classified under the following categories: optimization of the performance of existing plants, upgrade of existing plants, design of new facilities and development of new treatment concepts. Optimization of an existing facility represents the best opportunity for use of mathematical modeling approaches since the proposed model can be calibrated and validated using actual data from the facility. Recently, the use of simulation for educational purposes (teaching and staff training) has also been receiving more attention.

The IWA-family activated sludge models have been implemented in the computer software called simulation platforms. The most popular commercial simulators include ASIM (Switzerland), BioWin (Canada), GPS-X (Canada), SIMBA (Germany), STOAT (UK) and WEST (Belgium). This chapter deals with the role of simulators in predicting performance real systems, such as activated sludge bioreactors. The capabilities of the simulators are discussed with a special focus on unique utilities of individual programs.

5.1 INTRODUCTION

Shortly after the introduction of ASM1 (Henze *et al.* 1987), Grady (1989) listed five potential areas for the applications of mathematical modelling in the wastewater treatment practice:

"For the researcher, modelling serves as a conceptual framework upon which to build and test hypotheses, thereby extending knowledge. For the designer, modelling allows the exploration of the impact of a wide range of system variables, thereby greatly increasing his/her experimental space regarding a proposed process. Furthermore, when employed within the appropriate framework, modelling permits the development of near optimal designs in which the desired process objectives are obtained at minimal cost. For the operator, modelling allows the development of control strategies by facilitating the investigation of treatment system response to a wide range of inputs without jeopardizing actual system performance. For the regulator, modelling allows judgements to be made about the impact of new effluent requirements on treatment system design and cost. Finally, for the engineering educator, modelling provides a tool with which students can explore new ideas, thereby enriching their education by engaging them actively in the learning process."

In addition, in the following years, several authors (e.g. USEPA, 1993; Olsson and Newell 1999; Henze *et al.* 2000; Ferrer *et al.* 2004b; Gernaey *et al.* 2004; Makinia 2006; Henze *et al.* 2008) also identified another important model application which was diagnosis and optimization of the operation of existing WWTPs, i.e. predicting the effluent quality, detecting non-optimal performance and comparing modifications with regard to the plant configuration and operational strategies. Practical model applications, which have been reported in literature, can generally be classified under four categories: optimization of the performance of existing plants, upgrade of existing plants, design of new facilities and development of new treatment concepts (Makinia 2006). These issues are discussed in the following sections and the six general areas of model applications are presented in Figure 5.1. Henze *et al.* (2008) also noted that models are nowadays *"invaluable tools"* for the plant operator training and become a part of the university curricula for engineers and scientists.

Hauduc *et al.* (2009) presented the results of a world-wide survey on the use of activated sludge models. The main objectives identified for building and using a model were: optimisation (59%), design (42%) and prediction of future operations (21%), but the distribution of modelling tasks varied in terms of the organisation type using the model (Figure 5.2). It was emphasized that the majority of North-American and European modellers are using models in different ways. In Europe, models are primarily used by researchers for optimisation purposes, while most modellers in North America are employed by private companies and carry out

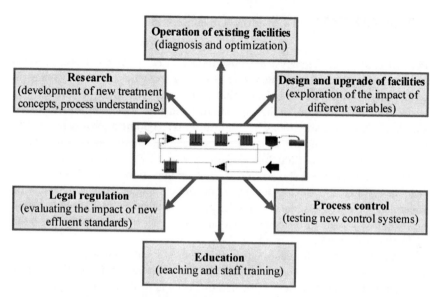

Figure 5.1 Potential practical applications of mathematical modelling and computer simulation of activated sludge systems.

design studies. The survey provided useful insights into the main limitations of modelling and the expectations of users for improvements with regard to:

- The complexity (apparent or actual) of the model theories and modelling procedures;
- The time consuming steps and therefore the cost of modelling;
- The modellers' appreciation of the reliability of the models.

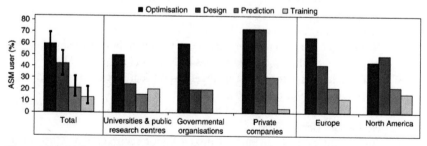

Figure 5.2 Main objectives of the simulation studies in terms of the organization type and geographical region (Hauduc *et al.* 2009).

5.2 OPTIMIZATION OF PROCESS PERFORMANCE

In the literature (USEPA 1993; Barker and Dold 1997a; Henze *et al.* 2000; Petersen *et al.* 2002; Gernaey *et al.* 2004), model-based optimization is understood as a scenario evaluation of alternative operating strategies, such as changes in wastewater or recycle flow rates and pollutant loads, etc. and physical modifications to the plant configuration. In this context, mathematical modelling and computer simulation have primarily found the application as an optimization tool focused on the optimal process performance neglecting the implicit constraints of maximum daily costs. Ferrer *et al.* (2004b) and Gernaey *et al.* (2004) differentiated off-line optimization (as defined above) from on-line optimization, in which simulations with the calibrated model are run in an optimization scheme, such as a plant-wide supervisory control system or a model predictive control (MPC) optimization algorithm. The issue of on-line simulation remains out of the scope of this book and will not further be discussed.

Some WWTPs can reach more stringent effluent standards by process optimization without extension of the plant. Optimization of an existing facility represents the best opportunity for use of mathematical modelling approaches since the proposed model can be calibrated and validated using actual data from the facility (USEPA 1993). Without interfering with the actual process performance, the potential effectiveness of various process modifications can be compared and the principal bottlenecks limiting treatment capacity can be identified (van Veldhuizen *et al.* 1999b; Bratby *et al.* 2001). The model can also be used on an ongoing basis by the successive interaction between predicting system responses and monitoring of actual system responses (USEPA 1993).

A number of different case studies focused on process optimization supported by modelling and simulation has been reported in the literature including:

- Integration of a dynamic process model within an optimization algorithm to optimize performance or operating costs of the oxidation ditch process (von Sperling and Lumbers 1991);
- Seeding the nitrifying bacteria from one treatment line to another lines for reducing the effluent concentrations of NH_4-N (Finnson 1993);
- Rearrangement of the size of nitrification and denitrification compartments (Coen *et al.* 1996; Leeuw *et al.* 1996);
- Implementation of a step-feed flow (Coen *et al.* 1996; Leeuw *et al.* 1996; Solley and Barr 1999);
- Implementation of an internal mixed liquor recirculation (Coen *et al.* 1996; Leeuw *et al.* 1996; Solley and Barr 1999; Ferrer *et al.* 2004a).
- Equalization of dosing a wastewater discharge from food industry (Leeuw *et al.* 1996);

- Adjustment of the phase lengths in alternating aeration and flow processes (Dupont and Sinkjaer 1994; Potter *et al.* 1996; Irizar *et al.* 2003; Fikar *et al.* 2005);
- DO control in the aerated zones (Leeuw *et al.* 1996; Solley and Barr 1999);
- Addition of the readily biodegradable substrate, such as fermentation products of the raw sludge from primary clarifiers (Funamizu *et al.* 1997);
- Shutting down one treatment line and overloading the other lines (Carucci *et al.* 1999);
- Reduction of the total nitrogen emission during stormwater events (Seggelke and Rosenwinkel 2002, Rosenwinkel *et al.* 2007b; Ahnert *et al.* 2009);
- Adjustment of the cycle range and the ratio between the nitrification and denitrification length in an intermittently aerated process (Andreottola *et al.* 2003);
- Adjustment of the mixed liquor recirculation between the nitrification and denitrification compartments (Moussa *et al.* 2004);
- Dosing ammonia rich substrate to the regeneration zone of the R-D-N process (Novak and Havrlikova 2004);
- Adjustment of production schedules in a chemical industry WWTP (van Hulle and Vanrolleghem 2004);
- Analysis of possible modifications in the layout of an oil refinery WWTP and reduction of operating costs (Pardo *et al.* 2007);
- Strategy of dosing external source to enhance denitrification (Phillips *et al.* 2009).

Many practical examples have proved that mathematical modelling and simulation are especially a valuable tool for searching optimal operating conditions in SBRs in terms of the cycle length, phase distribution and filling strategy (Oles and Wilderer 1991; Andreottola *et al.* 1997; Brenner 2000; Coelho *et al.* 2000; Hvala *et al.* 2001; Mikosz *et al.* 2001; Artan *et al.* 2002; Sousa *et al.* 2008).

Since the level of expertise required and the time to configure and calibrate the models may not be readily available at all plants, the results from simulation models can be converted into a series of convenient operating support charts (Bratby *et al.* 2001). Typical information which the charts provide to the operators include the following:

- Selection of the correct operating SRT for a given temperature and effluent ammonia concentration;
- Selection of the number of internal mixed liquor recycle pumps for given temperatures, RAS rates and required effluent nitrate concentrations;

- Prediction of the MLSS concentration at given SRTs, primary effluent loads, and temperatures;
- Prediction of the maximum allowable solids load to the final clarifiers to prevent failure.

5.3 EXPANSION AND UPGRADE OF EXISTING FACILITIES

In the case of upgrading of existing WWTPs, information on wastewater characteristics is available to facilitate model calibration and the existing wastewater and biomass may be used for some limited laboratory scale verification experiments (USEPA 1993). Using computer simulation, a number of plant upgrading options can be evaluated including formulation of different process retrofitting and expansion options as well as making best use of the available infrastructure (van Niekerk *et al.* 2000). Sometimes the principle of "*Optimize What You Have First*" can produce surprisingly high nutrient removal levels for a very modest capital expenditure (Solley and Barr 1999).

Several case studies have illustrated various aspects of WWTP upgrade supported by mathematical modelling and simulation:

- Estimation of a desired capacity for treatment processes (Coen *et al.* 1996; Coen *et al.* 1997; Cheng and Ribarova 1999; van Niekerk *et al.* 2000; Hanhan *et al.* 2002; Moussa *et al.* 2004);
- Efficiency comparison of different technologies for improving nitrogen removal (Hvala *et al.* 2002);
- Testing various operating strategies (Dupont and Sinkjaer 1994; Jobbagy *et al.* 2004);
- Splitting the wastewater flow between WWTPs (Ladiges *et al.* 1999; Ladiges *et al.* 2000; Ladiges *et al.* 2004) - this is one of the best examples illustrating benefits of using computer simulation in the wastewater industry. The cost of an intensive measurement campaign for model calibration was 150.000 Euro, but at least 20 million Euro were subsequently saved in the design phase;
- Installing an additional volume for nitrification and denitrification in a food industry WWTP (Vandekerckhove *et al.* 2008).

5.4 DESIGN OF NEW FACILITIES

During the design phase, models can assist in identifying and quantifying the key design parameters (Henze *et al.* 2008). However, if a complex mathematical model is to be used for optimizing the design of a new facility, model verification may be difficult to carry out because of lack of an appropriate data base. In such cases, significant data should be available based on experience with the particular system and/or the type of wastewater to be treated, otherwise an extensive pilot program has to be conducted to produce the data necessary to calibrate a sophisticated process model. In designing such a program, the issues of particular concern include pilot plant configuration, pilot plant operating conditions and experimental design for model calibration (USEPA 1993).

During the design phase for a new secondary treatment process, mathematical modelling and simulation can be used to (Buhr *et al.* 2001):

- Support the optimization and fine tune aeration basin design;
- Examine the effectiveness of alternative treatment strategies;
- Evaluate the impacts of future water quality standards;
- Identify the influence of dynamic process variations on daily composite effluent quality.

Several case studies have demonstrated the applicability of modelling and simulation for the design purposes ranging from simple small plants to complex large plants (Daigger and Nolasco 1995; Bingley and Upton 2000; Gabaldon *et al.* 2000; Larrea *et al.* 2001; Rivas *et al.* 2001; Oleszkiewicz *et al.* 2004; Seco *et al.* 2004). Simulation results can be used not only to make direct decisions on investments but also to verify the proposed design assumptions (Gorgun *et al.* 1996b; Coen *et al.* 1997).

In the design phase of activated sludge systems, steady-state models are suitable for a comparison of alternative process configurations or a preliminary estimation of the removal efficiency under constant flow and load (Koch *et al.* 2001b). Moreover, with the aid of such models, the impact of various parameters on the system response and internal relationships between processes can be examined (Wentzel and Ekama 1997). A number of design models for biological nutrient systems is available in the literature (e.g. Wentzel and Ekama 1997; Scheer and Seyfried 1997; Koch *et al.* 2001a; Koch *et al.* 2001b; Rosenwinkel *et al.* 2002).

5.5 DEVELOPMENT OF NEW TREATMENT CONCEPTS

In research applications, models can be used to test hypotheses in a consistent and integrated manner, which results in a better understanding of the fundamental behavioural patterns controlling the system response as well as developing new treatment concepts (Henze *et al.* 2008). For example, a proper model for the description of microbiological processes can be developed based on laboratory experiments. Coupling this model with the mass transport equations can provide the prediction of a large scale reactor and most sensitive process parameters can be further tested in a laboratory scale (Salem *et al.* 2002). Such an approach was successfully applied for scaling up the SHARON process from a 2 dm³ laboratory scale reactor to a 1500 m³ full-scale reactor (Hellinga *et al.* 1998).

The calibrated mathematical models (ASM2d and TUDP) were used at two Dutch WWTPs to evaluate the potential of bio-augmentation to maintain nitrification in high loaded activated sludge processes at low temperatures (Salem *et al.* 2002 and Salem *et al.* 2003). A new augmentation concept (Figure 5.3a), called the BABE process (Bio-Augmentation Batch Enhanced), consists of a nitrification reactor in the RAS line fed with the ammonia-rich water generated internally in WWTPs by sludge digestion and thickening (reject water). Simulation results revealed that implementing the BABE process can reduce the minimal SRT by approximately 50% with only 10% extra tank volume needed to be constructed. The full-scale application of this concept confirmed this capability by improving substantially (by almost 60%) the nitrification rate of the activated sludge in the main stream (Salem *et al.* 2004).

Another model-based approach was investigated based on the assumption of the reduced decay of autotrophic bacteria under anaerobic conditions (Yuan *et al.* 1998). The SRT in the main stream reactor could be reduced considerably (maintaining a similar nitrification capacity) by storing the waste activated sludge temporarily under anaerobic conditions in the additional tank with a retention time of a few days (Figure 5.3b) and returning the biomass to the main stream process when a shock nitrogen load or an inhibition/toxicity incident occurred. Experimental studies carried out at a pilot plant confirmed the simulation results (Yuan *et al.* 2000). The analysis also revealed that savings on the tank volume could reach 20%, but an increased sludge production is a potential negative side effect.

Siegriest *et al.* (1999) also demonstrated experimentally that the anaerobic and anoxic decay rate of autotrophic bacteria was reduced by over 50% compared to the decay rate under aerobic conditions. This hypothesis was incorporated in a biokinetic model (modified ASM2) and simulation results revealed that reduction of the DO supply by an intermittent aeration not only could save aeration energy but also could improve the nitrification capacity of the plant (through increasing the concentration of nitrifying bacteria).

Based on simulation studies, a concentration equalization was also proposed as a method to improve the effluent phosphorus quality in an A/O system (Filipe *et al.* 2001a). The improvements could be achieved by diminishing the potential for imbalances between phosphate release and uptake by avoiding sudden increases of VFA loading to the plant. Simulation results suggested that it might be possible to decrease the amount of phosphorus discharged by a factor as high as 4 through use of concentration equalization.

a)

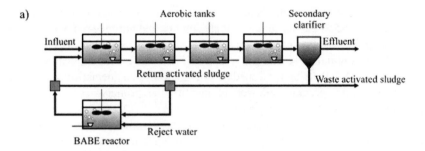

b)

c)

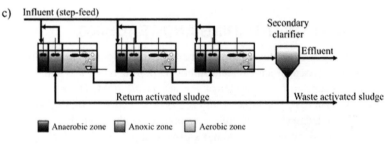

d)

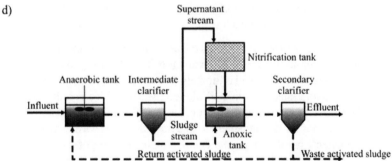

Figure 5.3 Model-based development of new treatment concepts: (a) BABE technology, (b) storage tank in the WAS line, (c) step-feed bio-P process, (d) A_2N process.

The model of Barker and Dold (1997a) was used to develop the Step Bio-P process (Figure 5.3c), a new biological phosphorus and nitrogen removal process with a step feed configuration (Nolasco *et al.* 1998). A pilot plant was operated to verify the simulation results and to optimize the process for application at the Lethbridge WWTP (Alberta, Canada). The combined mathematical modelling and pilot testing program allowed to identify the primary factors affecting performance of the Step Bio-P process.

An innovative concept process consisting of an anaerobic reactor and an anoxic reactor in a mainstream line with a nitrification reactor preferably with biofilm in a sidestream line was proposed by Kuba *et al.* (1996b) and referred to as an A_2N process. A schematic diagram of that process is presented in Figure 5.3d. Subsequently, the TUDP model (see: Section 1.2.3) was used to evaluate the process and simulation results revealed that an optimal split ratio between the mainstream and sidestream exists for certain influent characteristics (Hao *et al.* 2001a). The model was further used to demonstrate advantages of the A_2N process compared to the common UCT process in terms of performance, energy consumption and land occupation (Hao *et al.* 2001b).

5.6 EDUCATION (TRAINING AND TEACHING)

When teaching wastewater treatment, mathematical models can be used as an active tool to increase process understanding, exploring new ideas and developing a general conception of the system (Gernaey *et al.* 2004; Henze *et al.* 2008). Although the survey of Hauduc *et al.* (2009) identified self training as the main source of knowledge (over 80% of the respondents), the modelling courses have also been offered by universities and software suppliers. Different approaches (often combined) are possible for the use of mathematical models in teaching wastewater treatment (Morgenroth *et al.* 2002a; Hug *et al.* 2009):

- *"Teaching with models"* means the use of a predefined model (e.g. ASM1) with predefined parameters to evaluate different process configurations and operating strategies, and ultimately understand processes, their behaviour and interactions in activated sludge systems. This requires providing instructions for running a particular model on any simulator. Students should posses a fundamental understanding of wastewater treatment processes, but do not need a detailed understanding of the model behind the system nor the numerical procedures involved. The *"teaching with models"* approach could also be used for the staff training (improving knowledge and evaluating the operators own operational strategies). For example, the plant operator can investigate the effects of certain actions (e.g. changing SRT, MLRs or DO setpoints) or conditions (e.g. seasonal variations of the process temperature) on the performance of a WWTP without consequences to the actual operation of the plant.

- "*Teaching modelling*" means the use of a specialized simulation software to understand the mathematical models (including the model background) and steps involved in model building. The latter issue can be taught at three levels: (i) configuring a treatment system using predefined unit processes and parameters, (ii) calibrating the model with actual data, (iii) developing a new model (new set of equations) for processes that cannot be modelled with existing models. Furthermore, this concept also aims at understanding the "*responsible*" modelling. This requires familiarity with the structure of the models, as well as awareness of limitations due to data quality, calibration and validation procedures as well as parameter and model structure uncertainty.

- "*Teaching models*" is focused on providing knowledge about specific models. In most cases, this is a prerequisite for the two concepts outlined above (teaching with models and teaching modelling) rather than the ultimate goal of the teaching.

- Understanding basic numerical integration procedures low level programming languages (e.g. C++, Fortran, Pascal, Visual Basic, etc.) which, in turn, will help to predict numerical problems when using a simulator. In general, however, environmental engineers do not need to focus on developing their own numerical procedures as this can be facilitated by different kinds of advanced specialized software.

5.7 CHARACTERISTICS OF THE EXISTING SIMULATOR ENVIRONMENTS

Different types of a computer software can be used to implement mathematical models and subsequently predict the performance of WWTPs. These include spreadsheets (useful primarily for steady-state simulations and mass balancing), low level programming languages (see above), general-purpose simulators (e.g. MATLAB/Simulink, ACSL, Maple, Mathematica, Stella) and specific WWTP simulator environments (or simulation platforms).

The general-purpose simulators have a high flexibility, but the modeller has to supply the models for a specific WWTP configuration. As a consequence, this kind of software requires a skilled user who fully understands the implications resulting from each line of code in the models (Gernaey *et al.* 2004). In contrast, the WWTP simulator environments should offer a high level of flexibility with minimal user training and without the requirement of knowledge of programming Melcer *et al.* (2003).

The first specific simulator using ASM1 was SSSP (stands for Simulation of Single Sludge Processes), developed at Clemson University, South Carolina

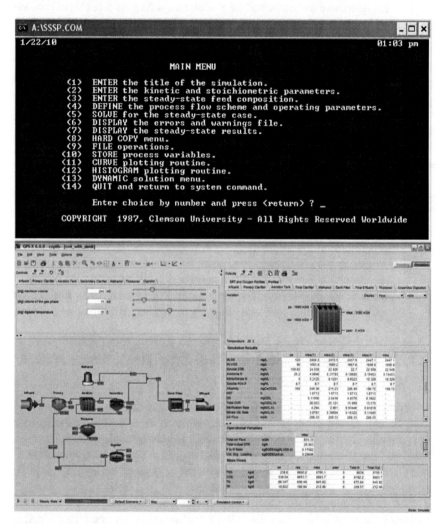

Figure 5.4 Main windows of the WWTP simulators: SSSP in 1987 (top) and GPS-X in 2009 (bottom) revealing some common features of the modern simulator environments.

(USA) and implemented in the DOS environment (Bidstrup and Grady, 1988). A chain (up to nine) completely mixed activated sludge bioreactors performing carbon oxidation, nitrification and denitrification can be simulated with ASM1, whereas the clarifier is modelled as a separation point (without any loss of biomass). The program may be used to predict both steady-state and dynamic performance. The user may specify plant configuration; reactor sizes; influent, recycle, and recirculation flow rates; the desired SRT; the kinetic and stoichiometric parameters; and the time-variable patterns of influent flows and concentrations of the model components (Figure 5.4).

Shortly after the SSSP release, the first version of ASIM was developed primarily as a teaching and research tool (see the description below).

All the common commercial simulation platforms (described below) were developed in the 1990s. The latest versions are modular, multi-purpose programs in which the studied process configuration can easily be laid out with a graphical user interface with a great variety of output options (Figure 5.4). The programs contain extended libraries of predefined process unit models and the model libraries are (usually) provided in an open format which allows to modify and extend the existing models using so-called model editors. Advanced tools for model calibration and process control are also available.

Earlier overviews of WWTP simulator environments were presented by Vanrolleghem and Jeppsson (1995), Olsson and Newell (1999), Copp (2002), Schütze et al. (2002), Melcer et al. (2003), Gernaey et al. (2004). In order to facilitate a rational (without favouring any simulator) selection, Olsson and Newell (1999) proposed a list of criteria which can be summarized (with some updates) as follows:

- A library of reliable models for various unit processes and equipment;
- Uncertainty, sensitivity analysis and parameter optimisation modules (automated procedures);
- Flexibility to build larger plant models (available models and modularity of the models);
- Graphical (block-oriented system) representation;
- Numerical methods (with more than one standard integration algorithm) to ensure a reliable and robust integration of the model equations;
- Possibility of being connected to a real-time environment (real-time simulation);
- Inclusion of measurement data as input to the simulator;
- A variety of output presentation formats and flexible reporting;
- Running under all common (modern) operating systems.

5.7.1 ASIM (ETH/EAWAG, Switzerland)

ASIM (stands for Activated Sludge SIMulation Program) is a simulation program, which allows for simulation of a variety of different activated sludge systems: The program has been developed under the supervision of Professor Willi Gujer at the Institute for Hydromechanics and Water Resources Management (ETH) in Zurich. Special licences for research and development can be obtained from ETH/EAWAG (www.asim.eawag.ch). A commercial distributor of the program is the company Holinger AG (www.holinger.com).

The studied plant layout may consist of up to ten different reactors in series (aerobic, anoxic, anaerobic), including RAS and internal MLR streams, batch reactors, chemostat reactors, etc. The biokinetic models may be freely defined, stored and edited by the user, however, several pre-defined models are available in a model library including ASM1 (adapted), ASM2d and ASM3. Control loops, using simple proportional and on/off type controllers, can also be implemented in the program.

Both steady-state and dynamic simulations can be run in ASIM. The dynamic input data, such as loads, process alteration, temperature and operational parameters (aeration intensity, excess sludge removal, recycle rates, etc.), are introduced from a "*variation*" file. Data analysis is supported by the possibility of comparing measured data with simulation results, which can be exported to spreadsheets for further treatment.

Examples of the application of ASIM are documented in the publications listed at the program web page (see above).

5.7.2 BioWin (EnviroSim, Canada)

Most types of wastewater treatment systems can be configured in BioWin using many process modules including a range of activated sludge bioreactor modules and various clarifier modules (primary, ideal and 1-D model clarifiers). BioWin uses the biokinetic "*supermodel*" approach to pass information between unit processes without having to convert state variables between process types. The "*supermodel*" includes biological carbon, nitrogen and phosphorus removal processes, fermentation and methane production, gas transfer modelling, and chemical equilibrium processes for modelling pH and precipitation reactions. Moreover, two-step nitrification, denitrification, and Anammox process are included to enable modelling sidestream treatment processes.

Both steady-state and dynamic simulations can be run from the main simulator window. BioWin uses an itinerary that allows to schedule different operational parameters, such as DO setpoints, air flowrates, and temperature. In order to display simulation results in an integrated form, BioWin incorporates an "*Album*" for this purpose. The album consists of a series of tabbed pages which may contain any or a combination of the following data display formats: a wide range of chart types, tables and element-specific information displays.

BioWin incorporates an automatic report generation feature. The type of information that appears in the report can be customized and include both the general information (e.g. plant layout, album pages, etc.) and specific information for each model element (e.g. physical data, local biokinetic model parameters, etc.).

The BioWin Controller is a separate Windows application (started within BioWin) which links to a BioWin configuration and allows specification of a range of process control features commonly used in activated sludge systems. Several types of controllers are available ranging from on/off controllers to proportional-integral-derivative (PID) feedback controllers.

A comprehensive list of the BioWin application references is available at the EnviroSim web page (www.envirosim.com).

5.7.3 GPS-X (Hydromantis, Canada)

GPS-X has a large selection of unit processes that can be used to describe a wide variety of wastewater treatment plant configurations. Over 50 plant layouts are pre-compiled, covering many popular BNR activated sludge systems with the integrated N and P removal (A/O, Bardenpho, VIP, Step Bio-P, contact stabilization, etc.). In activated sludge bioreactors, unlimited number of aerobic, anoxic and anaerobic zone combinations is available. There is also no limit with regard to the number of stages (passes) and internal recycle combinations (including step-feeding).

Standard biokinetic models include temperature dependant versions of ASM1, ASM2d, ASM3, NewGeneral, and two-step nitrification (2-step Mantis). The one-dimensional clarifier models can be reactive or non-reactive with the double exponential Tackacs settling function. All GPS-X models are open-code, and available to be edited. Additional model utilities assist in the characterization of influents (Influent Advisor) and edition of the biokinetic models in the Petersen matrix format (Model Developer).

Both steady-state and dynamic simulations can be run in GPS-X. The program provides flexible data input and graphical output features (graphs, data files, etc.) and communication with other programs (e.g. Excel, MATLAB). The link to MATLAB is used for design and simulation of advanced model based control systems. Powerful numerical methods comprise 8 integration routines, 2 steady-state routines and iterative flow-loop (recycle) solving and optimization.

GPS-X creates reports directly in the Excel format, including images, graphs, parameter settings, and simulation results. Either the default or customized report format can be used for each project.

A comprehensive list of the GPS-X application references is available at the Hydromantis web page (www.hydromantis.com).

5.7.4 SIMBA (IFAK, Germany)

The SIMBA simulation platform is based on MATLAB and Simulink and uses the mathematical functionality and the graphical capabilities of both tools. This ensures an open structure and high flexibility of the program (which can be used without any prior knowledge in programming). In a SIMBA library, the main components of activated sludge systems are available in the form of blocks. The library also contains other blocks, such as control and display blocks (e.g. Sankey diagrams) as well as some of the most frequently required SIMULINK blocks.

The model foundations used by SIMBA, e.g. the biokinetic models (ASM1, ASM2d, ASM3), are provided in open format, using a symbolic notation. These can be accessed and modified using a model editor. Each modelled fraction is described in the model editor in such a way that all related SIMBA blocks know the relevant properties of this fraction. Since SIMBA is an open modelling system, also new models and functions can be added by the user. The sensitivity analysis and calibration tool provides an automated model calibration procedure.

SIMBA supports the static design of WWTPs, e.g. according to the German A131 guidelines, by dedicated tools (activated sludge model, load scenarios, synthetic influent data), preparing for detailed planning with dynamic simulation. The SIMBA simulation platform allows for the holistic consideration of sewer system, wastewater treatment plant, sludge treatment and rivers.

A comprehensive list of the Simba application references is available at the program web page (simba.ifak.eu).

5.7.5 STOAT (WRc, UK)

In STOAT (stands for Sewage Treatment Operation Analysis over Time), models for all common wastewater treatment processes are available, both on the wastewater and sludge treatment sides. The biokinetic models include the common IWA models ASM1, ASM2d and ASM3, as well as various extensions of these models, and the Takacs model for the clarifier. Specific activated sludge systems, such as oxidation ditches and SBRs can be modelled. The program contains sensitivity analysis, model calibration and optimisation routines as well as two types of controllers (PID and ladder logic).

A basic model of the urban environment for studying the sewer-sewage works interaction is included in STOAT, but two external interfaces are also provided - one for the OpenMI communication protocol and a second for a proprietary, but documented, DHI interface. In addition, the program has support for file exchange with Wallingford Software's sewer and river models.

The technical references concerning the use of STOAT include Stokes *et al.* (1997), Smith *et al.* (1998), Stokes *et al.* (2000a, 2000b) and Ashley *et al.* (2002).

5.7.6 WEST (MOSTforWATER, Belgium)

WEST offers a platform for dynamic modelling and simulation of wastewater treatment plants and other water quality systems, such as rivers, sewers and urban catchments. The layout of a studied system (unlimited size and complexity) is set up in a graphical way by selecting the different units from a process library. Various control strategies can be implemented without additional coding by selecting and placing sensor and controller units within the layout.

An extensive model library is available including all the most common models for the activated sludge process and clarifiers. WEST uses the Model Specification Language (MSL), which enables the modeller to adapt and extend the available models or to implement new models. In addition, a Gujer (Petersen) matrix editor is available in order to facilitate modifications in the biokinetic models.

TAB-separated files (e.g. imported from Excel) can be used as input for the simulations. The plot environment offers flexibility in creating visually appealing plots of simulation results (a comparison with the experimental data is possible). For model calibration, a number of additional modules are available to automate several tasks, such as sensitivity analysis, parameter estimation, Monte Carlo simulation and scenario analysis. Furthermore, WEST has an open structure that allows for integration with supervisory and data management systems (SCADA, GIS, etc.), and ultimately implementation of an operator decision support system.

A list of publications describing WEST applications in design, optimization, automation and research is available on the web page of MOSTforWATER (www. mostforwater.com).

The following product suites are available:

- WESTforDESIGN - allows to validate of design options and evaluate of different plant layouts in dynamic conditions. This is done by running scenarios, e.g. for high load vs. low load conditions, and evaluating the effect of complex control strategies.
- WESTforOPERATORS - allows to perform both short-term (e.g. storm events) and long-term (e.g. consistent nutrient removal) evaluations of

WWTPs to assist operational decisions. The evaluations are done for bottleneck identification, running scenarios for specific influent and operational conditions, and for cost evaluation.

- This approach makes it possible, for operators, to re-use the modelling efforts of their consultants in a flexible and customizable tool. The tool is most useful to improve understanding of the WWTP and hence for operator training.
- WESTforOPTIMIZATION - allows to optimize the wastewater treatment processes thanks to a flexible and open model structure (any model in the model library can be changed without limitations) in combination with additional modules for easier calibration (uncertainty and sensitivity analysis, automatic parameter estimation, scenario analysis).
- WESTforAUTOMATION - allows to integrate modelling and simulation in custom applications, and link WEST with SCADA systems or other modelling software (e.g. Matlab, CFD, Mike Urban, etc.).

Table 5.1. Comparison of the features of most common simulation platforms

Feature	Simulation platform				
	BioWin	GPS-X	SIMBA	STOAT	WEST
Model building	Simple (based on a graphical interface).	Simple (based on a graphical interface).	Little more complicated (based on a graphical interface).	Simple (based on a graphical interface).	Simple (based on a graphical interface).
Activated sludge models	General biokinetic model and ASM family models.	ASM family, "New General".	ASM family, ASM3P.	ASM family, own model based on the BOD balance.	ASM family, ASM3P, TUDP.
Secondary settler models	Three kinds of models including the model of Takacs et al. (1991) (also reactive).	Two kinds of models including the model of Takacs et al. (1991) (also reactive).	Three kinds of models including the model of Takacs et al. (1991) (no reactive).	Model of Takacs et al. (1991) (no reactive).	Three kinds of models including the model of Takacs et al. (1991) (also reactive).
Primary settler models	Two kinds of models including a modified model of Takacs et al. (1991).	Two kinds of models including the model of Takacs et al. (1991) (also reactive).	Simplified model of Otterpohl and Freund (1992).	Model of Lessard and Beck (1993) with two biochemical processes.	Two kinds of models including the model of Lessard and Beck (1994) with two biochemical processes.
Anaerobic digestion models	General biokinetic model and ADM1.	Simplified and ADM1.	Two simplified models and ADM1.	Simple and more complex models for mesophillic digestion.	Three kinds including ADM1.
Reactor hydro-dynamic model	CSTR or a series of CSTRs.	CSTR or a series of CSTRs.	CSTR or a series of CSTRs.	CSTR or a series of CSTRs.	CSTR or a series of CSTRs.

(continued on next page)

Table 5.1 (cont.) Comparison of the features of most common simulation platforms

Feature	Simulation platform				
	BioWin	GPS-X	SIMBA	STOAT	WEST
Chemical P precipitation	In any point (the model is based on the chemical equilibrium).	In any point (the kinetic model is adopted from ASM2).	In any point (the kinetic model is adopted from ASM2).	In any point (the model is based on the chemical equilibrium and the kinetics is adopted from ASM2).	In any point (the model is based on the chemical equilibrium and the kinetics is adopted from ASM2).
Sewer system model	Not included, but can be linked with such programs.	Very simple.	Very complex model included.	Simplified, but can be linked with such programs.	Simplified model (KOSIM) included, can be linked with such programs.
Petersen matrix editor	Yes.	Yes.	Yes.	No.	Yes.
Introduction of input data	Directly in the program or import from MS Excel.	Directly in the program or text files (can be copied in the input data module).	Directly in the program, text files, import from MS Excel or data bases.	Directly in the program or import from MS Excel.	Directly in the program, import from MS Excel or data bases.

Table 5.1 (cont.) Comparison of the features of most common simulation platforms

Feature	Simulation platform				
	BioWin	GPS-X	SIMBA	STOAT	WEST
Presentation of output data	Several kinds of graphs including Sankey's diagrams. A special module for formatting graphs is included.	Several kinds of graphs. Formatting graphs is not possible.	One type of graphs (SIMBA Monitor), Sankey's diagrams, better visualisation is possible in MATLAB.	Tabular form or X-Y graphs.	Several kinds of graphs. A special module for formatting graphs is included.
Sensitivity analysis	No. The results of many simulations can be presented in one graph.	Yes.	Yes.	Yes.	Yes.
Parameter estimation	No.	Yes.	Yes.	Yes.	Yes.
Controllers	Several. Any measured variable can be controlled.	Several. Any measured variable can be controlled. An advanced control module is included.	Two kinds of controllers. Additional controllers can be designed in MATLAB.	Two kinds of controllers. No link to MATLAB or similar programs.	Several. User's own control alghortims can be designed.

References

Abbaspour, K.C., Schulin, R., Schläppi, E. and Flühler, H. (1996) A Bayesian approach for incorporating uncertainty and data worth in environmental projects. *Environ. Model. Assess.* **1**, 151–158.

Abrishamchi, A., Tajrishy, M. and Shafieian, P. (2005) Uncertainty analysis in QUAL2E model of Zayandeh-Rood River. *Water Environ. Res.* **77**, 279–286.

Abu-ghararah, Z.H. and Randall, C.W. (1991) The effect of organic compounds on biological phosphorus removal. *Water Sci. Technol.* **23**(4/6), 585–594.

Ahn Y. (2006) Sustainable nitrogen elimination biotechnologies: A review. *Process Biochem.* **41**, 1709–1721.

Ahnert, M., Blumensaat, F., Langergraber, G., Alex, J., Woerner, D., Frehmann, T., Halft, N., Hobus, I., Plattes, M., Spering, V. and Winkler, S. (2007) Goodness-of-fit measures for numerical modelling in urban water management – a summary to support practical applications. In *Proc. 10th IWA Specialized Conference on Design, Operation and Economics of Large Wastewater Treatment Plants (Posters)*, Vienna, 9–13 September, 69–72.

Ahnert, M., Tranckner, J., Gunther, N., Hoeft, S. and Krebs, P. (2009) Model-based comparison of two ways to enhance WWTP capacity under stormwater conditions. *Water Sci. Technol.* **60**(7), 1875–1883.

Akca, L., Kinaci, C. and Karpuzcu, M. (1993) A model for optimum design of activated sludge plants. *Water Res.* **27**, 1461–1468.

Albertson, O.E. and DiGregorio, D. (1975) Biologically mediated inconsistencies in aeration equipment performance. *J. Wat. Pollut. Cont. Fed.* **47**, 977–988.

Albertson, O.E. and Stensel, H.D. (1994) Aerated anoxic biological NdeN process. *Water Sci. Technol.* **29**(7), 167–176.

Alex, J., Kolisch, G. and Krause, K. (2002) Model structure identification for wastewater treatment simulation based on computational fluid dynamics. *Water Sci. Technol.* **45**(4–5), 325–334.

Alleman, J.E. and Prakasam, T.B.S. (1983) Reflections on seven decades of activated sludge history. *J. Water Pollut. Control Fed.* **55**, 436–443.

Andreasen, K. and Nielsen, P.H. (2000) Growth of Microthrix parvicella in nutrient removal activated sludge plants: Studies of in situ physiology. *Water Res.* **34**, 1559–1569.

Andreottola, G., Bortone, G. and Tilche, A. (1997) Experimental validation of a simulation and design model for nitrogen removal in sequencing batch reactors. *Water Sci. Technol.* **35**(1), 113–120.

Andreottola, G., Engl, K., Foladori, P. and Hilber, C. (2003) Optimisation of nitrogen removal in a full-scale intermittently aerated process. In *Proc. 9th IWA Specialized Conference on Design, Operation and Economics of Large Wastewater Treatment Plants*, part I, Prague, 1–4 September, 151–158.

Andrews, J.F. (1968) A mathematical model for the continuous culture of microorganisms utilizing inhibitory substrates. *Biotechnol. Bioeng.* **10**, 707–723.

Andrews, J.F. (1974) Dynamic models and control strategies for wastewater treatment processes. *Water Res.* **8**, 261–289.

Andrews, J.F. (1992) Mathematical modelling and computer simulation. In *Dynamics and Control of the Activated Sludge Process*, J.F. Andrews (ed.). Technomic Pub. Co., Lancaster, PA (USA), 23–66.

Andrews, J.F. (1993) Mathematical modeling and simulation of wastewater treatment processes. *Water Sci. Technol.* **28**(11/12), 141–150.

Anthonisen, A.C., Loehr, R.C., Prakasam, T.B. and Srinath, E.G. (1976) Inhibition of nitrification by ammonia and nitrous acid. *J. Water Pollut. Control Fed.* **48**, 835–852.

Antoniou, P., Hamilton, J., Koopman, B., Jain, R. and Holloway, B. (1990) Effect of temperature and pH on the effective maximum specific growth rate of nitrifying bacteria. *Water Res.* **24**, 97–102.

Ardern, E. and Lockett W.T. (1914) Experiments on the oxidation of sewage without the aid of Filters. *J. Soc. Chem. Ind.* **33**, 523–539.

Argaman, Y. and Adams Jr., C.E. (1977) Comprehensive temperature model for aerated biological systems. *Prog. Wat. Tech.* **9**, 397–409.

Aris, R. (1994) Mathematical Modelling Techniques. Courier Dover Publications, Mineola, NY (USA).

Aris, R. (1999) Mathematical Modeling: a Chemical Engineer's Perspective. Academic Press, New York.

Artan, N., Wilderer, P., Orhon, D., Tasli, R. and Morgenroth, E. (2002) Model evaluation and optimisation of nutrient removal potential for sequencing batch reactors. *Water SA* **28**, 423–432.

ASCE (1997) Standard Guidelines for In-Process Oxygen Transfer Testing. American Society of Civil Engineers, New York, NY.

Ashley, R.M., Dudley, J., Vollertsen, J., Saul, A.J., Jack, A. and Blanksby, J.R. (2002) The effect of extended in-sewer storage on wastewater treatment plant performance. *Water Sci. Technol.* **45**(3), 239–246.

Attir, U. and Denn, M. (1978) Dynamics and control of the activated sludge wastewater processes. *AIChE J.* **24**, 693–698.

Avcioglu, E., Karahan-Gul, O. and Orhon, D. (2003) Estimation of stoichiometric and kinetic coefficients of ASM3 under aerobic and anoxic conditions via respirometry. *Water Sci. Technol.* **48**(8), 185–194.

Babbitt, H.E. (1922) Sewerage And Sewage Treatment. John Wiley, New York.

Bailey, W., Tesfaye, A., Dakita, J., McGrath, M., Daigger, G., Benjamin, A. and Sadick, T. (1998) Large-scale nitrogen removal demonstration at the Blue Plains Wastewater Treatment Plant using post-denitrification with methanol. *Water Sci. Technol.* **38**(1), 79–86.

Balakrishnan S. and Eckenfelder, W.W. (1970) Nitrogen removal by modified activated sludge processes. *J. San. Eng. Div. ASCE* **96**, 501–512.

Barker P.S. and Dold P.L. (1995) COD and nitrogen mass balances in activated-sludge systems. *Water Res.* **29**, 633–643

Barker, P.S. and Dold, P.L. (1996a). Sludge production and oxygen demand in nutrient removal activated sludge systems. *Water Sci. Technol.* **34**(5/6), 43–50.

Barker, P.S. and Dold, P.L. (1996b). Denitrification behaviour in biological excess phosphorus removal activated sludge systems. *Water Res.* **30**, 769–780.

Barker, P.S. and Dold, P.L. (1997a) General model for biological nutrient removal activated-sludge systems: model presentation. *Water Environ. Res.* **69**, 969–984.

Barker, P.S. and Dold, P.L. (1997b) General model for biological nutrient removal activated-sludge systems: model application. *Water Environ. Res.* **69**, 985–991.

Barnard, J.L. (1973) Biological denitrification. *Wat. Pollut. Control* **72**, 705–720.

Barnard, J.L. (1974) Cut P and N without chemicals. *Water Wastes Eng.*, Part 1, **11**(7), 33–36; Part 2, **11**(8), 41–43.

Barnard, J.L. (1975) Biological nutrient removal without the addition of chemicals. *Water Res.* **9**, 485–490.

Barnard, J.L. (1976) A review of biological phosphorous removal in activated sludge process. *Water SA* **2**, 136–144.

Barnard, J.L. (1998) The development of nutrient-removal processes. *Water Environ. J.* **12**, 330–337.

Barnard, J.L. (2006) Biological nutrient removal: where we have been, where we are going? In *Proc. 79th WEF WEFTEC 2006*, Dallas (USA), 21–25 October, 1–25.

Barnard, J.L. and Abraham, K. (2006) Key features of successful BNR operation. *Water Sci. Technol.* **53**(12), 1–9.

Batchelor, B. (1982). Kinetic analysis of alternative configurations for single-sludge nitrification denitrification. *J. Water Pollut. Cont. Fed.* **54**, 1493–1504.

Batista, J.R., Becker, J.R., Unz, R.F. and Johnson, W. (2005) Phosphorus release in the secondary clarifier of a full-scale biological phosphorus removal system. In *Proc. 78th WEF WEFTEC 2005*, 533–547.

Beccari, M., Dionisi, D., Giuliani, A., Majone, M. and Ramadori, R. (2002) Effect of different carbon sources on aerobic storage by activated sludge. *Water Sci. Technol.* **45**(6), 157–168.

Beccari, M., Passino, R., Ramadori, R. and Tandoi, V. (1983) Kinetics of dissimilatory nitrate and nitrite reduction in suspended growth culture. *J. Water Pollut. Cont. Fed.* **55**, 58–64.

Beck, M.B. (1981) Operational estimation and prediction of nitrification dynamics in the activated sludge process. *Water Res.* **15**, 1313–1330.

Beck, M.B. (1989) System identification and control. In *Dynamic Modeling and Expert Systems in Wastewater Engineering*, G.G. Patry and D. Chapman (eds.), Lewis Publishers, Chelsea, MI (USA), 261–323.

Bennet, G.F., Kempe, L.L. (1965) Oxygen transfer in biological systems. In *Proc. 20th Purdue Ind. Waste Conf.*, Purdue University, 435–449.

Beun, J.J., Paletta, F., van Loosdrecht, M.C.M. and Heijnen, J.J. (2000a) Stoichiometry and kinetics of poly-beta-hydroxybutyrate metabolism in aerobic, slow growing, activated sludge cultures. *Biotechnol. Bioeng.* **67**, 379–389.

Beun, J.J., Verhoef, E.V., van Loosdrecht M.C.M. and Heijnen, J.J. (2000b) Stoichiometry and kinetics of poly-beta-hydroxybutyrate metabolism under denitrifying conditions in activated sludge cultures. *Biotechnol. Bioeng.* **68**, 496–507.

Bidstrup, S.M. and Grady, C.P.L. Jr. (1988) SSSP: Simulation of single-sludge processes. *J. Water Pollut. Control Fed.* **60**, 351–361.

Bingley, M. and Upton, J. (2000) The good, the bad and the ugly: process modeling for design. *Water Sci. Technol.* **41**(9), 171–178.

Black, E.B. and Phelps, E.B. (1914) Brooklyn sewage experiment station. *Eng. New-Rec.* **74**, 826.

Blok, J. and Struys, J. (1996) Measurement and validation of kinetic parameter values for prediction of biodegradation rates in sewage treatment. *Ecotoxicol. Environ. Safety* **33**, 217–227.

Bocken, S.M., Braae, M. and Dold, P.L. (1989) Dissolved oxygen control and oxygen utilization rate estimation: extension of the Holmberg/Olsson method. *Water Sci. Technol.* **21**(10/11), 1197–1208.

Boero, V.J., Eckenfelder, W.W. Jr. and Bowers, A.R. (1991) Soluble microbial product formation in biological systems. *Water Sci. Technol.* **23**(2), 1067–1076.

Bohnke, B. (1977) Das Adsorptions-Belebungsverfahren, *Korrespondenz Abwasser* **24**, 121–127. (in German).

Boon, B. and Laudelout, H. (1962) Kinetics of nitrite oxidation by Nitrobacter winogradskyi. *Biochem. J.* **85**, 440–447.

Bortone, G., Libelli, S.M., Tilche, A. and Wanner, J. (1999) Anoxic phosphate uptake in the DEPHANOX process. *Water Sci. Technol.* **40**(4/5), 177–185.

Bortone, G., Malaspina, F., Stante, L. and Tilche, A. (1994) Biological nitrogen and phosphorus removal in an anaerobic/anoxic sequencing batch reactor with separated biofilm nitrification. *Water Sci. Technol.* **30**(6), 303–313.

Bortone, G., Saltarelli, R., Alonso, V., Sorm, R., Wanner, J. and Tilche, A. (1996) Biological anoxic phosphorus removal – The DEPHANOX process. *Water Sci. Technol.* **34**(1/2), 119–128.

Boyle, W.C., Hellstrom, B.G. and Ewing, L. (1989) Oxygen transfer efficiency measurements using off-gas techniques. *Water Sci. Technol.* **21**(10/11), 1295–1300.

Brands, E., Liebeskind, M. and Dohmann, M. (1994) Parameters for dynamic simulation of wastewater treatment plants with high-rate and low-rate activated sludge tanks. *Water Sci. Technol.* **30**(4), 211–214.

Bratby, J., Johnson, L., Parker, D. and van Derveer, B. (2001) Are activated sludge simulation models useful to operators? In *Proc. 74th WEF WEFTEC 2001*, Atlanta (Georgia), 13–17 October [CD-Rom].

Bratby, J. and Parker, D. (2009) Accurately modeling the effect of dissolved oxygen on nitrification. In *Proc. WEF Spec. Conf. "Nutrient Removal 2009"*, Washington (USA), 28 June – 1 July, 868–879.

Bratley, P., Fox, B.L. and Schrage, L.E. (1987) A Guide to Simulation. 2nd Edition. Springer-Verlag, New York.

Brdjanovic, D., van Loosdrecht M.C.M., Versteeg, P., Hooijmans, C.M., Alaerts, G.J and Heijnen, J.J. (2000) Modeling COD, N and P removal in a full scale WWTP Haarlem Waarderpolder. *Water Res.* **34**, 846–858.

Brenner, A. (2000) Modeling of N and P transformations in an SBR treating municipal wastewater. *Water Sci. Technol.* **42**(1/2), 55–63.

Brezonik, P.L. (1994) Chemical Kinetics and Process Dynamics in Aquatic Systems. Lewis Publishers, Boca Raton, FL.

Brock, T.D. and Madigan, M.T. (1991) Biology of Microorganisms. Prentice Hall, New Jersey.

Broda, E. (1977) Two kinds of lithotrophs missing in nature. *Z. Allg. Mikrobiol.* **17**, 491–493.

Bround, S. and Scherfig, J. (1994) Dynamic temperature changes in nutrient removal plants. *Water Sci. Technol.* **30**(2), 205–208.

Brouwer, H., Klapwijk, A. and Keesman K.J. (1998) Identification of activated sludge and wastewater characteristics using respirometric batch-experiments. *Water Res.* **32**, 1240–1254.

Brun, R., Kuhni, M., Siegrist, H., Gujer, W. and Reichert, P. (2002) Practical identifiability of ASM2d parameters – systematic selection and tuning of parameter subsets. *Water Res.* **36**, 4113–4127.

Buhr, H., Lee, M., Narayanan, B., Leveque, E., Pai, P. Johnson, W. and Dold, P. (2001) Using computer simulation to optimize BNR design. In *Proc. 74th WEF WEFTEC 2001*, Atlanta (Georgia), 13–17 October [CD-Rom].

Bundgaard, E. and Petersen, J. (1991) Techniques for and experience with biological nutrient removal – Danish long-term operating and optimization experience. *Water Sci. Technol.* **24**(10), 211–216.

Bundgaard, E., Andersen, K.L. and Petersen, G. (1989) Bio-Denitro and Bio-Denipho systems – experiences and advanced model development – the Danish systems for biological N-removal and P-removal. *Water Sci. Technol.* **21**(12), 1727–1730.

Burger, R. and Wendland, W.L. (2001) Sedimentation and suspension flows: Historical perspective and some recent developments. *J. Eng. Math.* **41**, 101–116.

Burkhead, C.E. and McKinney, R.E. (1968) Application of complete-mixing activated sludge design equations to industrial wastes. *J. Wat. Pollut. Control Fed.* **40**, 557–570.

Busby, J.B. and Andrews, J.F. (1975) Dynamic modeling and control strategies for the activated sludge process. *J. Water Pollut. Control Fed.* **47**, 1055–1080.

Busch, A.W. and Kalinske, A.A. (1956) Utilization of the kinetics of the activated sludge process for process and equipment design. In *Biological treatment of sewage and industrial wastes*, J. McCabe and W.W. Eckenfelder (eds), Reinhold Publishing Corp., New York, **1**, 277–283.

Busch, A.W. (1984) Chemical reactor design theory and biological treatment of industrial-wastes – is there a gap. *J. Water Pollut. Control Fed.* **56**, 215–218.

Buswell, A.M. (1923) Activated sludge studies 1920-1922. Bulletin No. 18. State of Illinois, Division of the State Water Survey, Springfield.

Bye, C.M. and Dold, P.L. (1998) Sludge volume index settleability measures: Effects of solids characteristics and test parameters. *Water Environ. Res.* **70**, 87–93

Bye, C.M. and Dold, P.L. (1999) Evaluation of correlations for zone settling velocity parameters based on sludge volume index type measured and consequences in settling tank design. *Water Environ. Res.* **71**, 1333–1344.

Carpenter, W.T. and Horowitz, M.P. (1915) Some remarks on activated sludge. *Am. J. Public Health* **6**, 1218–1223.

Carrette, R., Bixio, D., Thoeye, C. and Ockier, P. (2001) Full-scale application of the IAWQ ASM No. 2d model. *Water Sci. Technol.* **44**(2/3), 17–24.

Carucci, A., Dionisi, D., Majone, M., Rolle, E. and Smurra, P. (2001) Aerobic storage by activated sludge on real wastewater. *Water Res.* **35**, 3833–3844.

Carucci, A., Rolle, E. and Smurra, P. (1999) Management optimisation of a large wastewater treatment plant. *Water. Sci. Technol.* **39**(4), 129–136.

Cech, J.S. and Hartman, P. (1993) Competition between polyphosphate and polysaccharide accumulating bacteria in enhanced biological phosphate removal systems. *Water Res.* **27**, 1219–1225.

Cech, J.S., Chudoba, J. and Grau, P. (1985) Determination of kinetic constants of activated sludge microorganisms. *Water Sci. Technol.* **17**(2/3), 259–272.

Chambers, B. and Jones, G.L. (1988) Optimisation and upgrading of activated sludge plants by efficient process design. *Water Sci. Technol.* **20**(4/5), 121–132.

Chandran, K. and Smets, B.F. (2000a) Applicability of two-step models in estimating nitrification kinetics from batch respirograms under different relative dynamics of ammonia and nitrate oxidation. *Biotechnol. Bioeng.* **68**, 54–64.

Chandran, K. and Smets, B.F. (2000b) Single-step nitrification models erroneously describe batch ammonia oxidation profiles when nitrate oxidation becomes rate limiting. *Biotechnol. Bioeng.* **68**, 396–406.

Chandran, K. and Smets, B.F. (2005) Optimizing experimental design to estimate ammonia and nitrite oxidation biokinetic parameters from batch respirograms. *Water Res.* **39**, 4969–4978.

Chapman, D.T. (1983) The influence of process variables on secondary clarification. *J. Water Pollut. Control Fed.* **55**, 1425–1434.

Charley, R.C., Hooper, D.G. and McLee, A.G. (1980) Nitrification kinetics in activated sludge at various temperatures and dissolved oxygen concentrations. *Water Res.* **14**, 1387–1396.

Chen, C.Y., Roth, J.A. and Eckenfelder, W.W. Jr. (1980) Response of dissolved oxygen to changes in influent organic loading to activated sludge systems. *Water Res.* **14**, 1449–1457.

Chen, G.H., Yip, W.K., Mo, H.K. and Liu, Y. (2004) Effect of sludge fasting/feasting on growth of activated sludge cultures. *Water Res.* **35**, 1029–1037.

Cheng, C.Y. and Ribarova, I. (1999) Activated sludge system modelling and simulations for improving the effluent water quality. *Water Sci. Technol.* **39**(8), 93–98.

Cho, S.H., Colin, F., Sardin, M. and Prost, C. (1993) Settling velocity model of activated sludge. *Water Res.* **27**, 1237–1242.

Choubert, J.-M., Marquot, A., Stricker, A.-E., Racault, Y., Gillot, S. and Heduit, A. (2009) Anoxic and aerobic values for the yield coefficient of the heterotrophic biomass: Determination at full-scale plants and consequences on simulations. *Water SA* **35**, 103–109.

Chudoba, P., Capdeville, B. and Chudoba, J. (1992) Explanation of biological meaning of the S_0/X_0 ratio in batch cultivation. *Water Sci. Technol.* **26**(3/4), 743–751.

Cinar, O., Daigger, G.T. and Graef, S.P. (1998) Evaluation of IAWQ Activated Sludge Model No.2 using steady-state data from four full-scale wastewater treatment plants. *Water Environ. Res.* **70**, 1216–1224.

Clara, M., Kreuzinger, N., Strenn, B., Gans, O. and Kroiss, H. (2005) The solids retention time – a suitable design parameter to evaluate the capacity of wastewater treatment plants to remove micropollutants. *Water Res.* **39**, 97–106.

Clayton, J.A., Ekama, G.A., Wentzel, M.C. and Marais, G.v.R. (1991) Denitrification kinetics in biological nitrogen and phosphorus removal activated sludge systems treating municipal wastewaters. *Water Sci. Technol.* **23**(4/6), 1025–1035.

Clifft, R.C. and Andrews, J.F. (1981) Predicting the dynamics of oxygen utilization in the activated sludge process. *J. Wat. Pollut. Cont. Fed.* **53**, 1219–1232.

Clifft, R.C. and Barnett, M.W. (1988) Gas transfer kinetics in oxygen activated sludge. *J. Env. Eng. ASCE* **114**, 415–432.

Coe, H.S. and Clevenger, G.H. (1916) Methods for determining the capacities of slime-settling tanks. *Trans. AIME* **55**, 356–384.

Coelho, M.A.Z., Russo, C. and Araujo, O.Q.F. (2000) Optimization of a sequencing batch reactor for biological nitrogen removal. *Water Res.* **34**, 2809–2817.

Coen, F., Petersen, B., Vanrolleghem, P.A., Vanderhaegen, B. and Henze, M. (1998) Model-based characterisation of hydraulic, kinetic and influent properties of an industrial WWTP. *Water Sci. Technol.* **37**(12), 317–326.

Coen, F., Vanderhaegen, B., Boonen, I., Vanrolleghem, P.A. and van Meenen, P. (1997) Improved design and control of industrial and municipal nutrient removal plants using dynamic models. *Water Sci. Technol.* **35**(10), 53–61.

Coen, F., Vanderhaegen, B., Boonen, I., Vanrolleghem, P.A., van Eyck, L. and van Meenen, P. (1996) Nitrogen removal upgrade of a wastewater treatment plant within existing reactor volumes: A simulation supported scenario analysis. *Water Sci. Technol.* **34**(3/4), 339–346.

Cokgor, E.U., Sozen, S., Orhon, D. and Henze, M. (1998) Respirometric analysis of activated sludge behaviour – part I. Assessment of the readily biodegradable substrate. *Water Res.* **32**, 461–475.

Comeau, Y., Hall, K.J., Hancock, R.E.W. and Oldham, W.K. (1986) Biochemical model for enhanced biological phosphorus removal. *Water Res.* **20**, 1511–1521.

Concha, F. and Bustos, M.C. (1987) A modification of the Kynch theory of sedimentation. *AICHE J.* **33**, 312–315.

Constantine, T., Shea, T. and Johnson, B. (2005) Newer approaches for treating return liquors from anaerobic digestion. In *Proc. IWA Spec. Conf. "Nutrient Managementin Wastewater Treatment Processes and Recycle Streams"*, Cracow (Poland), 19–21 September, 455–464.

Cooney, C.L. and Wang, D.I.C. (1971) Oxygen transfer and control. *Biotech. Bioeng. Symp. No. 2*, Wiley, New York, 63–75.

Cooper, P.F. (2001) Historical aspects of wastewater treatment. In: *Decentralised Sanitation and Reuse: Concepts, Systems and Implementation.* (Eds.) P. Lens, G. Zeeman and G. Lettinga, IWA Publishing, London, 11–38.

Cooper, P.F. and Downing, A.L. (1998) Milestones in the development of the activated-sludge process over the past eighty years. *Water Environ. J.* **12**, 303–313.

Copp, J.B. (ed.) (2002). The COST Simulation Benchmark – Description and Simulator Manual. Office for Official Publications of the European Communities, Luxembourg.

Copp, J.B. and Dold, P.L. (1998) Comparing sludge production under aerobic and anoxic conditions. *Water Sci. Technol.* **38**(1), 285–294.

Copp, J.B. and Murphy, K.L. (1995) Estimation of the active nitrifying biomass in activated sludge. *Water Res.* **29**, 1855–1862.

Crabtree, H.E. (1983) Some observations on denitrification in activated sludge final settlement tanks. *Water Pollut. Control* **82**, 315–329.

Czerwionka, K., Makinia, J., Pagilla, K. and Drewnowski, J. (2009) Transformations of dissolved and colloidal organic nitrogen in biological nutrient removal activated sludge systems. In *Proc. WEF Specialty Conference "Nutrient Removal 2009"*, Washington, DC, 28 June – 1 July, 1127–1139.

Daigger, G.T. (1995) Development of refined clarifier operating diagrams using an updated settling characteristics database. *Water Environ. Res.* **67**, 95–100.

Daigger, G.T. and Buttz, J.K. (1992) Upgrading Wastewater Treatment Plants. Technomic Pub. Co., Lancaster, PA.

Daigger, G.T. and Nolasco, D. (1995) Evaluation and design of full-scale wastewater treatment plants using biological process models. *Water Sci. Technol.* **31**(2), 245–255.

Daigger, G.T. and Roper, R.E. (1985) The relationship between SVI and activated sludge settling characteristics. *J. Water Pollut. Control Fed.* **57**, 859–866.

Daigger, G.T., Waltrip, G.D., Romm, E.D. and Morales, L.M. (1988) Enhanced secondary treatment incorporating biological nutrient removal. *J. Water Pollut. Contr. Fed.* **60**, 1833–1842.

Danckwerts, P.V. (1951) Significance of liquid film coefficients in gas absorption. *Ind. Eng. Chem.* **43**, 1460–1467.

Danckwerts, P.V. (1953) Continuous flow system – distribution residence times. *Chem. Eng. Sci.* **2**, 1–13.

Dapena-Mora, A., Van Hulle, S.W.H., Campos, J.L., Mendez, R., Vanrolleghem, P.A. and Jetten, M. (2004) Enrichment of Anammox biomass from municipal activated sludge: Experimental and modelling results. *J. Chem. Technol. Biotechnol.* **79**, 1421–1428.

De Clercq, J., Devisscher, M., Boonen, I., Vanrolleghem, P.A. and Defrancq, J. (2003). A new one-dimensional clarifier model – verification using full-scale experimental data. *Water Sci. Technol.* **47**(12), 105–112.

De Clercq, J., Nopens, I., Defrancq, J. and Vanrolleghem, P.A. (2008) Extending and calibrating a mechanistic hindered and compression settling model for activated sludge using in-depth batch experiments. *Water Res.* **42**, 781–791.

de la Sota, A., Larrea, L., Novak, L., Grau, P. and Henze, M. (1994) Performance and model calibration of R-D-N process in pilot plant. *Water Sci. Technol.* **30**(6), 355–364.

De Pauw, D.J.W. (2005) Optimal experimental design for calibration of bioprocess models: a validated software toolbox. Ph.D. Thesis, Ghent University.

DeClercq, B., Coen, F., Vanderhaegen, B. and Vanrolleghem, P.A. (1999) Calibrating simple models for mixing and flow propagation in waste water treatment plants. *Water Sci. Technol.* **39**(4), 61–69.

Dick R.I. and Young K. W. (1972) Analysis of thickening performance of final settling tanks. In *Proc. 27th Ind. Waste Conf.*, Purdue Univ. Lafayette, 25 May, 34 pp.

Dick, R.I. and Ewing, B.B. (1967) Evaluation of the activated sludge thickening theories. *J. San. Eng. ASCE* **93**, 9–29.

Diehl, S. (2001) Operating charts for continuous sedimentation I: Control of steady states. *J. Eng. Math.* **41**, 117–144.

Diehl, S. (2008) The solids-flux theory – Confirmation and extension by using partial differential equations. *Water Res.* **42**, 4976–4988.

Dionisi, D., Majone, M., Ramadori, R. and Beccari, M. (2001) The storage of acetate under anoxic conditions. *Water Res.* **35**, 2661–2668.

Dionisi, D., Renzi, V., Majone, M., Beccari, M. and Ramadori, R. (2004) Storage of substrate mixtures by activated sludges under dynamic conditions in anoxic or aerobic environments. *Water Res.* **38**, 2196–2206.

Dipankar S. and Randall, C.W. (1992) General activated sludge model for biological nitrogen and excess phosphorus removal. In *Design and Retrofit of Wastewater Treatment Plants for Nutrient Removal*. C.W. Randall, J.L. Barnard and H.D. Stensel (eds.). Technomic Publishing Co., Lancaster, PA (USA), 311–334.

Diplas, P. and Papanicolaou, A.N. (1997). Batch analysis of slurries in zone settling regime. *J. Env. Eng. ASCE* **123**, 659–667.

Dircks, K., Beun, J.J., van Loosdrecht, M.C.M., Heijnen, J.J. and Henze, M. (2001b) Glycogen metabolism in aerobic mixed cultures. *Biotechnol. Bioeng.* **73**, 85–94.

Dircks, K., Henze, M., van Loosdrecht, M.C.M., Mosbaek, H. and Aspegren, H. (2001a) Storage and degradation of poly-beta-hydroxybutyrate in activated sludge under aerobic conditions. *Water Res.* **35**, 2277–2285.

Dircks, K., Pind, P.F., Mosbaek, H. and Henze, M. (1999) Yield determination by respirometry. The possible influence of storage under aerobic conditions in activated sludge. *Water SA* **25**, 69–74

Dochain, D. and Vanrolleghem, P.A. (2001) Dynamical modelling and estimation in wastewater treatment processes. IWA Publishing, London.

Dold, P. and Fairlamb, M. (2001) Estimating oxygen transfer K_La, SOTE and air flow requirements in fine bubble diffused air systems. In *Proc. 74th WEF WEFTEC 2001*, Atlanta, 13–17 October [CD-Rom].

Dold, P.L. (1990) A general activated sludge model incorporating biological excess phosphorus removal. *Paper presented at CSCE Annual Conference*, Hamilton, Ontario (Canada), 16–18 May.

Dold, P.L. and Marais, G.v.R. (1986) Evaluation of the general Activated Sludge Model proposed by the IAWPRC Task Group. *Water Sci. Technol.* **18**(6), 63–89.

Dold, P.L., Ekama, G.A. and Marais, G.v.R. (1980) The activated sludge process. Part 1 - A general model for the activated sludge process. *Prog. Water Technol.* **12**(6), 47–77.

Dold, P.L., Wentzel, M.C., Billing, A.E., Ekama, G.A. and Marais, G.v.R. (1991) Activated sludge simulation programs: Nitrification and nitrification/denitrification systems (Version 1.0). Water Research Commission, Pretoria.

Downing, A.L. and Cooper, P.F. (1998) European Practices. In *Activated sludge process design and control: theory and practice*. (eds.) W.W. Eckenfelder and P. Grau, Technomic Pub. Co., Lancaster, PA (USA), 99–334.

Downing, A.L., Painter, H.A. and Knowles, G. (1964) Nitrification in the activated sludge process. *J. Proc. Inst. Sew. Purif.* **63**, 130–158.

Drtil, M., Nemeth, P. and Bodik, I. (1993) Kinetic constants of nitrification. *Water Res.* **27**, 35–39.

Dueholm, T.D., Andreasen, K. and Nielsen, P.H. (2001) Transformation of lipids in activated sludge. *Water Sci. Technol.* **43**(1), 165–172.

Dupont, R. and Henze, M. (1992) Modelling of the secondary clarifier combined with the Activated Sludge Model No. 1. *Water Sci. Technol.* **25**(6), 285–300.

Dupont, R. and Sinkjaer, O. (1994) Optimization of waste-water treatment plants by means of computer models. *Water Sci. Technol.* **30**(4), 181–190.

Eckenfelder, W.W. (1966) Industrial Water Pollution Control. 1st ed., McGraw-Hill, New York.

Eckenfelder, W.W. (1989) Industrial Water Pollution Control. 2nd ed., McGraw-Hill, New York.

Eckenfelder, W.W. and Ford, D.L. (1970) Water Pollution Control. The Pemberton Press, Austin, Texas.

Eckenfelder, W.W. and O'Connor, D.J. (1954) Aerobic biological treatment of organic wastes. In *Proc. 9th Ind. Waste Conf.*, Purdue Univ. Lafayette, 512–530.

Eckenfelder, W.W. Jr. (1961) Theory and practice of activated-sludge process modifications. *Water Sewage Works* **108**, 145–150.

Eckenfelder, W.W. Jr. (1966) Industrial Water Pollution Control. McGraw-Hill, New York.

Eckenfelder, W.W. Jr. and O'Connor, D.J. (1961) Biological Waste Treatment. Pergamon Press, London.

Eckenfelder, W.W., Goronszy, M.C. and Watkin, A.T. (1985) Comprehensive activated sludge design. In *Mathematical Models in Biological Waste Water Treatment*. S.E. Jorgensen and M.J. Gromiec (eds.), *Developments in Environmental Modeling No. 7*, Elsevier, Amsterdam, 95–132.

Eckenfelder, WW, Jr. and Porges, N. (1957) Activity of microorganisms in organic waste disposal. IV. Bio-calculations. *Appl. Microbiol.* **5**, 180–187.

Edmondson, H. and Goodrich, R. (1943) The cyclo-nitrifying filter. *J. Inst. Sew. Purif.* **1**, 57–92.

Edmondson, H. and Goodrich, R. (1947) Experimental work leading to the increased efficiency in the bio-aeration process of sewage purification and further experiments on nitrification and re-circulation in percolating filters. *J. Inst. Sew. Purif.* **2**, 17–43.

Edwards, V.H. (1970) The influence of high substrate concentrations on microbial kinetics. *Biotechnol. Bioeng.* **12**, 679–712.

Egli, K., Fanger, U., Alvarezz, P.J.J., Siegrist, H., van der Meer, J.R. and Zehnder, A.J.B. (2001) Enrichment and characterization of an anammox bacterium from a rotating biological contactor treating ammonium rich leachate. *Arch. Microbiol.* **175**, 198– 207.

Ekama G.A. and Marais, G.v.R. (1979) Dynamic behavior of the activated-sludge process. *J. Wat. Pollut. Cont. Fed.* **51**, 534–556.

Ekama, G.A., Dold, P.L. and Marais, G.v.R. (1986) Procedures for determining influent COD fractions and the maximum specific growth rate of heterotrophs in activated sludge systems. *Water Sci. Technol.* **18**(6), 91–114.

Ekama, G.A. and Marais, G.V. (1986) Sludge settleability and secondary settling-tank design procedures. *Water Pollut. Control* **85**, 101–113.

Ekama, G.A., Barnard, JL., Gunthert, F.W., Krebs, P., McCorquodale, J.A., Parker, D.S. and Wahlberg, E.J. (1997) Secondary settling tanks: theory, modelling design and operation. IWAQ Scientific and Technical Report No. 6, IWA Publishing, London.

Ekama, G.A. and Wentzel, M.C. (1999a) Difficulties and developments in biological nutrient removal technology and modelling. *Water Sci. Technol.* **39**(6), 1–11.

Ekama, G.A. and Wentzel, M.C. (1999b) Denitrification kinetics in biological N and P removal activated sludge systems treating municipal wastewaters. *Water Sci. Technol.* **39**(6), 69–77.

Ekama G.A. and Marais, P. (2004) Assessing the applicability of 1D flux theory to full-scale secondary settling tank design with a 2D hydrodynamic model. *Water Res.* **38**, 495–506.

Ewing, L., Redmon, D.T. and Wren, J.D. (1979) Testing and data analysis of diffused aeration equipment. *J. Water Pollut. Control Fed.* **51**, 2384–2401.

Fall, C. and Loaiza-Navia, J.L. (2007) Design of a tracer test experience and dynamic calibration of the hydraulic model for a full-scale wastewater treatment plant by use of AQUASIM. *Water Environ. Res.* **79**, 893–900.

Ferrer J., Morenilla J.J., Bouzas A. and Garcia-Usach F. (2004a) Calibration and simulation of two large wastewater treatment plants operated for nutrient removal. *Water Sci. Technol.* **50**(6), 87–94.

Ferrer, J., Seco, A., Garcia-Usach, F., Bouzas, A. and Barat, R. (2004b) Simulation of full-scale plants: benefits and drawbacks. In *Water Environ. Manag. Series*, M. van Loosrecht and J. Clement (eds.), IWA Publishing, London, 155–163.

Fikar, M., Chachuat, B. and Latifi, M.A. (2005) Optimal operation of alternating activated sludge processes. *Cont. Eng. Pract.* **13**, 853–861.

Filipe, C.D.M. and Daigger, G.T. (1998) Development of a revised metabolic model for the growth of phosphorus-accumulating organisms. *Water Environ. Res.* **70**, 67–79.

Filipe, C.D.M. and Daigger, G.T. (1999) Evaluation of the capacity of phosphorus accumulating organisms to use nitrate and oxygen as final electron acceptors: A theoretical study on population dynamics. *Water Environ. Res.* **71**, 1140–1150.

Filipe, C.D.M., Daigger, G.T. and Grady, C.P.L. (2001a) Stoichiometry and kinetics of acetate uptake under anaerobic conditions by an enriched culture of phosphorus-accumulating organisms at different pHs. *Biotech. Bioeng.* **76**, 32–43.

Filipe, C.D.M., Daigger, G.T. and Grady, C.P.L. (2001b) pH as a key factor in the competition between glycogen accumulating organisms and phosphorus accumulating organisms. *Water Environ. Res.* **73**, 223–232.

Filipe, C.D.M., Daigger, G.T. and Grady, C.P.L. (2001c) A metabolic model for acetate uptake under anaerobic conditions by glycogen accumulating organisms: stoichiometry, kinetics, and the effect of pH. *Biotechnol. Bioeng.* **76**, 17–31.

Filipe, C.D.M., Meinhold, J., Jorgensen, S.B., Daigger, G.T. and Grady, C.P.L. (2001d) Evaluation of the potential effects of equalization on the performance of biological phosphorus removal systems. *Water Env. Res.* **73**, 276–285.

Fillos, J., Diyamandoglu, V., Carrio, L.A. and Robinson, L. (1996) Full-scale evaluation of biological nitrogen removal in the step-feed activated sludge process. *Water Environ. Res.* **68**, 132–142.

Finnson, A. (1993) Simulation of a strategy to start up nitrification at Bromma sewage plant using a model-based on the IAWPRC Model No. 1. *Water Sci. Technol.* **28**(11/12), 185–195.

Fitch, B. (1983) Kynch theory and compression zones. *AICHE J.* **29**, 940–947.

Ford, D.L., Chih, S.S. and Sobosta, E.C. (1972) Temperature prediction in activated sludge basins using mechanical screens. In *Proc. 27th Ind. Waste Conf.*, Purdue Univ. Lafayette, 587.

Fowler, G.J. (1934) An introduction to the biochemistry of nitrogen conservation. Edward Arnold, London.

Freitas, F., Temudo, M., Almeida, J.S. and Reis, M.A.M. (2004) Flux of carbon, nitrogen and phosphorus in a biological nutrient removal process at different operating conditions. *Water Environ. Manag. Series* (Young Researchers 2004), IWA Publishing, London, 79–88.

French, R.H. (1985) Open channel hydraulics. McGraw-Hill, New York.

Fujie, K., Sekizawa, T. and Kubota, H. (1983) Liquid mixing in activated sludge aeration tank. *J. Ferment. Technol.* **61**, 295–304.

Funamizu, N., Yamamoto, S., Kitagawa, Y. and Takakuwa, T. (1997) Simulation of the operational conditions of the full-scale municipal wastewater treatment plant to improve the performance of nutrient removal. *Water Sci. Technol.* **36**(12), 9–17.

Furumai, H., Kazmi, A.A., Fujita, M., Furuya, Y. and Sasaki, K. (1999) Modeling long term nutrient removal in a sequencing batch reactor. *Water Res.* **33**, 2708–2714.

Fux, C. and Siegrist, H. (2004) Nitrogen removal from sludge digester liquids by nitrification/denitrification or partial nitritation/anammox: environmental and economical considerations. *Water Sci. Technol.* **50**(10), 19–26.

Fux, C., Lange, K., Faessler, A., Huber, P., Grueniger, B. and Siegrist, H. (2003) Nitrogen removal from digester supernatant via nitrite – SBR or SHARON? *Water Sci. Technol.* **48**(8), 9–18.

Fux, C., Velten, S., Carozzi, V., Solley, D. and Keller, J. (2006) Efficient and stable nitritation and denitritation of ammonium-rich sludge dewatering liquor using an SBR with continuous loading. Water Res. **40**, 2765–2775.

Gabaldon, C., Ferrer, J., Seco, A. and Marzal, P. (2000) A steady-state model for the design of biological wastewater treatment facilities. *Environ. Technol.* **21**, 733–744.

Gali, A., Dosta, J., Mace, S. and Mata-Alvarez, J. (2007) Comparison of reject water treatment with nitrification/denitrification via nitrite in SBR and SHARON chemostat process. *Environ. Technol.* **28**, 173–176.

Garrett, M.T. (1958) Hydraulic control of activated sludge growth rate. *Sew. Ind. Wastes* **30**, 252–261.

Garrett, M.T. and Sawyer, C.N. (1952) Kinetics of removal of soluble BOD by activated sludge. In *Proc. 7th Annual Ind. Wastes Conf.*, Purdue University, Lafayette, IN, 7–9 May, 51–77.

Gaudy, A.F. Jr., Bhatla, M.N. and Gaudy, E.T. (1964) Use of chemical oxygen demand values of bacterial cells in waste-water purification. *Appl. Microbiol.* **12**, 254–260.

Gaudy, A.F. Jr., Engelbrecht, R.S. and De Moss, R.D. (1960) Laboratory scale activated sludge unit. *Appl. Environ. Microbiol.* **8**, 298–304.

Gaudy, A.F. Jr. and Kincannon, D.F. (1977) Comparing design models for activated sludge. *Water Sewage Works* **124**, 66–69.

Gee, C.S., Suidan, M.T. and Pfeffer, J.T. (1990) Modeling of nitrification under substrate-inhibiting conditions. *J. Environ. Eng. ASCE.* **116**, 18–31.

Gellman, I. and Heukelekian, H (1953) Studies of biochemical oxidation by direct methods - III. Oxidation and purification of industrial wastes by activated sludge. *Sewage Ind. Wastes* **25**, 1196–1209.

Gernaey, K.V., Jeppsson, U., Batstone, D.J. and Ingildsen, P. (2006) Impact of reactive settler models on simulated WWTP performance. *Water Sci. Technol.* **53**(1), 159–167.

Gernaey, K.V., van Loosdrecht, M.C.M., Henze, M., Lind, M. and Jorgensen, S.B. (2004) Activated sludge wastewater treatment plant modelling and simulation: state of the art. *Environ. Model. Softw.* **19**, 763–783.

Gillot, S. and Heduit, A. (2007) Prediction of alpha factor values for fine pore aeration systems. In *Proc. 10th IWA Specialized Conference on Design, Operation and Economics of Large Wastewater Treatment Plants*, Vienna, 9–13 September, 89–96.

Gillot, S. and Vanrolleghem, P.A. (2003) Equilibrium temperature in aerated basins – comparison of two prediction models. *Water Res.* **37**, 3742–3748.

Gillot, S., Ohtsuki, T., Rieger, L., Shaw, A., Takacs, I. and Winkler, S. (2009) Development of a unified protocol for good modeling practice in activated sludge modeling. *Influents* **4** (Spring 2009), 70–72.

Ginestet, P., Maisonnier, A. and Sperandio, M. (2002) Wastewater COD characterization: biodegradability of physico-chemical fractions. *Water Sci. Technol.* **45**(6), 89–97.

Giokas, D.L., Daigger, G.T., von Sperling, M, Kim, Y. and Paraskevas, P.A. (2003) Comparison and evaluation of empirical zone settling velocity parameters based on sludge volume index using a unified settling characteristics database. *Water Res.* **37**, 3821–3836.

Goel, R., Mino, T., Satoh, H. and Matsuo, T. (1998) Intracellular storage compounds, oxygen uptake rates and biomass yield with readily and slowly degradable substrates. *Water Sci. Technol.* **38**(8/9), 85–93.

Goel, R., Mino, T., Satoh, H. and Matsuo, T. (1999) Modeling hydrolysis processes considering intracellular storage. *Water Sci. Technol.* **39**(1), 97–105.

Gokcay, C.F. and Sin, G. (2004) Modeling of a large-scale wastewater treatment plant for efficient operation. *Water. Sci. Technol.* **50**(7), 123–130.

Goodman, B.L. and Englande, A.J. Jr. (1974) A unified model of the activated sludge process. *J. Water Pollut. Contr. Fed.* **46**, 312–332

Gorgun, E., Artan, N., Orhon, D. and Sozen, S. (1996a) Evaluation of nitrogen removal by step feeding in large treatment plants. *Water Sci. Technol.* **34**(1/2), 253–260.

Gorgun, E., Artan, N., Orhon, D. and Sozen, S. (1996b) Simulation of nitrogen removal by step feeding for Istanbul wastewaters. *Water Sci. Technol.* **33**(12), 259–264.

Goto, M. and Andrews, J.F. (1985) On-line estimation of oxygen uptake rate in the activated sludge process. In *Instrumentation and Control of Water and Wastewater Treatment and Transport Systems*, R. Dake (ed.), Pergamon Press, London, 465–472.

Gottman, J. M. and Kumar Roy, A. (1990) Sequential analysis. A guide for behavioral researchers. Cambridge University Press, Cambridge (UK).

Gould, R.H. (1953) Sewage aeration practice in New York City. In *Proc. Am. Soc. Civil Engrs.* **79**, 1–11.

Grady, C.P.L. (1989) Dynamic modeling of suspended growth biological wastewater treatment processes. In *Dynamic Modeling and Expert Systems in Wastewater Engineering*, G.G. Patry and D. Chapman (eds.), Lewis Publishers, Chelsea, MI (USA), 1–38.

Grady, C.P.L. and Lim, H.C. (1980) Biological Wastewater Treatment: Theory and Applications. M. Dekker, New York (USA).

Grady, C.P.L. Jr. (1989) Dynamic modeling of suspended growth biological wastewater treatment processes. In *Dynamic Modeling and Expert Systems in Wastewater Engineering*, G.G. Patry and D. Chapman (eds.), Lewis Publishers, Chelsea, MI (USA), 1–38.

Grady, C.P.L. Jr., Daigger G.T. and Lim H.C. (1999) Biological Wastewater Treatment. Second Edition, Revised and Expanded. Marcel Dekker, New York (USA).

Grady, C.P.L., Gujer, W., Henze, M., Marais, G.v.R. and Matsuo, T. (1986) A model for single sludge wastewater treatment systems. *Water Sci. Technol.* **18**(6), 47–61.

Gray, N.F. (1989) Biology of Wastewater Treatment. Oxford University Press, Oxford (UK).

Greenburg, A.E., Levin, G. and Kauffmann, J. (1955) Effect of phosphorus removal on the activated-sludge process. *Sew. Ind. Wastes* **27**, 277–282.

Griborio, A., McCorquodale, J.A., Pitt, P., Vinci, P. and Wang, T. (2007) Uncertainty analysis in wastewater engineering modeling: Understanding the limitations. In *Proc. 80th WEF WEFTEC*, San Diego, CA, 13–17 October, 1584–1596.

Grijspeerdt, K., Vanrollenghem, P. and Verstraete, W. (1995) Selection of one-dimensional sedimentation models for on-line use. *Water Sci. Technol.* **31**(2), 193–204.

Guimet, V., Savoye, P. Audic, J. and Do-Quang, Z. (2004) Advanced CFD tool for wastewater: today complex modelling and tomorrow easy-to-use interface. In *Proc. 2nd IWA Leading-Edge Conference on Water and Wastewater Treatment Technologies*, IWA Publishing, London, 165–172.

Gujer, W. (1977) Design of a nitrifying activated-sludge process with the aid of dynamic simulation. *Prog. Water Technol.* **9**(2), 323–336.

Gujer, W. (2006) Activated sludge modelling: Past, present and future. *Water Sci. Technol.* **53**(3), 111–119.

Gujer, W. (2010) Nitrification and me - A subjective review. *Water Res.* **44**, 1–19.

Gujer, W. and Henze, M. (1991) Activated sludge modelling and simulation. *Water Sci. Technol.* **23**(4/6), 1011–1023.

Gujer, W. and Jenkins, D. (1975) A nitrification model for the contact stabilization activated sludge process. *Water Res.* **9**, 561–566.

Gujer, W. and Larsen, T.A. (1995) Implementation of biokinetics and conservation principles in ASIM. *Water Sci. Technol.*, **31**(2), 257–266.

Gujer, W., Henze, M., Mino, T. and van Loosdrecht, M.C.M. (1999) Activated Sludge Model No. 3. *Water Sci. Technol.* **39**(1), 183–193.

Güven, D., Dapena, A., Kartal, B., Schmid, M. C., Maas, B., van de Pas-Schoonen, K., Sozen, S., Mendez, R., Op den Camp, H.J.M., Jetten, M.S.M., Strous, M. and Schmidt, I. (2005) Propionate oxidation by and methanol oxidation of anaerobic ammonium-oxidizing bacteria. *Appl. Environ. Microbiol.* **71**, 1066–1071.

Hall, E.R. and Murphy, K.L. (1980) Estimation of nitrifying biomass and kinetics in wastewater. *Water Res.* **14**, 297–304.

Hamilton, J., Jain, R., Antoniou, P., Svoronos, S.A. and Koopman, B. (1992) Modeling and pilot-scale experimental verification for pre-denitrification process. *J. Env. Eng. ASCE* **118**, 38–55.

Hanada, S., Satoh, H. and Mino, T. (2002) Measurement of microorganisms with PHA production capability in activated sludge and its implication in Activated Sludge Model No. 3. *Water Sci. Technol.* **45**(6), 107–113.

Hanhan, O., Artan, N. and Orhon, D. (2002) Retrofitting activated sludge systems to intermittent aeration for nitrogen removal. *Water Sci. Technol.* **46**(8), 75–82.

Hao, X., Heijnen, J.J. and van Loosdrecht, M.C.M. (2002a) Sensitivity analysis of a biofilm model describing a one-stage completely autotrophic nitrogen removal (CANON) process. *Biotechnol. Bioeng.* **77**, 266–277.

Hao, X., Heijnen, J.J. and van Loosdrecht, M.C.M. (2002b) Model-based evaluation of temperature and inflow variations on a partial nitrification-ANAMMOX biofilm process. *Water Res.* **36**, 4839–4849.

Hao, X., van Loosdrecht, M.C.M., Meijer, S.C.F., Heijnen, J.J. and Qian, Y. (2001a) Model-based evaluation of denitrifying P removal in a two-sludge system. *J. Environ. Eng. ASCE* **127**, 112–118.

Hao, X., van Loosdrecht, M.C.M., Meijer, S.C.F. and Qian, Y. (2001b) Model-based evaluation of two BNR processes - UCT and A₂N. *Water Res.* **35**, 2851–2860.

Harremoes, P. (1979) Dimensionless analysis of circulation, mixing and oxygenation in aeration tanks. *Prog. Water Technol.* **11**(3), 49–57.

Hartel, L. and Popel, H.J. (1992) A dynamic secondary clarifier model including processes of sludge thickening. *Water Sci. Technol.* **25**(6), 267–284.

Hauduc, H. Gillot, S. Rieger, L. Ohtsuki, T., Shaw, A., Takacs, I. and Winkler, S. (2009) Activated sludge modelling in practice: an international survey. *Water Sci. Technol.* **60**(8), 1943–1951.

Head, M.A. and Oleszkiewicz, J.A. (2004) Bioaugmentation for nitrification at cold temperatures. *Water Res.*, **38**, 523–530.

Hellinga, C. (1992) Macrobal 2.02. Delft University of Technology.

Hellinga, C., Schellen, A.A.J.C., Mulder, J.W., van Loosdrecht, M.C.M. and Heijnen, J.J. (1998) The SHARON process: an innovative method for nitrogen removal from ammonium-rich wastewater. *Water Sci. Technol.* **37**(9), 35–42.

Hellinga, C., van Loosdrecht, M.C.M. and Heijnen, J.J. (1999) Model based design of a novel process for nitrogen removal from concentrated flows. *Math. Comp. Model. Dyn. Syst.* **5**, 351–371.

Helmer-Madhok, C., Schmid, M., Filipov, E., Gaul, T., Hippen, A., Rosenwinkel, K.-H., Seyfried, C.F., Wagner, M. and Kunst, S. (2002) Deammonification in biofilm systems: population structure and function. *Water Sci. Technol.* **46**(1/2), 223–231.

Henze, M. (1986) Nitrate versus oxygen utilization rates in wastewater and activated sludge systems. *Water Sci. Technol.* **18**(6), 115–122.

Henze, M. (1992) Characterisation of wastewater for modelling of activated sludge processes. *Water Sci. Technol.* **25**(6), 1–15.

Henze, M. and Harremoes, P. (1992) Characterization of wastewater: the effect of chemical precipitation on the wastewater composition and its consequences for biological denitrification. In *Chemical Water and Wastewater Treatment II*, R. Klute, H.H. Hahn (eds.), Springer-Verlag, Heidelberg, 299–311.

Henze, M. and Mladenovski, C. (1991) Hydrolysis of particulate substrate by activated sludge under aerobic, anoxic and anaerobic conditions. *Water Res.* **25**, 61–64.

Henze, M., Dupont, R., Grau, P. and de la Sota, A. (1993) Rising sludge in secondary settlers due to denitrification. *Water Res.* **27**, 231–236.

Henze, M., Grady, C.P.L. Jr, Gujer, W., Marais, G.v.R. and Matsuo, T. (1987) Activated Sludge Model No. 1. Scientific and Technical Report No.1, IAWPRC, London.

Henze, M., Gujer W., Mino T. and van Loosdrecht M. (eds.) (2000) Activated Sludge Models ASM1, ASM2d and ASM3. Scientific and Technical Report No. 9. IWA, London.

Henze, M., Gujer, W., Mino, T., Matsuo, T., Wentzel, M. C. and Marais, G.v.R. (1995a) Activated Sludge Model No. 2. *Scientific and Technical Report No. 3*, IAWQ, London.

Henze, M., Gujer, W., Mino, T., Matsuo, T., Wentzel, M. C., Marais, G.v.R. and van Loosdrecht, M. (1999) Activated Sludge Model No. 2d. *Water Sci. Technol.* **39**(1), 165–182.

Henze, M., Harremoës, P., la Cour Jansen, J. and Arvin, E. (1995b) Wastewater treatment. Biological and chemical processes. Springer-Verlag, Berlin.

Henze, M., Kristensen, G.H. and Strube, R. (1994) Rate-capacity characterization of wastewater for nutrient removal processes. *Water Sci. Technol.* **29**(7), 101–107.

Henze, M., van Loosdrecht, M.C.M., Ekama, G.A. and Brdjanovic, D. (eds.) (2008) Biological Wastewater Treatment Principles, Modelling and Design. IWA Publishing, London.

Herbert, D. (1958) Continuous cultivation of microorganisms: some theoretical aspects. In *Continuous Cultivation of Microorganisms (Proc. 2nd Symposium)*, 45–52.

Herbert, D., Elsworth, R. and Telling, R.C. (1956) The continuous culture of bacteria; a theoretical and experimental study. *J. Gen. Microbiol.* **14**, 601–622.

Hiatt, W.C. and Grady, C.P.L. (2008) An updated process model for carbon oxidation, nitrification, and denitrification. *Water Environ. Res.* **80**, 2145–2156.

Higbie, R. (1935) The rate of absorption of a pure gas into a still liquid during short periods of exposure. *Trans. Am. Inst. Chem. Engrs.* **35**, 365–389.

Hinshelwood, C.N. (1946) Influence of temperature on the growth of bacteria. In *The Chemical Kinetics of the Bacterial Cell*, Clarendon Press, Oxford, 254–257.

Holmberg, U. (1986) Adaptive dissolved oxygen control and on-line estimation of oxygen transfer and respiration rates. In *Proc. AIChE's Annual Meeting*, Miami Beach, 7 November.

Holmberg, U., Olsson, G. and Andersson, B. (1989) Simultaneous DO control and respiration estimation. *Water Sci. Technol.* **21**(10/11), 1185–1195.

Hoover, S.R. and Proges, N. (1952) Assimilation of dairy wastes by activated sludge: II. The equations of synthesis and rate of oxygen utilization. *Sewage Ind. Wastes* **24**, 306–312.

Horan, N.J. (1990) Biological Wastewater Treatment Systems: Theory and Operation. Wiley, Chichester (UK).

Hu, Z.Q., Chandran, K., Smets, B.F. and Grasso, D. (2002a) Evaluation of a rapid physical-chemical method for the determination of extant soluble COD. *Water Res.* **36**, 617–624.

Hu, Z-r., Sötemann, S., Moodley, R., Wentzel, M. C. and Ekama, G. A. (2003a) Experimental investigation of the external nitrification biological nutrient removal activated sludge (ENBNRAS) system. *Biotechnol. Bioeng.* **83**, 260–273.

Hu, Z-r., Wentzel, M.C. and Ekama G.A. (2000) External nitrification in biological nutrient removal activated sludge system. *Water SA* **26**, 225–238.

Hu, Z-r., Wentzel, M.C. and Ekama, G.A. (2002b) Anoxic growth of phosphate accumulating organisms (PAOs) in biological nutrient removal activated sludge systems. *Water Res.* **36**, 4927–4937.

Hu, Z-r., Wentzel, M.C. and Ekama, G.A. (2002c) The significance of denitrifying polyphosphate accumulating organisms in biological nutrient removal activated sludge systems. *Water Sci. Technol.* **46**(1/2), 129–138.

Hu, Z-r., Wentzel, M.C. and Ekama, G.A. (2003b) Modelling biological nutrient removal activated sludge systems – a review. *Water Res.* **37**, 3430–3444.

Hu, Z-r., Wentzel, M.C. and Ekama, G.A. (2007) A general kinetic model for biological nutrient removal activated sludge systems: Model development. *Biotechnol. Bioeng.* **98**, 1242–1258.

Hug, T., Benedetti, L., Hall, E.R., Johnson, B.R., Morgenroth, E., Nopens, I., Rieger, L., Shaw, A. and Vanrolleghem, P.A. (2009) Wastewater treatment models in teaching and training: the mismatch between education and requirements for jobs. *Water Sci. Technol.* **59**(4), 745–753.

Hug, T., Gujer, W. and Siegrist, H. (2005) Rapid quantification of bacteria in activated sludge using fluorescence in situ hybridization and epifluorescence microscopy. *Water Res.* **39**, 3837–3848.

Hug, T., Gujer, W. and Siegrist, H. (2006) Modelling seasonal dynamics of "Microthrix parvicella". *Water Sci. Technol.* **54**(1), 189–198.

Hughes, T., Oswalt, B., Chapman, J., Swartzlander, D., Benisch, M. and Neethling, J.B. (2004) Teaming up to meet startup goals at Las Vegas' BNR facility. In *Proc. 77th WEF WEFTEC 2004*, New Orleans, 2–6 October, 427–444.

Hulsbeek, J.J.W., Kruit, J., Roeleveld, P.J. and van Loosdrecht, M.C.M. (2002) A practical protocol for dynamic modelling of activated sludge systems. *Water Sci. Technol.* **45**(6), 127–136.

Hvala, N., Vrecko, D., Burica, O., Strazar, M. and Levstek, M. (2002) Simulation study supporting wastewater treatment plant upgrading. *Water Sci. Technol.* **46**(4/5), 325–332.

Hvala, N., Zec, M., Ros, M. and Strmcnik, S. (2001) Design of a sequencing batch reactor sequence with an input load partition in a simulation-based experimental environment. *Water Env. Res.* **73**, 146–153.

Hydromantis, Inc. (2006). GPS-X 5.0 - User's Guide and Technical Reference. Hydromantis, Inc., Hamilton, Ontario (Canada).

Iacopozzi, I., Innocenti, V., Marsili-Libelli, S. and Giusti, E. (2007) A modified Activated Sludge Model No. 3 (ASM3) with two-step nitrification-denitrification. *Environ. Model. Softw.* **22**, 847–861.

Iida, Y. (1988) Performance analysis of the aeration tanks in the activated sludge system. *Water Sci. Technol.* **20**(4/5), 109–120.

Imhoff, K. (1955) Two-Stage Operation of Activated Sludge Plants. *Sew. Ind. Wastes* **27**, 431–433.

Insel, G., Karahan-Gul, O., Orhon, D., Vanrolleghem, P.A. and Henze, M. (2002) Important limitations in the modeling of activated sludge: biased calibration of the hydrolysis process. *Water Sci. Technol.* **45**(12), 23–36.

Insel, G., Orhon, D. and Vanrolleghem, P.A. (2003) Identification and modelling of aerobic hydrolysis - application of optimal experimental design. *J. Chem. Technol. Biotechnol.* **78**, 437–445.

Irizar, I., Suescun, J., Plaza, F. and Larrea, L. (2003) Optimizing nitrogen removal in the BioDenitro process. *Water Sci. Technol.* **48**(11/12), 429–436.

Irvine, R.L. and Busch, A.W. (1979) Sequencing batch biological reactors - An overview. *J. Water Pollut. Control Fed.* **51**, 235–243.

Irvine, R.L. and Davis, W.B. (1971) Use of sequencing batch reactors for waste treatment - CPC International, Corpus Christi, Texas. In *Proc. 26th Annual Purdue Ind. Waste Conf.*, Purdue University, West Lafayette, IN, 450–462.

Irvine, R.L., Allemen, J.E., Miller, G. and Dennis, R.W. (1980) Stoichiometry and kinetics of biological waste treatment. *J. Water Pollut. Control Fed.* **52**, 1997–2006.

Itokawa, H., Inoki, H. and Murakami, T. (2008) JS Protocol: A practical guidance for the use of activated sludge modelling in Japan. In *Proc. the IWA World Water Congress and Exhibition*, Vienna, 7–12 October [CD-Rom].

Izquierdo, J., Perez, R. and Iglesias, P. L. (2004) Mathematical models and methods in the water industry. *Math. Comp. Model.* **39**, 1353–1374.

Jakeman, A.J., Letcher, R.A. and Norton, J.P. (2006) Ten iterative steps in development and evaluation of environmental models. *Environ. Model. Softw.* **21**, 602–614.

Japan Sewage Works Agency (2006) Technical Evaluation of the Practical Use of Activated Sludge Models. Report No. 05-004, Research and Technology Development Division, Japan Sewage Works Agency, Toda (in Japanese).

Jardin, N. (2006) Urban wastewater treatment in Europe: from operational problems to plant design. *Manag. Environ. Quality* **17**, 642–653.

Jenkins, D. (2008) From total suspended solids to molecular biology tools – a personal view of biological wastewater treatment process population dynamics. *Water Environ. Res.* **80**, 677–687.

Jeppsson, U. (1996) Modelling aspects of wastewater treatment processes. Ph.D. Thesis, Lund Institute of Technology, Lund. (http://www.iea.lth.se/publications)

Jeppsson, U. and Diehl, S. (1996) An evaluation of a dynamic model of the secondary clarifier. *Water Sci. Technol.* **34**(5/6), 19–26.

Jetten, M.S.M.,Wagner, M., Fuerst, J., van Loosdrecht, M.C.M., Kuenen, G. and Strous, M. (2001) Microbiology and application of the anaerobic ammonium oxidation ('anammox') process. *Curr. Opin. Microbiol.* **12**, 283–288.

Jobbagy, A., Simon, M.J. and Plosz, B. (2000) The impact of oxygen penetration on the estimation of denitrification rates in anoxic processes. *Water Res.* **34**, 2606–2609.

Jobbagy, A., Tardy, G.M. and Literathy, B. (2004) Enhanced nitrogen removal in the combined activated sludge-biofilter system of the Southpest Wastewater Treatment Plant. *Water Sci. Technol.* **50**(7), 1–8.

Johansson, P., Carlsson, H. and Jönsson, K. (1996) Modelling of the anaerobic reactor in a biological phosphate removal process. *Water Sci. Technol.* **34**(1/2), 49–55.

Johnson, B.R. (2009) How to use simulators in the design and operation of wastewater treatment facilities. WEF Water Quality Training (Modeling 101), webcast, 25 February.

Johnson, B.R., Goodwin, S., Daigger, G.T. and Crawford, G.V. (2005) A comparison between the theory and reality of full-scale step-feed nutrient removal systems. *Water Sci. Technol.* **52**(10/11), 587–596.

Jones, R.M., Dold, P.L., Takacs, I., Chapman, K., Wett, B., Murthy, S. and O'Shaughnessy, M. (2007) Simulation for operation and control of reject water treatment processes. In *Proc. 80th WEF WEFTEC 2007*, San Diego, CA, 13–17 October, 4357–4372.

Jones, R.M., Bye, C.M. and Dold, P.L. (2005) Nitrification parameter measurement for plant design: Experience and experimental issues with new methods. *Water Sci. Technol.* **52**(10/11), 461–468.

Kaelin, D., Manser, R., Rieger, L., Eugster, J., Rottermann, K. and Siegrist, H. (2009) Extension of ASM3 for two-step nitrification and denitrification and its calibration and validation with batch tests and pilot scale data. *Water Res.* **43**, 1680–1692.

Kampschreur, M.J., Picioreanu, C., Tan, N., Kleerebezem, R., Jetten, M.S.M. and van Loosdrecht, M.C.M. (2007) Unraveling the source of nitric oxide emission during nitrification. *Water Environ. Res.* **79**, 2499–2509.

Kappe, S.E. (1957) Digester supernatant: problems, characteristics, and treatment. *Sewage Ind. Wastes* **30**, 937–952.

Kappeler, J. and Gujer, W. (1992) Estimation of kinetic parameters of heterotrophic biomass under aerobic conditions and characterization of wastewater for activated sludge modelling. *Water Sci. Technol.* **25**(6), 125–140.

Karahan-Gul, O., Artan, N., Orhon, D., Henze, M. and van Loosdrecht, M.C.M. (2002b) Experimental assessment of bacterial storage yield. *J. Environ. Eng. ASCE* **128**, 1030–1035.

Karahan-Gul, O., van Loosdrecht, M.C.M. and Orhon, D. (2003) Modification of Activated Sludge Model No. 3 considering direct growth on primary substrate. *Water Sci. Technol.* **47**(11), 219–225.

Kawase, Y. and Moo-Young, M. (1990) Mathematical models for design of bioreactors: applications of Kolmogoroff's theory of isotropic turbulence. *Chem. Eng. J.* **43**, B19–B41.

Kayser, R., Stobbe, G. and Werner, M. (1992) Operational results of the Wolfsburg wastewater treatment plant. *Water Sci. Technol.* **25**(4/5), 203–209.

Keinath, T.M. (1985) Operational dynamics and control of secondary clarifiers. *J. Water Pollut. Control Fed.* **57**, 770–776.

Keinath, T.M., Ryckman, M.D., Dana, C.H. and Hofer, D.A. (1977) Activated sludge – unified system design and operation. *J. Env. Eng. ASCE*, **103**, 829–849.

Keller, J. (2005) Sequencing batch reactor processes for biological nutrient removal. In *Proc. 1st IWA Specialty Conf. "Nutrient Management in Wastewater Treatment Processes and Recycle Streams"*, Cracow, 19–21 September, 245–256.

Kerrn-Jespersen, J.P. and Henze, M. (1993) Biological phosphorus uptake under anoxic and oxic condition. *Water Res.* **27**, 617–624.

Kershaw, J.H. and Finch, J. (1936) The treatment of diluted sewage by bioaeration in the presence of nitric and nitrous nitrogen. *J. Inst. Sew. Purif.* **2**, 397.

Khudenko, B.M. and Shpirt, E. (1986) Hydrodynamic parameters of diffused air systems. *Water Res.* **20**, 905–915.

Kim, H., Hao, O.J. and McAvoy, T.J. (2001) SBR system for phosphorus removal: ASM2 and simplified linear model. *J. Env. Eng. ASCE* **127**, 98–104.

Knowles, G., Downing, A.L. and Barrett, M.J. (1965) Determination of kinetic constants for nitrifying bacteria in mixed culture, with the aid of an electronic computer. *J. Gen. Microbiol.* **38**, 263–278.

Koch, G., Egli, K., Van Der Meer, J.R. and Siegrist, H. (2000a) Mathematical modeling of autotrophic denitrification in a nitrifying biofilm of a rotating biological contactor. *Water Sci. Technol.* **41**(4/5), 191–198.

Koch, G., Kuhni, M., Gujer, W. and Siegrist, H. (2000b) Calibration and validation of Activated Sludge Model No. 3 for Swiss municipal wastewater. *Water Res.* **34**, 3580–3590.

Koch, G., Kuhni, M. and Siegrist, H. (2001a) Calibration and validation of an ASM3-based steady-state model for activated sludge systems. Part I: Prediction of nitrogen removal and sludge production. *Water Res.* **35**, 2235–2245.

Koch, G., Kuhni, M. and Siegrist, H. (2001b) Calibration and validation of an ASM3-based steady-state model for activated sludge systems. Part II: Prediction of phosphorus removal. *Water Res.* **35**, 2246–2255.

Koch, G., Pianta, R., Krebs, P. and Siegrist, H. (1999) Potential of denitrification and solids removal in the rectangular clarifier. *Water Res.* **33**, 309–318.

Koopman, B. and Cadee, K. (1983) Prediction of thickening capacity using diluted sludge volume index. *Water Res.* **17**, 1427–1431.

Kops, S., Vangheluwe, H., Claeys, F., Vanrolleghem, P., Yuan, Z. and Vansteenkiste, G. (1999) The process of model building and simulation of ill-defined systems: application to wastewater treatment. *Math. Comp. Model. Dyn. Syst.* **5**, 298–312.

Kos, P., Head, M.A., Oleszkiewicz, J. and Warakomski, A. (2000) Demonstration of low temperature nitrification with a short SRT. In *Proc. 74th WEF WEFTEC 2000*, Alexandria, 14–18 October, 338–347.

Kovarova-Kovar, K. and Egli, T. (1998) Growth kinetics of suspended microbial cells: from single-substrate-controlled growth to mixed-substrate kinetics. *Microbiol. Mol. Biol. Rev.* **62**, 646–666.

Krebs, P. (1995) Success and shortcomings of clarifier modelling. *Water Sci. Technol.* **31**(2), 181–192.

Krhutkova, O., Novak, L., Pachmanova, L., Wanner J. and Kos, M. (2006) In situ bioaugmentation of nitrification in the regeneration zone: practical application and experiences at full-scale plants. *Water Sci. Technol.*, **53**(12), 39–46.

Krishna, C. and van Loosdrecht, M.C.M. (1999a) Substrate flux into storage and growth in relation to activated sludge modeling. *Water Res.* **33**, 3149–3161.

Kristensen, G.H., Jorgensen, P.E. and Henze, M. (1992) Characterization of functional microorganism groups and substrate in activated sludge and wastewater by AUR, NUR and OUR. *Water Sci. Technol.* **25**(6), 43–57.

Kristensen, G.H., la Cour Jansen, J. and Jorgensen, P.E. (1998) Batch test procedures as tools for calibration of the activated sludge model - a pilot scale demonstration. *Water Sci. Technol.* **37**(4/5), 235–242.

Kuba, T., Murnleitner, E., van Loosdrecht, M.C.M. and Heijnen, J.J. (1996a) An integrated metabolic model for the aerobic and denitrifying biological phosphorus removal. *J. Env. Eng. ASCE* **54**, 434–450.

Kuba, T., van Loosdrecht, M.C.M. and Heijnen, J.J. (1996b) Phosphorus and nitrogen removal with minimal COD requirement by integration of denitrifying dephosphatation and nitrification in a two-sludge system. *Water Res.* **30**, 1702–1710.

Kynch, G. J. (1952). A theory of sedimentation. *Trans. Faraday Soc.* **48**, 166–176.

la Cour Jansen, J., Kristensen, G.H. and Laursen, K.D. (1992) Activated sludge nitrification in temperature climate. *Water Sci. Technol.* **25**(4/5), 177–184.

Ladiges, G. and Bertram, N.-P. (2004) Optimisation of Hamburg's WWTPs - three years of experience with the new concept. *Water Sci. Technol.* **50**(7), 45–48.

Ladiges, G., Bertram, N.-P. and Otterpohl, R. (2000) Concept development for the optimisation of the Hamburg Wastewater Treatment Plants. *Water Sci. Technol.* **41**(9), 89–96.

Ladiges, G., Gunner, C. and Otterpohl, R. (1999) Optimisation of the Hamburg wastewater treatment plants by dynamic simulation. *Water Sci. Technol.* **39**(4), 37–44.

Langergraber, G., Rieger, L., Winkler, S., Alex, J., Wiese, J., Owerdieck, C., Ahnert, M., Simon, J. and Maurer, M. (2004) A guideline for simulation studies of wastewater treatment plants. *Water Sci. Technol.* **50**(7), 131–138.

Laquidara, V.D. and Keinath, T.M. (1983) Mechanism of clarification failure. *J. Water Pollut. Control Fed.* **55**, 54–57.

Larrea, L., Irizar, I. and Hildago, M.E. (2002) Improving the predictions of ASM2d through modelling in practice. *Water Sci. Technol.* **45**(6), 199–208.

Larrea, L., Larrea, A., Ayesa, E., Rodrigo, J.C., Lopez-Carrasco M.D. and Cortacans, J.A. (2001) Development and verification of design and operation criteria for the step feed process with nitrogen removal. *Water Sci. Technol.* **43**(1), 261–268.

Larsen, P. (1977) On the hydraulics of rectangular settling basins, experimental and theoretical studies, Report No.1001, Dept. of Water Resources Engineering, Lund Institute of Technology, Lund, Sweden.

Latimer, R.J., Pitt, P.A. and Rohrbacher, J. (2007) BNR ENR model calibration experience indicates deviations in key modeling parameters. In *Proc. WEF/IWA Specialty Conf. "Nutrient Removal 2007"*, Baltimore, 4–6 March, 1215–1223.

Law, A.M. and Kelton, W.D. (2000) Simulation Modeling and Analysis. 3rd Edition. McGraw Hill, New York.

Lawrence, A.W. and McCarthy, P.L. (1970) Unified basis for biological treatment design and operation. *J. San. Eng. ASCE* **96**, 757–778.

Lebek, M. (2003) Bekämpfungsmaßnahmen von Blähschlamm verursacht durch Microthrix parvicella. *Veröffentlichungen des Institutes für Siedlungs-wasserwirtschaft und Abfalltechnik der Universität Hannover*, Hannover, Heft 125. (in German)

Lee, T.T., Wang, F.Y. and Newell, R.B. (1999) Distributed parameter approach to the dynamics of complex biological processes. *AIChE J.* **45**, 2245–2268.

Leeuw, E.J., Kramer, J.F., Bult, B.A. and Wijcherson, M.H. (1996) Optimization of nutrient removal with on-line monitoring and dynamic simulation. *Water Sci. Technol.* **33**(1), 203–209.

Lemos, P.C., Viana, C., Salgueiro, E.N., Ramos, A.M., Crespo, J.P.S.G. and Reis, M.A.M. (1998) Effect of carbon source on the formation of polyhydroxyalkanoates (PHA) by a phosphate-accumulating mixed culture. *Enz. Microb. Technol.* **22**, 662–671.

Lemoullec, Y., Potier, O., Gentric, C. and Leclerc, J.P. (2008) A general correlation to predict axial dispersion coefficients in aerated channel reactors. *Water Res.* **42**, 1767–1777.

Lesouef, A., Payraudeau, M., Rogalla, F. and Kleiber, B. (1992) Optimizing nitrogen removal reactor configurations by on-site calibration of the IAWPRC Activated Sludge Model. *Water Sci. Technol.* **25**(6), 105–123.

Lessard, P. and Beck, M.B. (1991) Dynamic modeling of wastewater treatment processes. Its current status. *Environ. Sci. Technol.* **25**, 30–39.

Lessard, P. and Beck, M.B. (1993) Dynamic modelling of the activated sludge process: a case study. *Water Res.* **27**, 963–978.

Lester, J.N. and Birkett, J.W. (1999) Microbiology and Chemistry for Environmental Scientists and Engineers. Taylor & Francis, London.

Levenspiel, O. (1972) Chemical Reaction Engineering, 2nd Edition. Wiley, New York.

Levin, G.V. (1970) US. Patent No. 3654147. U S . Patent Office, Washington, DC

Levin, G.V. and Shapiro, J. (1965) Metabolic uptake of phosphorus by wastewater organisms. *J. Water Pollut. Cont. Fed.* **37**, 800–821.

Lewis, W.K. and Whitman, W.C. (1924) Principles of gas absorption. *Ind. Eng. Chem.* **16**, 1215–1224.

Lijklema, L. (1973) Model for nitrification in activated sludge process. *Environ. Sci. Technol.* **7**, 428–433.

Lippi, S., Rosso, D., Lubello, C., Canziani, R. and Stenstrom, M.K. (2009) Temperature modelling and prediction for activated sludge systems. *Water Sci. Technol.* **59**(1), 125–131.

Littleton, H.X., Daigger, G.T., Strom, P.F. and Cowan, R.A. (2003) Simultaneous biological nutrient removal: evaluation of autotrophic denitrification, heterotrophic nitrification, and biological phosphorus removal in full-scale systems. *Water Environ. Res.* **75**, 138–150.

Liu, W.T., Mino, T., Nakamura, K. and Matsuo, T. (1996) Glycogen accumulating population and its anaerobic substrate uptake in anaerobic-aerobic activated sludge without biological phosphorus removal. *Water Res.* **30**, 75–82.

Lopez-Vazquez, C.M., Hooijmans, C.M., Brdjanovic, D., Gijzen, H.J. and van Loosdrecht, M.C.M. (2008) Factors affecting the microbial populations at full-scale enhanced biological phosphorus removal (EBPR) wastewater treatment plants in the Netherlands. *Water Res.* **42**, 2349–2360.

Lopez-Vazquez, C.M., Oehmen, A., Hooijmans, C.M., Brdjanovic, D., Gijzen, H.J., Yuan, Z. and van Loosdrecht, M.C.M. (2009) Modeling the PAO-GAO competition: Effects of carbon source, pH and temperature. *Water Res.* **43**, 450–462.

Ludzack, F.J. and Ettinger, M.B. (1962) Controlled operation to minimize activated sludge effluent nitrogen. *J. Water Pollut. Cont. Fed.* **34**, 920–931.

Luong, J.H.T. (1987) Generalization of Monod kinetics for analysis of growth data with substrate inhibition. *Biotechnol. Bioeng.* **29**, 242–248.

Mace, S. and Mata-Alvarez, J. (2002) Utilization of SBR technology for wastewater treatment: An overview. *Ind. Eng. Chem. Res.* **41**, 5539–5553.

Machado, V.C., Tapia, G., Gabriel, D., Lafuente, J. and Baeza, J.A. (2009) Systematic identifiability study based on the Fisher Information Matrix for reducing the number of parameters calibration of an activated sludge model. *Environ. Model. Softw.* **24**, 1274–1284.

Majone, M., Dircks, K. and Beun, J.J. (1999) Aerobic storage under dynamic conditions in activated sludge processes. The state of the art. *Water Sci. Technol.* **39**(1), 61–73.

Majone, M., Massanisso, P. and Ramadori, R. (1998) Comparison of carbon storage under aerobic and anoxic conditions. *Water Sci. Technol.* **38**(8/9), 77–84.

Makinia, J. (1998) Mathematical modeling of the activated sludge reactor with dispersive flow. Ph.D. dissertation, Department of Civil Engineering, Portland State University, Portland, OR (USA).

Makinia, J. (2006) Performance Prediction of Full-Scale Biological Nutrient Removal Systems Using Complex Activated Sludge Models. *Veröffentlichungen des Institutes für Siedlungswasserwirtschaft und Abfalltechnik der Universität Hannover*, Hannover, Heft 135.

Makinia, J. and Wells, S.A. (2000) A general model of the activated sludge reactor with dispersive flow (part I): model development and parameter estimation. *Water Res.* **34**, 3987–3996.

Makinia, J. and Wells, S.A. (2005) Evaluation of empirical formulae for estimation of the longitudinal dispersion in activated sludge reactors. *Water Res.* **39**, 1533–1542.

Makinia, J. Drewnowski, J., Swinarski, M. and Czerwionka, K. (2009) Internal vs. external (alternative) carbon sources for denitrification and EBPR accomplished by a full-scale process biomass. In *Proc. WEF Specialty Conference "Nutrient Removal 2009"*, Washington, DC, 28 June – 1 July, 16–30.

Makinia, J. Rosenwinkel, K.-H. and Spering, V. (2005a) Long-term simulation of the activated sludge process at the Hanover-Gümmerwald pilot WWTP. *Water Res.* **39**, 1489–1502.

Makinia, J., Rosenwinkel, K.-H. and Phan, L.-C. (2006a) Modification of ASM3 for the determination of biomass adsorption capacity in bulking sludge control. *Water Sci. Technol.* **53**(3), 91–99.

Makinia, J., Rosenwinkel, K.-H. and Spering, V. (2006b) A comparison of two model concepts for simulation of nitrogen removal at a full-scale BNR pilot plant. *J. Environ. Eng. ASCE* **132**, 476–487.

Makinia, J., Rosenwinkel, K.-H., Swinarski, M. and Dobiegala, E. (2006c) Experimental and model-based evaluation of the role of denitrifying PAO at two large scale WWTPs in northern Poland. *Water Sci. Technol.* **54**(8), 73–81.

Makinia, J., Swinarski, M. and Dobiegala, E. (2002) Experiences with computer simulation at two large wastewater treatment plants in northern Poland. *Water Sci. Technol.* **45**(6), 209–218.

Makinia, J., Wells, S.A. and Zima, P. (2005b) Temperature modeling in activated sludge systems: A case study. *Water Environ. Res.* **77**, 525–532.

Mamais, D., Jenkins, D. and Pitt, P. (1993) A rapid physical chemical method for the determination of readily biodegradable soluble COD in municipal wastewater. *Water Res.* **27**, 195–197.

Manga, J., Ferrer, J., Garcia-Usach, F. and Seco, A. (2001) A modification to the Activated Sludge Model No. 2 based on the competition between phosphorus-accumulating organisms and glycogen-accumulating organisms. *Water Sci. Technol.* **43**(11), 161–171.

Manser, R., Muche, K., Gujer, W. and Siegrist, H. (2005) A rapid method to quantify nitrifiers in activated sludge. *Water Res.* **39**, 1585–1593.

Marais, G. v. R. (1973) The activated sludge process at long sludge ages. Research Report No. W 3, Dept. of Civil Engineering, Univ. of Cape Town, South Africa.

Marais, G.v.R. and Ekama, G.A. (1976) Activated sludge process (part 1). Steady state behaviour. *Water SA* **2**, 164–200.

Martin, A.J. (1927) The Activated Sludge Process. MacDonald and Evans, London.

Martins, A.M.P., Pagilla, K., Heijnen, J.J. and van Loosdrecht, M.C.M. (2004) Filamentous bulking sludge – a critical review. *Water Res.* **38**, 793–817.

Mauer, M. and Gujer, W. (1998) Dynamic modelling of enhanced biological phosphorus and nitrogen removal in activated sludge systems. *Water Sci. Technol.* **38**(1), 203–210.

McHaney, R. (1991) Computer simulation: a practical perspective. Academic Press, San Diego, CA.

McKinney, R.E. (1970) Design and operation of complete-mixing activated sludge systems. EPCS-Reports, 1, 3, 1

McKinney, R.E. (1957) Activity of microorganisms in organic waste disposal - II. Aerobic processes. *Appl. Microbiol.* **5**, 167–174.

McKinney, R.E. (1962) Mathematics of complete mixing activated sludge, *J. San. Eng. Div. ASCE* **88(SA3)**, 87–113.

McKinney, R.E. and Ooten, R.J. (1969) Concepts of complete mixing activated. sludge. In *Proc. Trans. 19th Sanit. Eng. Conf.*, Univ. Kansas, Lawrence, 32–59.

Meijer, S.C.F. (2004) Theoretical and practical aspects of modelling activated sludge processes. Ph.D. Thesis, Delft University of Technology, Delft (the Netherlands).

Meijer, S.C.F., van der Spoel, H., Susanti, S., Heijnen, J.J. and van Loosdrecht, M.C.M. (2002a) Error diagnostics and data reconciliation for activated sludge modelling using mass balances. *Water Sci. Technol.* **45**(6), 145–156.

Meijer, S.C.F., van Loosdrecht, M.C.M. and Heijnen, J.J. (2001) Metabolic modeling of full-scale biological nitrogen and phosphorus removing WWTP's. *Water Res.* **35**, 2711–2723.

Meijer, S.C.F., van Loosdrecht, M.C.M. and Heijnen, J.J. (2002b) Modelling the start-up of a full-scale biological phosphorous and nitrogen removing WWTP. *Water Res.* **36**, 4667–4682.

Meinhold, J., Arnold, E. and Isaacs, S. (1999) Effect of nitrite on anoxic phosphate uptake in biological phosphorus removal activated sludge. *Water Res.* **33**, 1871–1883.

Melcer, H., Dold, P.L., Jones, R.M., Bye, C.M., Takacs, I., Stensel, H.D., Wilson, A.W., Sun, P. and Bury, S. (2003) Methods for wastewater characterization in activated sludge modeling. Report no. 99-WWF-3, Water Environment Research Foundation (WERF), Alexandria, VA (USA).

Melcer., H. (1999) Full scale experience with biological process models – calibration issues. *Water Sci. Technol.* **39**(1), 245–252.

Metcalf and Eddy, Inc. (1922) Sewerage and sewage disposal. McGraw-Hill, New York.

Metcalf and Eddy, Inc. (1991) Wastewater Engineering, 3rd Edition. New York.

Metcalf and Eddy, Inc. (2003) Wastewater Engineering, 4th Edition. McGraw-Hill, New York.

Michaelis, L. and Menten, M. (1913) Die kinetik der invertinwirkung, *Biochemistry Zeitung* **49**, 333–369. (in German)

Mikola, A. Rautiainen, J. and Vahala, R. (2009) Secondary clarifier conditions conducting to secondary phosphorus release in a BNR plant. *Water Sci. Technol.* **60**(9), 2413–2418.

Mikosz, J., Plaza, E. and Kurbiel, A. (2001) Use of computer simulation for cycle length adjustment in sequencing batch reactor. *Water Sci. Technol.* **43**(3), 61–68.

Miller, A.P. (1927) Public Health Engineering. *Am. J. Public Health* **17**, 974–977.

Miller, A.P. (1930) Public Health Engineering. *Am. J. Public Health* **20**, 1352–1356.

Mines, R.O. (1997) Design and modeling of post-denitrification single-sludge activated sludge processes. *Water Air Soil Pollut.* **100**, 79–88.

Mines, R.O. and Sherrard, J.H. (1987) Biological enhancement of oxygen transfer in the activated sludge process. *J. Water Pollut. Control Fed.* **59**, 19–24.

Mines, R.O., Vilagos, J.L., Echelberger, W.F. and Murphy, R.J. (2001) Conventional and AWT mixed-liquor settling characteristics. *J. Environ. Eng. ASCE* **127**, 249–258.

Mino, T., Liu, W.T., Kurisu, F. and Matsuo, T. (1995a) Modelling glycogen storage and denitrification capability of microorganism in enhanced biological phosphate removal processes. *Water Sci. Technol.* **31**(2), 25–34.

Mino, T., San Pedro, D.C. and Matsuo, T. (1995b) Estimation of the rate of slowly biodegradable COD (SBCOD) hydrolysis under anaerobic, anoxic and aerobic conditions by experiments using starch as model substrate. *Water Sci. Technol.* **31**(2), 95–103.

Mino, T., Tsuzuki, Y. and Matsuo, T. (1987) Effect of phosphorus accumulation on acetate metabolism in the biological phosphorus removal process. In *Proc. the IAWPRC Int. Conf. on Biol. Phosphate Removal from Wastewaters*, R. Ramadori (ed.), Rome, 28–30 September, 27–38.

Mino, T., van Loosdrecht, M.C.M. and Heijnen, J.J. (1998) Review paper: Microbiology and biochemistry of the enhanced biological phosphate removal processes. *Water Res.* **32**, 3193–3207.

Miyaji, Y., Iwasaki, M. and Sekigawa, Y. (1980) Biological nitrogen removal by step-feed process. *Prog. Water Technol.* **12**(6), 193–202.

M'Kendrick, A.G. and Pai, M.K. (1911) The rate of multiplication of microorganisms: A mathematical study. In *Proc. Roy. Soc. Edin.* **31**, 649–655.

Mohlman, F.W. (1917) The activated-sludge method of sewage treatment. *Univ. Illinois Bullet.* **15**(11), 75–113.

Mohlman, F.W. (1938) Twenty five years of activated sludge, Modern Sewage Disposal. Anniversary book of the Federation of Sewage Works Association (USA), Chapter VI, p. 38.

Monod, J. (1942) Recherches sur la Croissance des Cultures Bactériennes. Hermann & Co., Paris. (in French)

Monod, J. (1949) The growth of bacterial cultures. *Annual Rev. Microbiol.* **3**, 371–394.

Monod, J. (1950) La technique de culture continue, théorie et applications. *Ann Inst Pasteur* **79**, 390–410. (in French)

Montgomery, Inc. (1985) Water Treatment Principles and Design. Wiley, New York.

Morgenroth, E., Arvin, E. and Vanrolleghem, P.A. (2002a) The use of mathematical models in teaching wastewater treatment engineering. *Water Sci. Technol.* **45**(6), 229–233.

Morgenroth, E., Kommedal, R. and Harremoës P. (2002b) Processes and modeling of hydrolysis of particulate organic matter in aerobic wastewater treatment – a review. *Water Sci. Technol.* **45**(6), 25–40.

Moser-Engeler, R., Udert, K.M., Wild, D. and Siegrist, H. (1998) Products from primary sludge fermentation and their suitability for nutrient removal. *Water Sci. Technol.* **38**(1), 265–273.

Moussa, M.S., Hooijmans, C.M., Lubberding, H.J., Gijzen, H.J. and van Loosdrecht, M.C.M. (2005) Modelling nitrification, heterotrophic growth and predation in activated sludge. *Water Res.* **39**, 5080–5098.

Moussa, M.S., Rojas, A.R., Hooijmans, C.M., Gijzen, H.J. and van Loosdrecht, M.C.M. (2004) Model-based evaluation of nitrogen removal in a tannery wastewater treatment plant. *Water Sci. Technol.* **50**(6), 251–260.

Mueller, J.A. and Boyle, W.C. (1988) Oxygen transfer under process conditions. *J. Water Pollut. Control Fed.* **60**, 332–341.

Mueller, J.A., Boyle, W.C. and Popel, H.J. (2002) Aeration: Principles and Practice. CRC Press, Boca Raton (USA).

Mulder, A., Vandegraaf, A.A., Robertson, L.A. and Kuenen, J.G. (1995) Anaerobic ammonium oxidation discovered in a denitrifying fluidized-bed reactor. *FEMS Microbiol. Ecol.* **16**, 177–183.

Mulder, J.W., van Loosdrecht, M.C.M., Hellinga, C. and van Kempen, R. (2001) Full-scale application of the SHARON process for treatment of rejection water of digested sludge dewatering. *Water Sci. Technol.* **43**(11), 127–134.

Mulkerrins, D., Dobson, A.D.W. and Colleranb, E. (2004) Parameters affecting biological phosphate removal from wastewaters. *Environ. International* **30**, 249–259.

Mulkerrins, D., Jordan, C., McMahon, S. and Colleran, E. (2000) Evaluation of the parameters affecting nitrogen and phosphorus removal in anaerobic/anoxic/oxic (A/A/O) biological nutrient removal systems. *J. Chem. Technol. Biotechnol.* **75**, 261–268.

Muller, A., Wentzel, M.C., Loewenthal, R.E. and Ekama G.A. (2003) Heterotroph anoxic yield in anoxic aerobic activated sludge systems treating municipal wastewater. *Water Res.* **37**, 2435–2441.

Müller-Rechberger, H., Wandl, G., Winkler, S., Svardal, K. and Matsche, N. (2001) Comparison of different operational modes of a two-stage activated sludge pilot plant for the extension of the Vienna STP. *Water Sci. Technol.* **44**(1), 137–144.

Murnleitner, E., Kuba, T., van Loosdrecht, M.C. and Heijnen, J.J. (1997) An integrated metabolic model for the aerobic and denitrifying biological phosphorus removal. *J. Environ. Eng. ASCE* **54**, 434–450.

Murphy, K.L. and Boyko, B.I. (1970) Longitudinal mixing in spiral flow aeration tanks. *J. San. Eng. ASCE* **96**(2), 211–221.

Murphy, K.L. and Timpany, P.L. (1967) Design and analysis of mixing for an aeration tank. *J. San. Eng. ASCE* **93**(5), 1–15.

Murphy, K.L., Sutton, P.M. and Jank, B.E. (1977) Dynamic nature of nitrifying biological suspended growth systems. *Prog. Water Technol.* **9**(2), 279–290.

Naidoo, V., Urbain, V. and Buckley, C.A. (1998) Characterization of wastewater and activated sludge from European municipal wastewater treatment plants using the NUR test. *Water Sci. Technol.* **38**(1), 303–310.

NASA (1995) Systems Engineering Handbook. The National Aeronautics and Space Administration Program/Project Management Initiative (PPMI), NASA Headquarters, Washington, DC.

Ni, B.J. and Yu, H.Q. (2007) A new kinetic approach to microbial storage process. *Appl. Microbiol. Biotechnol.* **76**, 1431–1438.

Nicholls H.A., Osborn D.W. and Pitman A.R. (1987) Improvement to the stability of the biological phosphate removal process at the Johannesburg Northern Works. In *Advances in Water Pollution Control: Biological Phosphate Removal from Wastewaters*, Pergamon Press, Oxford, 261–272.

Nielsen, M.K. (2001) Control of wastewater systems in practice. Scientific and Technical Report (part 3) presented at the 1st IWA Specialized Conference on "*Instrumentation, Control and Automation*", 3–7 June 2001, Malmo (Sweden), 82 pp. (unpublished)

Nielsen, P.H., Roslev, P., Dueholm, T. and Nielsen, J.L. (2002) Microthrix parvicella, a specialized lipid consumer in anaerobic – aerobic activated sludge plants. *Water Sci. Technol.* **46**(1/2), 73–80.

Nolasco, D.A., Daigger, G.T., Stafford, D.R., Kaupp, D.M. and Stephenson, J.P. (1998) The use of mathematical modeling and pilot plant testing to develop a new biological phosphorus and nitrogen removal process. *Water Environ. Res.* **70**, 1205–1215.

Novak, L. and Havrlikova, D. (2004) Performance intensification of Prague wastewater treatment plant. *Water Sci. Technol.* **50**(7), 139–146.

Novak, L., Larrea, L. and Wanner, J. (1994) Estimation of maximum specific growth rate of heterotrophic and autotrophic biomass: A combined technique of mathematical modelling and batch cultivations. *Water Sci. Technol.* **30**(11), 171–180.

Novak, L., Larrea, L. and Wanner, J. (1995) Mathematical model for soluble carbonaceous substrate biosorption. *Water Sci. Technol.* **31**(2), 67–77.

Novick, A. and Szilard, L. (1950) Experiments with the chemostat on spontaneous mutations of bacteria. In *Proc. Nat. Acad. Sci.* **36**, 708–719.

Novotny, V. and Krenkel, P.A. (1973) Evaporation and heat balance in aerated basins. *AIChE Symposium Series* **70**, 150–159.

Novotny, V., Jones, H., Feng, X. and Capodaglio, A. (1991) Time series analysis models of activated sludge plants. *Water Sci. Technol.* **23**(4/6), 1107–1116.

Nowak, O., Franz, A., Svardal, K., Muller, V. and Kuhn, V. (1999) Parameter estimation for activated sludge models with the help of mass balances. *Water Sci. Technol.* **39**(4), 113–120.

Nowak, O., Schweighofer, P. and Svardal, K. (1994) Nitrification inhibition a method for the estimation of actual maximum autotrophic growth rates in activated sludge systems. *Water Sci. Technol.* **30**(6), 9–19.

Nyberg, U., Aspegren, H., Andersson, B., Jansen la, J.C. and Villadsen, I.S. (1992) Full–scale application of nitrogen removal with methanol as carbon source. *Water Sci. Technol.* **26**(5/6), 1077–1086.

Oda, T., Yano, T. and Niboshi, Y. (2006) Development and exploitation of a multipurpose CFD tool for optimisation of microbial reaction and sludge flow. *Water Sci. Technol.* **53**(3), 101–110.

Oh, J. and Silverstein, J. (1999) Oxygen inhibition of activated sludge denitrification. *Water Res.* **33**, 1925–1937.

Oles, J. and Wilderer, P.A. (1991) Computer aided design of sequencing batch reactors based on the IAWPRC Activated Sludge Model. *Water Sci. Technol.* **23**(4/6), 1087–1095.

Oleszkiewicz, J.A., Kalinowska, E., Dold, P., Barnard, J.L., Bieniowski, M., Ferenc, Z., Jones, R., Rypina, A. and Sudol, J. (2004) Feasibility studies and pre-design simulation of Warsaw's new wastewater treatment plant. *Environ. Technol.* **25**, 1405–1411.

Olsson, G. and Andrews, J.F. (1978) The dissolved oxygen profile – a valuable tool for control of the activated sludge process. *Water Res.* **12**, 985–1004.

Olsson, G. and Newell, B. (1999) Wastewater Treatment Systems. Modelling, Diagnosis and Control. IWA Publishing, London.

Orhon, D., Karahan, O. and Sozen, S. (1999) The effect of residual microbial products on the experimental assessment of the particulate inert COD in wastewaters. *Water Res.* **33**, 3191–3203.

Orhon, D. and Cokgor, E.U. (1997) COD fractionation in wastewater characterization – The state of the art. *J. Chem. Technol. Biotechnol.* **68**, 283–293.

Orhon, D. and Artan, N. (1994) Modelling of Activated Sludge Systems. Technomic Pub. Co., Lancaster, PA (USA).

Orhon, D., Artan, N. and Ates, E. (1994a) A description of three methods for the determination of the initial inert particulate chemical oxygen-demand of wastewater. *J. Chem. Technol. Biotechnol.* **61**, 73–80.

Orhon, D., Sozen, S. and Ubayo, E. (1994b) Assessment of nitrification-denitrification potential of Istanbul domestic wastewaters. *Water Sci. Technol.* **30**(6), 21–30.

Orhon, D., Artan, N. and Cimsit, Y. (1989) The concept of soluble residual product formation in the modelling of activated sludge. *Water Sci. Technol.* **21**(4/5), 339–350.

Orhon, D., Artan, N., Buyukmurat, S. and Gorgun, E. (1992) The effect of residual COD on the biological treatability of textile wastewaters. *Water Sci. Technol.* **26**(3/4), 815–825.

Orhon, D., Ates, E., Sozen, S. and Cokgor, E.U. (1997) Characterization and COD fractionation of domestic wastewaters. *Environ. Pollut.* **95**, 191–204.

Orhon, D., Cokgor, E.U. and Sozen, S. (1999) Experimental basis for the hydrolysis of slowly biodegradable substrate in different wastewaters. *Water Sci. Technol.* **39**(1), 87–95.

Orhon, D., Okutman, D. and Insel, G. (2002) Characterisation and biodegradation of settleable organic matter for domestic wastewater. *Water SA* **28**, 299–306.

Orhon, D., Sozen, S. and Artan, N. (1996) The effect of heterotrophic yield on the assessment of the correction factor for anoxic growth. *Water Sci. Technol.* **34**(5/6), 67–74.

Ossenbruggen, P.J., Spanjers, H. and Klapwijk, A. (1996) Assesment of a two-step nitrification model for activated sludge. *Water Res.* **30**, 939–953.

Ottengraf, S.P.P. and Rietema, K. (1969) The influence of mixing on the activated sludge process in industrial aeration basins. *J. Water Pollut. Control Fed.* **41**, R282-R293.

Otterpohl, R. and Freund, M. (1992) Dynamic models for clarifiers of activated sludge plants with dry and wet weather flows. *Water Sci. Technol.* **26**(5/6), 1391–1400.

Ozinsky, A.E. and Ekama, G.A. (1995) Secondary settling tank modelling and design Part 1: review of theoretical and practical developments. *Water SA* **21**, 325–332.

Ozinsky, A.E., Ekama, G.A. and Reddy, B.D. (1994) Mathematical simulation of dynamic behaviour of secondary settling tanks. Research report W85. Department of Civil Engineering, University of Cape Town, South Africa.

Pagilla, K.R., Czerwionka, K., Urgun-Demirtas, M. and Makinia, J. (2008) Nitrogen speciation in wastewater treatment plant influents and effluents - the US and Polish case studies. *Water Sci. Technol.* **57**(10), 1511–1517.

Painter, H.A. (1970) A review of the literature on inorganic nitrogen metabolism. *Water Res.* **4**, 393–450.

Panikov, N.S. (1995) Microbial Growth Kinetics. Chapman & Hall, New York.

Papp, M. and Zelenka, S. (2007) Wastewater treatment and sludge disposal in Vienna. In *Proc. 10th IWA Specialized Conference on Design, Operation and Economics of Large Wastewater Treatment Plants*, Vienna, 9–13 September, 1–8.

Pardo, A.L.P., Brdjanovic, D., Moussa, M.S., Lopez-Vazquez, C.M., Meijer, S.C.F., Vanstraten, H.H.A., Janssen, A.J.H., Amy, G. and van Loosdrecht, M.C.M. (2007) Modelling of an oil refinery wastewater treatment plant. *Environ. Technol.* **28**, 1273–1284.

Parker, D.S. and Wanner, J. (2007) Improving nitrification through bioaugmentation. In *Proc. WEF/IWA Spec. Conf. "Nutrient Removal 2007. The State of the Art"*, Baltimore (USA), 4–7 March, 740–765.

Pasveer, A. (1959) A contribution to the development in activated-sludge treatment. *J. Proc. Inst. Sew. Purif.* **4**, 436–465.

Patry, G.G. and Chapman, D. (eds.) (1989) Dynamic Modeling and Expert Systems in Wastewater Engineering. Lewis Publishers, Chelsea, MI (USA).

Pedersen, J. and Sinkjaer, O. (1992) Test of the activated sludge model's capabilities as a prognostic tool on a pilot scale wastewater treatment plant. *Water Sci. Technol.* **25**(6), 185–194.

Penfold, W.J. and Norris, D. (1912) The relation of concentration of food supply to the generation time for bacteria. *J. Hyg.* **12**, 527–531.

Penya-Roja, J.M., Seco, A., Ferrer, J. and Serralta, J. (2002) Calibration and validation of Activated Sludge Model No.2d for Spanish municipal wastewater. *Environ. Technol.* **23**, 849–862.

Petersen, B., Gernaey, K., Henze, M. and Vanrolleghem, P.A. (2002) Evaluation of an ASM1 model calibration procedure on a municipal–industrial wastewater treatment plant. *J. Hydroinformatics* **4**(1), 15–38.

Petersen, E.E. (1965) Chemical Reaction Analysis. Prentice-Hall, Engelwood Cliffs, NJ (USA).

Pflanz, P. (1969) Performance of activated sludge secondary sedimentation basins. In *Adv. Water Pollut. Res., Proc. 4th Int. Conf.*, (ed.) S.H. Jenkins, Pergamon Press, New York, 569–581.

Phan, L.-C. and Rosenwinkel, K.-H. (2004) Adsorption ability of activated sludge and ist application to bulking sludge control with selector. In *Water Environ. Manag. Series (Young Researchers 2004)*, P. Lens and R. Stuetz (eds), IWA Publishing, London, 11–20.

Phillips, H.M., Sahlstedt, K.E., Frank, K., Bratby, J., Brennan, W., Rogowski, S., Pier, D., Anderson, W., Mulas, M., Copp, J.B. and Shirodkar, N. (2009) Wastewater treatment modelling in practice: a collaborative discussion of the state of the art. *Water Sci. Technol.* **59**(4), 695–704.

Pijuan, M., Saunders, A.M., Guisasola, A., Baeza, J.A., Casas, C. and Blackall, L.L. (2004) Enhanced biological phosphorus removal in a sequencing batch reactor using propionate as the sole carbon source. *Biotechnol. Bioeng.* **85**, 56–67.

Pitman, A.R. (1984) Settling of nutrient removal activated sludges. *Water Sci. Technol.* **17**(4/5), 493–504.

Platt, H.L. (2004) "Clever microbes:" bacteriology and sanitary technology in Manchester and Chicago during the progressive age. *Osiris* **19**, 149–166.

Plosz, B., Jobbagy, A. and Grady, C.P.L. Jr. (2003) Factors influencing deterioration of denitrification by oxygen entering an anoxic reactor through the surface. *Water Res.* **37**, 853–863.

Plosz, B.G., Weiss, M., Printemps, C., Essemiani, K. and Meinhold, J. (2007). One-dimensional modelling of the secondary clarifier-factors affecting simulation in the clarification zone and the assessment of the thickening flow dependence. *Water Res.* **41**, 3359–3371.

Pochana, K. and Keller, J. (1999) Study of factors affecting simultaneous nitrification and denitrification (SND). *Water Sci. Technol.* **39**(6), 61–68.

Pochana, K., Keller, J. and Lant, P. (1999) Model development for simultaneous nitrification and denitrification. *Water Sci. Technol.* **39**(1), 235–243.

Poduska, R.A. and Andrews, J.F. (1975) Dynamics of nitrification in the activated sludge process. *J. Water Pollut. Control Fed.* **47**, 2599–2619.

Popel, H.J. and Wagner, M. (1994) Modelling of oxygen transfer in deep diffused-aeration tanks and comparison with full-scale plant data. *Water Sci. Technol.* **30**(4), 71–80.

Porter, J.E. (1921) The activated sludge process of sewage treatment. General Filtration Co., Rochester, NY.

Potier, O., Leclerc, J.P. and Pons, M.N. (2005) Influence of geometrical and operational parameters on the axial dispersion in an aerated channel reactor. *Water Res.* **39**, 4454–4462.

Potter, T.G., Koopman, B. and Svoronos, S.A. (1996) Optimization of a periodic biological process for nitrogen removal from wastewater. *Water Res.* **30**, 142–152.

Prosser, J.I. (1990) Mathematical modeling of nitrification processes. *Adv. Microb. Ecol.* **11**, 263–304.

Puig, S., van Loosdrecht, M.C.M., Colprim, J. and Meijer, S.C.F. (2008) Data evaluation of full-scale wastewater treatment plants by mass balance. *Water Res.* **42**, 4645–4655.

Rabinowitz, B. and Marais, G.v.R. (1980) Chemical and biological phosphorus removal in the activated sludge. Research Report W32, Dept. of Civil Engineering, University of Cape Town, Rondeboesch, 7700, South Africa.

Ramanathan, M. and Gaudy, A.F. Jr. (1971) Steady state model for activated sludge with constant recycle sludge concentration. *Biotechnol. Bioeng.* **13**, 125–145.

Randall, A.A. and Liu, Y. (2002) Polyhydroxyalkanoates form potentially a key aspect of aerobic phosphorus uptake in enhanced biological phosphorus removal. *Water Res.* **36**, 3473–3478.

Randall, C.W. and Buth, D. (1984) Nitrite build-up in activated sludge resulting from temperature effects. *J. Water Pollut. Control Fed.* **56**, 1039–1044.

Randall, C.W., Benefield, L.D. and Buth, D. (1982) The effects of temperature on the biochemical reaction rates of the activated sludge process. *Water Sci. Technol.* **14**(1/2), 413–430.

Raphael, J.M. (1962) Prediction of temperatures in rivers and reservoirs. *J. Power* **88**, 157–181.

Ratkowsky, D.A., Lowry, R.K., McMeekin, T.A., Stokes, A.N. and Chandler, R.E. (1983) Model for bacterial culture-growth rate throughout the entire biokinetic temperature-range. *J. Bacteriol.* **154**, 1222–1226.

Ratkowsky, D.A., Olley, J., McMeekin, T.A. and Ball, A. (1982) Relationship between temperature and growth-rate of bacterial cultures. *J. Bacteriol.* **149**, 1–5.

Redmon, D., Boyle, W.C. and Ewing, L. (1983) Oxygen transfer efficiency measurements in mixed liquor using off-gas techniques. *J. Water Pollut. Cont. Fed.* **55**, 1338–1347.

Reinius, L.G. and Hultgren, J. (1988) Evaluation of the efficiency of a new aeration system at Henriksdal Sewage Treatment Plant. *Water Sci. Technol.* **20**(4/5), 85–92.

Richardson J.F. and Zaki, W.N. (1954) The sedimentation of a suspension of uniform spheres under conditions of viscous flow. Chem. Eng. Sci. **3**, 65–73.

Rieger, L., Comeau, Y. and Siegrist, H. (2005) Planning of measuring campaigns for simulation studies – a procedure to obtain high quality data. *Presented at the "Modeling workshop at the WEF WEFTEC 2005"*, Washington, DC, 3–5 November.

Rieger, L., Koch, G., Kuhni, M., Gujer, W. and Siegrist, H. (2001) The EAWAG Bio-P module for Activated Sludge Model No. 3. *Water Res.* **35**, 3887–3903.

Rivas, A., Ayesa, E., Galarza, A. and Salterain, A. (2001) Application of mathematical tools to improve the design and operation of activated sludge plants. Case study: the new WWTP of Galindo-Bilbao. Part I: Optimum design. *Water Sci. Technol.* **43**(7), 157–165.

Roberts, N., Andersen, D., Deal, R., Garet, M. and Shaffer, W. (1983) Introduction to computer simulation: A system dynamics approach. Productivity Press, Portland, OR.

Roeleveld P. J. and Kruit J. (1998) Guidelines for wastewater characterization in the Netherlands. *Korrespondenz Abwasser* **45**, 465–468 (in German).

Roeleveld, P.J. and van Loosdrecht, M.C.M. (2002) Experience with guidelines for wastewater characterisation in the Netherlands. *Water Sci. Technol.* **45**(6), 77–87.

Rosen, B. and Huljbregsen C. (2003) The ScanDeNi (R) process could turn an existing under-performing activated sludge plant into an asset. *Water Sci. Technol.* **47**(11), 31–36.

Rosenwinkel, K.-H., Beier, M., Phan, L.-C. and Hartwig, P. (2007a) Conventional and advanced technologies for biological nitrogen removal in Europe. In *Proc. 10th IWA Specialized Conference on Design, Operation and Economics of Large Wastewater Treatment Plants*, Vienna, 9–13 September, 41–44.

Rosenwinkel, K.-H., Makinia, J. and Pabst, M. (2007b) Experimental and model-based evaluation of the maximum allowable flows in activated sludge systems under storm conditions. In *Proc. the WEF-IWA Spec. Conf. "Nutrient Removal 2007: State of the Art"*, Baltimore, 4–6 March [CD-Rom].

Rosenwinkel, K.-H., Wichern, M., Lippert, C., Arnold, B. and Fengler, T. (2002) DENIKAplus – program for dimensioning and optimisation of biological wastewater treatment plants (Manual). Institute of Sanitary Engineering and Waste Management, Hanover (Germany).

Rossetti, S. Tomei, M.C., Nielsen, P.H. and Tandoi, V. (2005) "Microthrix parvicella", a filamentous bacterium causing bulking and foaming in activated sludge systems: a review of current knowledge. *FEMS Microbiol. Rev.* **29**, 49–64.

Rossetti, S., Tomei, M.C., Levantesi, C., Ramadori, R. and Tandoi, V. (2002) "Microthrix parvicella": a new approach for kinetic and physiological characterization. *Water Sci. Technol.* **46**(1/2), 65–72.

Rosso, D. (2005) Mass transfer at contaminated bubble interfaces. Ph.D. dissertation, Univ. of California, Los Angeles.

Rosso, D. and Stenstrom, M.K. (2006) Surfactant effects on α-factors in aeration systems. *Water Res.* **40**, 1397–1404.

Rosso, D., Larson, L.E. and Stenstrom, M.K. (2008) Aeration of large-scale municipal wastewater treatment plants: state of the art. *Water Sci. Technol.* **57**(7), 973–978.

Rozich, A.F. and Castens, D.J. (1986) Inhibition kinetics of nitrification in continuous-flow reactors. *J. Water Pollut. Control Fed.* **58**, 220–226.

Ruano, M.V., Ribes, J., De Pauw, D.J.W. and Sin, G. (2007) Parameter subset selection for the dynamic calibration of activated sludge models (ASMs): experience versus systems analysis. *Water Sci. Technol.* **56**(8), 107–115.

Ruel, S.M., Ginestet, P, Akerman, A. and Audic, J.M. (2005) Nitrite production over digester supernatant and landfill leachate at low cost operating conditions. In: *Proc. the IWA Spec. Conf. "Nutrient Management in Wastewater Treatment Processes and Recycle Streams"*, Cracow, 19–21 September, 445–453.

Sahlstedt, K.E., Aurola, A.M. and Fred, T. (2003) Practical modelling of a large activated DN-process with ASM3. In *Proc. 9th IWA Specialized Conference on Design, Operation and Economics of Large Wastewater Treatment Plants*, I. Ruzickova and J. Wanner (eds), Prague, 1–4 September, 141–148.

Salem, S., Berends, D., Heijnen, J.J. and van Loosdrecht, M.C.M. (2002) Model-based evaluation of a new upgrading concept for N-removal. *Water Sci. Technol.* **45**(6), 169–176.

Salem, S., Berends, D., Heijnen, J.J. and van Loosdrecht, M.C.M. (2003) Bio-augmentation by nitrification with return sludge. *Water Res.* **37**, 1794–1804.

Salem, S., Berends, D.H.J.G., van der Roest, H.F., van der Kuij, R.J. and van Loosdrecht, M.C.M. (2004) Full-scale application of the BABE technology. *Water Sci. Technol.* **50**(7), 87–96.

San, H.A. (1994) Impact of dispersion and reaction kinetics on performance by biological reactors – solution by "S" series. *Water Res.* **28**, 1639–1651.

Sarioglu, M. and Horan, N.J. (1996) An equation for the empirical design of anoxic zones used to eliminate rising sludges at nitrifying activated sludge plants. *Water Sci. Technol.* **33**(3), 185–194.

Satoh, H., Mino, T. and Matsuo, T. (1992) Uptake of organic substrates and accumulation of polyhydroxyalkanoates linked with glycolysis of intracellular carbohydrates under anaerobic conditions in the biological excess phosphate removal process. *Water Sci. Technol.* **26**(5/6), 933–942.

Satoh, H., Okuda, E., Mino, T. and Matsuo, T. (2000) Calibration of kinetic parameters in the IAWQ Activated Sludge Model: a pilot scale experience. *Water Sci. Technol.* **42**(3/4), 29–34.

Saunders, A.M., Oehmen, A., Blackall, L.L., Yuan, Z. and Keller, J. (2003) The effect of GAOs (glycogen accumulating organisms) on anaerobic carbon requirements in fullscale Australian EBPR (enhanced biological phosphorus removal) plants. *Water Sci. Technol.* **47**(11), 37–43.

Sawyer, C.N. and Bradney, L. (1945) Rising of activated sludge in final settling tanks. *Sew. Works J.* **17**, 1191–1209.

Sawyer, C.N. (1965) Milestones in the development of the activated sludge process. *J. Water Pollut. Control Fed.* **37**, 151–162.

Sayigh, B.A. and Malina, J.F. Jr (1978) Temperature effects on the activated sludge process. *J. Wat. Pollut. Cont. Fed.* **50**, 678–687.

Scheer, H. and Seyfried, C.F. (1997) Enhanced biological phosphate removal: Modelling and design in theory and practice. *Water Sci. Technol.* **35**(10), 43–52.

Scherfig, J., Schleisner, L., Brond, S. and Kilde, N. (1996) Dynamic temperature changes in wastewater treatment plants. *Water Environ. Res.* **68**, 143–151.

Schlegel, S. (1992) Operational results of wastewater treatment plants with biological N and P elimination. *Water Sci. Technol.* **25**(4/5), 105–112.

Schmid, M., Twachtmann, U., Klein, M., Strous, M., Juretschko, S., Jetten, M.S.M., Metzger, J., Schleifer, K.H. and Wagner, M. (2000) Molecular evidence for genus level diversity of bacteria capable of catalyzing anaerobic ammonium oxidation. *Syst. Appl. Microbiol.* **23**, 93–106.

Schmidt, I., Sliekers, O., Schmid, M., Bock, E., Fuerst, J., Kuenen, J.G., Jetten, M.S.M. and Strous, M. (2003) New concepts of microbial treatment processes for the nitrogen removal in wastewater. *FEMS Microbiol. Rev.* **27**, 481–492.

Schonberger, R. (1990) Conversion of existing primary clarifiers according to the EASC process for biological phosphorus removal. *Water Sci. Technol.* **22**(7/8), 45–51.

Schoolfield, R.M., Sharpe, P.J.H. and Magnuson, C.E. (1981) Non-linear regression of biological temperature-dependent rate models based on absolute reaction-rate theory. *J. Theor. Biol.* **88**, 719–731.

Schroeder, E.D. (1977) Water and Wastewater Treatment. McGraw-Hill, New York.

Schuler, A.J. and Jang, H. (2007) Density effects on activated sludge zone settling velocities. *Water Res.* **41**, 1814–1822.

Schulze, K.L. (1956) The effect of phosphorus supply on the rate of growth and fat formation in yeasts. *Appl. Microbiol.* **4**, 207–210.

Schunn, C.D. and Wallach, D. (2005) Evaluating goodness-of-fit in comparison of models to data. In *Psychologie der Kognition: Reden and Vorträge anlässlich der Emeritierung von Werner Tack*, W. Tack (Ed.), University of Saarland Press, Saarbruecken, 115–154.

Schütze, M.R. Butler, D. and Beck, M.B. (2002) Modelling, simulation and control of urban wastewater systems. Springer-Verlag, London.

Scott, K.J. (1966) Mathematical models of mechanism of thickening. *Ind. Eng. Chem. Fundamentals*, **2**, 203–211.

Scott, K.J. (1968) Theory of thickening: factors affecting settling rate of solids in flocculated pulps. *Trans. Inst. Min. Met. (section C)* **77**, 85–97.

Seco, A., Ferrer, J., Serralta, J., Ribes, J., Barat, R. and Bouzas, A. (2004) Use of biological and sedimentation models for designing Peniscola WWTP. *Environ. Technol.* **25**, 681–687.

Sedlak, R. (ed.) (1991) Phosphorus and Nitrogen Removal from Municipal Wastewater. Principles and Practice. Lewis Publisher, Chelsea, MI.

Sedory, P.E. and Stenstrom, M.K. (1995) Dynamic Prediction of wastewater aeration basin temperature. *J. Environ. Eng. ASCE* **121**, 609–618.

Seeger, H. (1999) The history of German waste water treatment. *Europ. Water. Manag.* **2**(5), 51-56.

Seggelke., K. and Rosenwinkel, K.-H. (2002) Online-simulation of the WWTP to minimise the total emission of WWTP and sewer system. *Water Sci. Technol.* **45**(3), 101–108.

Seicz, G., Endrodi, G. and Tajchman, S. (1969) Aerodynamic and surface factors in evaporation. *Water Resour. Res.* **5**, 380–393.

Shah, D.B. and Coulman, G.A. (1978) Kinetics of nitrification and denitrification reactions. *Biotechnol. Bioeng.* **20**, 43–72.

Shannon, P.T., Stroupe E. and Tory E.M. (1963) Batch and continuous thickening. *Ind. Eng. Chem.* **2**, 203–211.

Sharma, B. and Ahlert, R.C. (1997) Nitrification and nitrogen removal. *Water Res.* **11**, 897–925.

Shoji, T., Satoh, H. and Mino, T. (2003) Quantitative estimation of the role of denitrifying phosphate accumulating organisms in nutrient removal. *Water Sci. Technol.* **47**(11), 23–29.

Siebritz, I.P., Ekama, G.A. and Marais, C.v.R. (1980) Excess biological phosphorus removal In the activated sludge process. In *Proc. Int. Symp. on Waste Treatment and Utilization (Vol. 2)*, Pergamon Press, Toronto, 233–251.

Siegrist, H. and Tschui, M. (1992) Interpretation of experimental data with regard to the Activated Sludge Model No. 1 and calibration of the model for municipal wastewater treatment plants. *Water Sci. Technol.* **25**(6), 167–183.

Siegrist, H., Brunner, I., Koch, G., Phan, L.C. and van Chieu, L. (1999) Reduction of biomass decay rate under anoxic and anaerobic conditions. *Water Sci. Technol.* **39**(1), 129–137.

Siegrist, H., Krebs, P., Buhler, R., Purtschert, I. and Rock, C. (1995) Denitrification in secondary clarifiers. *Water Sci. Technol.* **31**(2), 205–214.

Sin, G., Vanrolleghem, P.A. (2006) Evolution of an ASM2d-like model structure due to operational changes of an SBR process. *Water Sci. Technol.* **53**(12), 237–245.

Sin, G., van Hulle, S.W.H., de Pauw, D.J.W., van Griensven, A. and Vanrolleghem, P.A. (2005a) A critical comparison of systematic calibration protocols for activated sludge models: A SWOT analysis. *Water Res.* **39**, 2459–2474.

Sin, G., Guisasola, A., De Pauw, D.J.W., Baeza, J.A., Carrera, J. and Vanrolleghem, P.A. (2005b) A new approach for modelling simultaneous storage and growth processes for activated sludge systems under aerobic conditions. *Biotechnol. Bioeng.* **92**, 600–613.

Sin, G., De Pauw, D.J.W., Weijers, S. and Vanrolleghem, P.A. (2008a) An efficient approach to automate the manual trial and error calibration of activated sludge models. *Biotechnol. Bioeng.* **100**, 516–528.

Sin, G., Kaelin, D., Kampschreur, M.J., Takacs, I., Wett, B., Gernaey, K.V., Rieger, L., Siegrist, H. and van Loosdrecht, M. (2008b) Modelling nitrite in wastewater treatment systems: a discussion of different modelling concepts. *Water Sci. Technol.* **58**(6), 1155–1171.

Sin, G., Gernaey, K.V., Neumann, M.B., van Loosdrecht, M.C.M. and Gujer, W. (2009a). Uncertainty analysis in WWTP model applications: A critical discussion using an example from design. *Water Res.* **43**, 2894–2906.

Sin, G., Gernaey, K.V. and Lantz, A.E. (2009b) Good modeling practice for PAT applications: Propagation of input uncertainty and sensitivity analysis. *Biotechnol. Prog.* **25**, 1043–1053.

Sinkjaer, O., Yndgaard, L., Harremoes, P. and Hansen, J.L. (1994) Characterisation of the nitrification process for design puroposes. *Water Sci. Technol.* **30**(4), 47–56.

Skinner, F.A. and Walker, N. (1961) Growth of *Nitrosomonas europea* in batch and continuous culture. *Arch. Microbiol.* **38**, 339–349.

Slijkhuis, H. (1983) The physiology of the filamentous bacterium M. parvicella. Ph.D. Thesis, Agriculture College of Wageningen, the Netherlands.

Smith, M., Cooper, P., McMurchie, J., Stevenson, D., Mann, B., Stocker, D., Bayes, C. and Clark, D. (1998) Nitrification trials at Dunnswood sewage treatment works and process modelling using WRc STOAT. *Water Environ. J.* **12**, 157–162.

Smith, R.C. and Oerther, D.B. (2007) The impact of side-stream reactor configuration on functional stability in activated sludge. In *Proc. WEF/IWA Spec. Conf. "Nutrient Removal 2007. The State of the Art"*, Baltimore, 4–7 March, 455–469.

Smith, R.D. (1998) Essential techniques for military modeling and simulation. In *Proc. the 1998 Winter Simulation Conference "Simulation in the 21st Century"*, Washington, DC, 13–16 December, 805–812.

Smith, R.D. (1999) Simulation: the engine behind the virtual world. *Simulation 2000 Series* **1**, 1–24.

Smolders, G.J.F., van der Meij, J., van Loosdrecht, M.C.M. and Heijnen, J.J. (1994a) Model of anaerobic metabolism of the biological phosphorus removal process: stoichiometry and pH influence. *Biotechnol. Bioeng.* **42**, 461–470.

Smolders, G.J.F., van Loosdrecht, M.C.M. and Heijnen, J.J. (1994b) Stoichiometric model of the aerobic metabolism of the biological phosphorus removal process. *Biotechnol. Bioeng.* **43**, 837–848.

Smolders, G.J.F., van der Meij, J., van Loosdrecht, M.C.M. and Heijnen, J.J. (1995a) A structured metabolic model for the anaerobic and aerobic stoichiometry and kinetics of the biological phosphorus removal process. *Biotechnol. Bioeng.* **47**, 277–287.

Smolders, G.J.F., van der Meij, J., van Loosdrecht, M.C.M. and Heijnen, J.J. (1995b) A metabolic model for the biological phosphorus removal process: Effect of the sludge retention time. *Biotechnol. Bioeng.* **48**, 222–233.

Solley, D. and Barr, K. (1999) Optimise what you have first! Low cost upgrading of plants for improved nutrient removal. *Water Sci. Technol.* **39**(6), 127–134.

Sollfrank, U. and Gujer, W. (1991) Characterisation of domestic wastewater for mathematical modelling of the activated sludge process. *Water Sci. Technol.* **23**(4/6), 1057–1066.

Sollfrank, U., Kappeler, J. and Gujer, W. (1992) Temperature effects on wastewater characterization and the release of soluble inert organic material. *Water Sci. Technol.* **25**(6), 33–41.

Sorm, R., Bortone, G., Saltarelli, R., Jenicek, P., Wanner, J. and Tilche, A. (1996) Phosphate uptake under anoxic conditions and fixed film nitrification in nutrient removal activated sludge system. *Water Res.* **30**, 1573–1584.

Sorm, R., Wanner, J., Saltarelli, R., Bortone, G. and Tilche, A. (1997) Verification of anoxic phosphate uptake as the main biochemical mechanism of the DEPHANOX process. *Water Sci. Technol.* **35**(10), 87–94.

Souza, S.M., Araujo, O. and Coelho, M.A.Z. (2008) Model-based optimization of a sequencing batch reactor for biological nitrogen removal. *Biores. Technol.* **99**, 3213–3223.

Sozen, S., Orhon, D. and San, H.A. (1996) A new approach for the evaluation of the maximum specific growth rate in nitrification. *Water Res.* **30**, 1661–1669.

Spagni, A., Stante, L. and Bortone, G. (2002) Modelling population dynamics of denitrifying phosphorus accumulating organisms in activated sludge. *Water Sci. Technol.* **46**(1/2), 323–326.

Spanjers, H. and Vanrollenghem, P. (1995) Respirometry as a tool for rapid characterization of wastewater and activated sludge. *Water Sci. Technol.* **31**(2), 105–114.

Spanjers, H., Olsson, G. and Klapwijk, A. (1993) Determining influent short-term biochemical oxygen-demand by combined respirometry and estimation. *Water Sci. Technol.* **28**(11/12), 401–414.

Spanjers, H., Takacs. I. and Brouwer, H. (1999) Direct parameter extraction from respirograms for wastewater and biomass characterization. *Water Sci. Technol.* **39**(4), 137–145.

Spanjers, H., Vanrolleghem, P., Nguyen, K., Vanhooren, H. and Patry, G.G. (1998) Towards a simulation-benchmark for evaluating respirometry-based control strategies. *Water Sci. Technol.* **50**(12), 219–226.

Sperandio, M. and Paul, E. (2000) Estimation of wastewater biodegradable COD fractions by combining respirometric experiments in various S_0/X_0 ratios. *Water Res.* **34**, 1233–1246.

Spering, V., Makinia, J. and Rosenwinkel, K.-H. (2008) Analyzing and improving a mechanistic model for the Microthrix parvicella behaviour in activated sludge systems. In *Proc. 1st IWA/WEF Wastewater Treatment Modelling Seminar*, Mont Sainte Anne (Canada), 1-3 June, 79–88.

Spering, V., Schlösser, N., Märker, S. and Rosenwinkel, K.-H. (2007) Modelling biokinetics of Microthrix parvicella as a module for the ASM3. In *Proc. 10th IWA Spec. Conf. on Design, Operation and Economics of Large Wastewater Treatment Plants (Posters)*, Vienna, 9–13 September, 105–108.

Srinath, E.G., Loehr, R.C. and Prakasam, T.B.S. (1976) Nitrifying organism concentration and activity. *J. Environ. Eng. ASCE* **102**, 449–463.

Srinath, E.G., Sastry, C.A. and Pillai, S.C. (1959) Rapid removal of phosphorus from sewage by activated sludge. *Experientia* **15**, 339–340.

Stamou, A.I. (1994) Modeling oxidation ditches using the IAWPRC Activated Sludge Model with hydrodynamic effects. *Water Sci. Technol.* **30**(2), 185–192.

Stamou, A.I. (1995) Modelling of settling tanks – a critical review. *Trans. Ecol. Environ.*, L. C. Wrobel and P. Latinopoulos (eds.) **7**, 305–312.

Stamou, A.I., Katsiri, A., Mantziaras, I., Boshnakov, K., Koumanova, B. and Stoyanov, S. (1999) Modelling of an alternating oxidation ditch system. *Water Sci. Technol.* **39**(4), 169–176.

Standard Methods (1992) Standard Methods for the Examination of Water and Wastewater. 18th Edition, APHA, Washington, DC.

Steinour, H. H. (1944) Rate of sedimentation, nonflocculated suspensions of uniform spheres. *Ind. Eng. Chem.* **36**, 618–624.

Stensel, H.D. (1992) Pilot plant testing of BNR systems. In *Design and Retrofit of Wastewater Treatment Plants for Biological Nutrient Removal*. C.W. Randall, J.L. Barnard, H.D. Stensel (eds.). Technomic Pub. Co., Lancaster, PA.

Stensel, H.D. (2001) Biological nutrient removal: merging engineering innovation and science. In *Proc. 74th WEF WEFTEC 2001*, Atlanta, 13–17 October [CD-Rom].

Stensel, H.D., *et. al.* (2008) Dissolved organic nitrogen (DON) in biological nutrient removal wastewater treatment processes. Water Environment Research Foundation. http:www.werf.org/nutrients/LOTDissolvedOrganicNitrogen.

Stenstrom, M.K. (1975) A dynamic model and computer compatible control strategies for wastewater treatment plant. Ph.D. dissertation, Clemson University. Clemson, SC (USA).

Stenstrom, M.K. and Gilbert, R.G. (1981) Effects of alpha, beta and theta factor upon the design, specification and operation of aeration systems. *Water Res.* **15**, 643–654.

Stenstrom, M.K. and Poduska, R.A. (1980) The effect of dissolved oxygen concentration on nitrification. *Water Res.* **14**, 643–649.

Stenstrom, M.K. and Rosso, D. (2008) Aeration. In *Biological Wastewater Treatment. Principles, Modelling and Design.* M. Henze, M. C. M. van Loosdrecht, G.A. Ekama and D. Brdjanovic (eds.), IWA Publishing, London.

Stephenson, R.V. and Luker, M. (1994) Full-scale demonstration of nitrogen control by operationally modified plug-flow and step-feed activated sludge process. In Proc. 67th WEF WEFTEC 1994, Chicago, 15–19 October, **1**, 563–571.

Stokes, L., Tackacs, I., Watson, B. and Watts, J.B. (1993) Dynamic modelling of an ASP sewage works – a case study. *Water Sci. Technol.* **28**(11/12), 151–162.

Stokes, A.J., West, J.R., Forster, C.F., Kruger, R.C.A., de Bel, M. and Davies, W.J. (1997) Improvements to a stoat model of a full scale wastewater treatment works through the use of detailed mechanistic studies. *Water Sci. Technol.* **36**(5), 277–284.

Stokes, A.J., West, J.R., Forster, C.F. and Davies, W.J. (2000a) Understanding some of the differences between the COD- and BOD-based models offered in STOAT. *Water Res.* **34**, 1296–1306.

Stokes, A.J., Forster, C.F., West, J.R. and de Bel, M. (2000b) Process modelling at Oldham sewage treatment works using WRc STOAT. *Water Environ. J.* **14**, 15–21.

Stricker, A.-E., Takacs, I. and Marquot, A. (2007) Hindered and compression settling: parameter measurement and modelling. *Water Sci. Technol.* **56**(12), 101–110.

Strous, M., Van Gerven, E., Kuenen, J.G. and Jetten, M.S.M. (1997) Effects of aerobic and microaerobic conditions on anaerobi ammonium-oxidizing (anammox) sludge. *Appl. Environ. Microbiol.* **63**, 2446–2448.

Strous, M., Heijnen, J.J., Kuenen, J.G. and Jetten, M.S.M. (1998) The sequencing batch reactor as a powerful tool for the study of slowly growing anaerobic ammonium-oxidizing microorganisms. *Appl. Microbiol. Biotechnol.* **50**, 589–596.

Strous, M., Fuerst, J.A., Kramer, E.H.M., Logemann, S., Muyzer, G., van de Pas-Schoonen, K.T., Webb, R., Kuenen, J.G. and Jetten, M.S.M (1999a) Missing lithotroph identified as new Planctomycete. *Nature* **400**, 446–449.

Strous, M., Kuenen, J.G. and Jetten, M.S.M. (1999b) Key physiology of anaerobic ammonium oxidation. *Appl. Environ. Microbiol.* **65**, 3248–3250.

Sundstrom, D.W. and Klei, H.E. (1979) Wastewater Treatment. Prentice-Hall, Englewood Cliffs (USA)

Symons, J.M. and McKinney, R.E. (1958) The biochemistry of nitrogen in the synthesis of activated sludge. *Sew. Ind. Wastes* **30**, 874–890.

Takacs, I. (2008) Experiments in activated sludge modelling. Ph.D. Thesis, Ghent University.

Takacs, I., Patry, G.G. and Nolasco, D. (1991) A dynamic model of the clarification-thickening process. *Water Res.* **25**, 1263–1271.

Talati, S.N. and Stenstrom, M.K. (1990) Aeration basin heat loss. *J. Environ. Eng. ASCE* **116**, 70–86.

Tan, K.H. (1996) Soil sampling, preparation and analysis. Marcel Dekker, New York.

Tandoi, V., Rossetti, S., Blackall, L.L. and Majone, M. (1998) Some physiological properties of an Italian isolate of "Microthrix parvicella". *Water Sci. Technol.* **37**(4/5), 1–8.

Tanner, F. and Wallace, G. (1925) Relation of temperature to the growth of thermophilic bacteria. *J. Bact.* **10**, 421–437.

Temmink, H., Petersen, B., Isaacs, S. and Henze, M. (1996) Recovery of biological phosphorus removal after periods of low organic loading. *Water Sci. Technol.* **34**(1/2), 1–8.

Thackston, E.L. and Parker, F.L. (1972) Geographical influence on cooling ponds. *J. Wat. Pollut. Cont. Fed.* **44**, 1334–1351.

Thomas, D.G. (1963) Transport characteristics of suspension. Relation of hindered settling floc characteristics to rheological parameters. *AIChE J.* **9**, 310–316.

Thomas, H.A. and McKee, J.E. (1944) Longitudinal mixing in aeration tanks. *Sew. Works J.* **16**, 42–55.

Thomas, V.K., Chambers, B. and Dunn, W. (1989) Optimisation of aeration efficiency: a design procedure for secondary treatment using a hybrid aeration system. *Water Sci. Technol.* **21**(4), 1403–1419.

Tiller, F.M. (1981) Revision of Kynch sedimentation theory. *AICHE J.* **27**, 823–829.

Tsao, G.T. (1968) Simultaneous gas-liquid interfacial oxygen absorption and biochemical oxidation. *Biotech. Bioeng.* **10**, 765–785.

Tsivoglou, E.C., Cohen, J.B. and Shearer, S.D. (1968) Tracer measurements of atmospheric reaeration – II. Field Studies. *J. Water Pollut. Cont. Fed.* **40**, 285–305.

Turian, R.M., Fox, G.E. and Rice, P.A. (1975) The dispersed flow model for a biological reactor as applied to the activated sludge process. *The Can. J. Chem. Eng.* **53**, 431–437.

U.S. EPA (1983) Development of Standard Procedures for Evaluating Oxygen Transfer Devices. EPA/600/2-83/002, Municipal Environ. Research Lab., Cincinnati, OH.

U.S.EPA (1987a) Design Manual. Phosphorus Removal. EPA/625/I-87/001, Center for Environmental Research Information, Cincinnati, OH.

U.S.EPA (1987b) QUAL2E – the enhanced stream water quality model. EPA/823/B-95/003, Environmental Research Laboratory, Athens, GA.

U.S. EPA (1993) Manual Nitrogen Control. EPA/625/R-93/010, U.S. EPA, Washington, DC.

Ubukata, Y. (2005) Role of particulate organic matter and acetic acid for phosphate removal in anaerobic/aerobic activated sludge process. In *Proc. IWA Spec. Conf. "Nutrient Management in Wastewater Treatment Processes and Recycle Streams"*, Cracow, 19–21 September, 183–192.

USGS (1982) Measurement of time of travel and dispersion in streams by dye tracing. United States Government Printing Office, Washington, DC.

USGS (1986) Fluorometric procedures for dye tracing. United States Government Printing Office, Washington, DC.

Vaerenbergh, E.V. (1980) Numerical computation of secondary settler area using batch settling data. *Trib. Cebedeau* **33**, 369–374.

van de Graaf A.A., De Bruijn, P., Robertson, L.A., Jetten, M.S.M. and Kuenen, J.G. (1996) Autotrophic growth of anaerobic ammonium-oxidizing micro-organisms in a fluidized bed reactor. *Microbiol.–UK* **142**, 2187–2196.

van de Graaf, A.A., Mulder, A., de Bruijn, P., Jetten, M.S., Robertson, L.A. and Kuenen, J.G. (1995) Anaerobic oxidation of ammonium is a biologically mediated process. *Appl. Environ. Microbiol.* **61**, 1246–1251.

van der Graf, J.H.J.M. (1976) Laten biologische zuiveringsprocessen zich naar temperatuur optimaliseren. H_2O **9**, 87–93 (in Dutch)

van der Star, W.R.L. Abma, W.R., Blommers, D., Mulder, J.W., Tokutomi, T., Strous, M., Picioreanu, C. and van Loosdrecht, M.C.M. (2007) Experiences from the first full-scale anammox reactor in Rotterdam. *Water Res.* **41**, 4149–4163.

van Dongen, U., Jetten, M.S.M., van Loosdrecht, M.C.M. (2001) The SHARON-Anammox process for treatment of ammonium rich wastewater. *Water Sci. Technol.* **44**(1), 153–160.

van Haandel, A.C., Ekama, G.A. and Marais, G.v.R. (1981) The activated sludge process 3 – single sludge denitrification. *Water Res.* **15**, 1135–1152.

van Hulle, S.W.H. and Vanrolleghem, P.A. (2004) Modelling and optimisation of a chemical industry wastewater treatment plant subjected to varying production schedules. *J. Chem. Technol. Biotechnol.* **79**, 1084–1091.

van Hulle S.W.H. (2005) Modelling, simulation and optimization of autotrophic nitrogen removal processes. Ph.D. Thesis. Faculty of Bioengineering Sciences, Ghent University.

van Hulle S.W.H., Volcke, E.I.P., Teruel, J.L., Donckels, B., van Loosdrecht, M.C.M. and Vanrolleghem, P.A. (2007) Influence of temperature and pH on the kinetics of the Sharon nitritation process. *J. Chem. Technol. Biotechnol.* **82**, 471–480.

van Loosdrecht, M.C.M. and Henze, M. (1999) Maintenance, endogeneous respiration, lysis, decay and predation. *Water Sci. Technol.* **39**(1), 107–117.

van Loosdrecht, M.C.M. and Salem, S. (2006) Biological treatment of sludge digester liquids. *Water Sci. Technol.* **53**(12), 11–20.

van Loosdrecht, M.C.M., Brandse, F.A. and de Vries, A.C. (1998) Upgrading of wastewater treatment processes for integrated nutrient removal – The BCFS process. *Water Sci Technol.* **37**(9), 209–217.

van Loosdrecht, M.C.M., Pot, M.A. and Heijnen J.J. (1997) Importance of bacterial storage polymers in bioprocesses. *Water Sci. Technol.* **35**(1), 41–47.

van Niekerk, A., Pitt, P. and Prentice, M. (2000) Rating of Henrico County BNR plant treatment capacity using a calibrated activated sludge model. In *Proc. 73rd WEF WEFTEC 2000*, Ahaheim (California), 15–18 October [CD-Rom].

van Veldhuizen, H.M., van Loosdrecht M.C.M. and Heijnen, J.J. (1999a) Modelling biological phosphorus and nitrogen removal in a full scale activated sludge process. *Water Res.* **33**, 3459–3468.

van Veldhuizen, H.M., van Loosdrecht, M.C.M. and Brandse, F.A. (1999b) Model based evaluation of plant improvement strategies for biological nutrient removal. *Water Sci. Technol.* **39**(4), 45–53.

Vand, V. (1948) Design of prototype thickeners from batch settling tests. *Water Sew. Works* **20**, 302–307.

Vandekerckhove, A., Moerman, W. and van Hulle, S.W.H. (2008) Full-scale modelling of a food industry wastewater treatment plant in view of process upgrade. *Chem. Eng. J.* **135**, 185–194.

Vanderhasselt, A. and Vanrolleghem, P.A. (2000) Estimation of sludge sedimentation parameters from single batch settling curves. *Water Res.* **34**, 395–406.

Vangheluwe, H., de Lara, J. and Mosterman, P. J. (2002) An Introduction to Multi-Paradigm Modelling and Simulation. In *Proc. AI Simulation and Planning in High Autonomy System*, Lisbon, 7–10 April, 9–20.

Vanhooren, H., Meirlaen, J., Amerlinck, Y., Claeys, F., Vangheluwe, H. and Vanrolleghem, P.A. (2003) WEST: modelling biological wastewater treatment. *J. Hydroinformatics* **5**(1), 27–50.

Vanrolleghem, P.A. and Coen, F. (1995) Optimal design in-sensor-experiments for on-line modeling of nitrogen removal processes. *Water Sci. Technol.* **31**(2), 149–160.

Vanrolleghem, P.A. and Jeppsson, U. (1995) Simulators for Modelling of WWTP. In *Optimizing the Design and Operation of Biological Wastewater Treatment Plants through the use of Computer Programmes based on Dynamic Modelling of the Process*, D. Dochain, P.A. Vanrolleghem and M. Henze (eds), European Commission, Directorate-General XII: Science, Research and Development, Luxembourg, 67–78.

Vanrolleghem, P.A., Insel, G., Petersen, B., Sin, G., de Pauw, D.J.W., Nopens, I., Weijers, S. and Gernaey, K. (2003) A comprehensive model calibration procedure for activated sludge models. In: *Proc. 76th WEF WEFTEC 2003*, Los Angeles, 11–15 October [CD-Rom].

Vanrolleghem, P.A., Spanjers, H., Petersen, B., Ginestet, P. and Takacs, I. (1999) Estimating (combinations of) Activated Sludge Model No. 1 parameters and components by respirometry. *Water Sci. Technol.* **39**(1), 195–214.

Vanrolleghem, P.A., van Daele, M. and Dochain, D. (1995) Practical identifiability of a biokinetic model of activated sludge respiration. *Water Res.* **29**, 2561–2570.

Vanrollenghem P.A. and Keesman, K.J. (1996) Identification of biodegradation models under model and data uncertainty. *Water Sci. Technol.* **33**(2), 91–105.

Vavilin, V.A. (1982) The theory and design of aerobic biological treatment. *Biotechnol. Bioeng.* **24**, 1721–1747.

Verdickt, L.B., Voitovich, T. V., Vandewalle, S., Lust, K., Smets, I.Y. and van Impe, J. F. (2006) Role of the diffusion coefficient in one-dimensional convection-diffusion models for sedimentation/thickening in secondary settling tanks. *Math. Comp. Model. Dyn. Syst.* **12**, 455–468.

Vesilind, P.A. (1968) Design of prototype thickener from batch settling tests. *Water Sew. Works* **115**, 302–307.

Vesilind, P.A. (2003) Wastewater Treatment Plant Design. IWA Publishing, London.

Vitasovic, Z. (1989) Continuous settler operation: a dynamic model. In *Dynamic Modeling and Expert Systems in Wastewater Engineering*, G.G. Patry and D.T. Chapman (eds), Lewis Publishers, Chelsea (USA), 59–81.

Volcke, E.I.P. (2006) Modelling, analysis and control of a SHARON reactor in view of its coupling with an Anammox process. Ph.D. Thesis, Faculty of Bioengineering Sciences, Ghent University.

von Munch, E. and Pollard, P.C. (1997) Measuring bacterial biomass-COD in wastewater containing particulate matter. *Water Res.* **31**, 2550–2556.

von Munch, E., Lant, P. and Keller, J. (1996) Simultaneous nitrification and denitrification in bench-scale sequencing batch reactors. *Water Res.* **30**, 277–284.

von Schulthess, R. and Gujer, W. (1996) Release of nitrous oxide (N_2O) from denitrifying activated sludge: verification and application of a mathematical model. *Water Res.* **30**, 521–530.

von Sperling, M. and Lumbers, J.P. (1991) Optimization of the operation of the oxidation ditch process incorporating a dynamic model. *Water Sci. Technol.* **24**(6), 225–233.

Wachtmeister, A., Kuba, T., van Loosdrecht, M.C.M. and Heijnen, J.J. (1997) A sludge characterization assay for aerobic and denitrifying phosphorus removing sludge. *Water Res.* **31**, 471–478.

Wagner, M. and Popel, H.J. (1996) Surface active agents and their influence on oxygen transfer. *Water Sci. Technol.* **34**(3/4), 249–256.

Wahlberg, E.J. and Keinath, T.M. (1988) Development of settling flux curves using SVI. *J. Water Pollut. Control Fed.* **60**, 2095–2100.

Wahlberg, E.J. and Keinath, T.M. (1995) Development of settling flux curves using SVI: an addendum. *Water Environ. Res.* **67**, 872–874.

Wall, D.J. and Petersen, G. (1986) Model for winter heat loss in uncovered clarifiers. *J. Environ. Eng. ASCE* **112**, 123–138.

Wanner, J. (1994) Activated sludge population dynamics. *Water Sci. Technol.* **30**(11), 159–170.

Wanner, J. (1998) Process theory: biochemistry, microbiology, kinetics, and activated sludge quality control. In *Activated Sludge Process Design and Control: Theory and Practice*. W. W. Eckenfelder, P. Grau (eds), Technomic Pub. Co., Lancaster, PA, 1–55.

Wanner, J., Cech, J.S. and Kos, M. (1992) New process design for biological nutrient removal. *Water Sci. Technol.* **25**(4/5), 445–448.

Wanner, J., Kos, M. and Grau, P. (1990) An innovative technology for upgrading nutrient removal activated sludge plants. *Water Sci. Technol.*, **22**(7/8), 9–20.

Wanner, J., Ruzickova, I., Krhutkova, O. and Pribyl, M. (2000) Activated sludge population dynamics and wastewater treatment plant design and operation. *Water Sci. Technol.* **41**(9), 217–225.

Water Research Commision (1984) Theory, Design and Operation of Nutrient Removal Activated Sludge Processes. Water Research Commission, Pretoria.

Watts, R.W., Svoronos, S.A. and Koopman, B. (1996) One-dimensional modeling of secondary clarifiers using a concentration and feed velocity-dependent dispersion coefficient. *Water Res.* **30**, 2112–2124.

Weber, W.J. (1972) Physicochemical Processes for Water Quality Control. Wiley, New York.

Wehrner, J.F. and Wilhelm, R.H. (1956) Boundary condition of flow reactors. *Chem. Eng. Sci.* **6**, 89–96.

Weijers, S.R. and Vanrolleghem, P.A. (1997) A procedure for selecting the most important parameters in calibrating the Activated Sludge Model No.1 with full-scale plant data. *Water Sci. Technol.* **36**(5), 69–79.

Wells, S.A. (1990) Effect of winter heat loss on treatment plant efficiency. *J. Wat. Pollut. Control Fed.* **62**, 34–39.

Wentzel, M. C., Loetter, L. H., Loewenthal, R. E. and Marais, G.v.R. (1986) Metabolic behaviour of Acinetobacter spp. in enhanced biological phosphorus removal – a biochemical model. *Water SA* **12**, 209–224.

Wentzel, M.C. and Ekama, G.A. (1997) Principles in the design of single-sludge activated-sludge systems for biological removal of carbon, nitrogen, and phosphorus. *Water Environ. Res.* **69**, 1222–1231.

Wentzel, M.C., Dold, P.L., Ekama, G.A. and Marais, G.v.R. (1985) Kinetics of biological phosphorus release. *Water Sci. Technol.* **17**(11/12), 57–71.

Wentzel, M.C., Dold, P.L., Ekama, G.A. and Marais, G.v.R. (1989) Enhanced polyphosphate organism cultures in activated sludge systems. Part III: Kinetic model. *Water SA* **15**, 89–102.

Wentzel, M.C., Dold, P.L., Ekama, G.A. and Marais, G.v.R. (1989a) Enhanced polyphosphate organism cultures in activated sludge systems. Part II: Experimental behaaviour. *Water SA* **15**, 71–88.

Wentzel, M.C., Dold, P.L., Ekama, G.A. and Marais, G.v.R. (1989b) Enhanced polyphosphate organism cultures in activated sludge systems. Part III: Kinetic model. *Water SA* **15**, 89–102.

Wentzel, M.C., Ekama, G.A. and Marais, G.v.R. (1992) Processes and modelling of nitrification denitrification biological excess phosphorus removal systems – a review. *Water Sci. Technol.* **25**(6), 59–82.

Wentzel, M.C., Loewenthal, R.E., Ekama, G.A. and Marais, G.v.R. (1988) Enhanced polyphosphate organism cultures in activated sludge systems. Part I: Enhanced culture development. *Water SA* **14**, 81–92.

Wentzel, M.C., Lötter, L.H., Ekama, G.A., Loewenthal, R.E. and Marais, G.v.R. (1991) Evaluation of biochemical models for biological excess phosphorus removal. *Water Sci. Technol.* **23**(4/6), 567–576.

Wentzel, M.C., Mbewe, A. and Ekama, G.A. (1995) Batch test for measurement of readily biodegradable COD and active organism concentrations in municipal waste-waters. *Water SA* **21**, 117–124.

Wentzel, M.C., Ekama, G.A. and Marais, G.v.R. (1992) Processes and modelling of nitrification denitrification biological excess phosphorus removal systems – a review. *Water Sci. Technol.* **25**(6), 59–82.

Wett, B. (2002) A straight interpretation of the solids flux theory for a three-layer sedimentation model. *Water Res.* **36**, 2949–2958.

Wett, B. and Rauch, W. (2003). The role of inorganic carbon limitation in biological nitrogen removal of extremely ammonia concentrated wastewater. *Water Res.* **37**, 1100–1110.

Whang, L.M., Filipe, C.D.M. and Park, J.K. (2007) Model-based evaluation of competition between polyphosphate- and glycogen-accumulating organisms. *Water Res.* **41**, 1312–1324.

Wichern, M., Lübken, M., Blömer, R. and Rosenwinkel, K.-H. (2003) Efficiency of the Activated Sludge Model No. 3 for German wastewater on six different WWTPs. *Water Sci. Technol.* **47**(11), 211–218.

Wichern, M., Obenaus F., Wulf, P. and Rosenwinkel, K.-H. (2001) Modelling of full-scale wastewater treatment plants with different treatment processes using the Activated Sludge Model No. 3. *Water Sci. Technol.* **44**(1), 49–56.

Wild, D., von Schulthess, R. and Gujer, W. (1995) Structured modelling of denitrification intermediates. *Water Sci. Technol.* **31**(2), 45–54.

Wilderer, P.A., Bungartz, H.-J., Lemmer, H., Wagner, M., Keller, J. and Wuertz, S. (2002) Modern scientific methods and their potential in wastewater science and technology. *Water Res.* **36**, 370–393.

Wilderer, P.A., Irvine, R.L. and Goronszy, M.C. (2001) Sequencing Batch Reactor Technology. IWA Publishing, London.

Wilderer, P.A., Warren. L.J. and Dau, U. (1987) Competition in denitrification systems affecting reduction rate and accumulation of nitrite. *Water Res.* **21**, 239–245.

Williamson, K.J. and McCarthy, P.L. (1975) Rapid measurement of Monod half-velocity coefficients for bacterial kinetics. *Biotechnol. Bioeng.* **14**, 915–924.

Wilson, E.M. (1974) Engineering Hydrology. MacMillan Press, London.

Winkler, M. (1981). Biological Treatment of Wastewater. Ellis Horwood, Chichester (UK).

Winkler, S., Muller-Rechberger, H., Nowak, O., Svardal, K. and Wandl, G. (2001) A new approach towards modelling of the carbon degradation cycle at two-stage activated sludge plants. *Water Sci. Technol.* **43**(7), 19–27.

Witteborg, A., van der Last, A., Hamming, R. and Hemmers, I. (1996) Respirometry for determination of the influent S_s concentration. *Water Sci. Technol.* **33**(1), 311–323.

Wolman, A. (1924) Sanitary engineering. *Am. J. Public Health* **14**, 450–452.

Wouters-Wasiak, K., Heduit, A. and Audic, J.M. (1996) Consequences of an occasional secondary phosphorus release on enhanced biological phosphorus removal. *Water SA* **22**, 91–96.

Wuhrmann, K. (1962) Nitrogen removal in the sewage treatment process. In *Proc. 15th Int. Congress of Limnology*, Madison, 22 August.

Wuhrmann, K. (1964) Nitrogen removal in sewage treatment process. *Verh. Int. Ver. Limnol.* **15**, 580–596.

Xu, S. and Hultman, B. (1996) Experiences in wastewater characterization and model calibration for the activated sludge process. *Water Sci. Technol.* **33**(12), 89–98.

Xu, S.L. and Hasselblad, S. (1997). A simple biological method to estimate the readily biodegradable organic matter in wastewater. *Water Res.* **30**, 1023–1025.

Yagci, N., Insel, G. and Orhon, D. (2004) Modelling and calibration of phosphate and glycogen accumulating organism competition for acetate uptake in a sequencing batch reactor, *Water Sci. Technol.* **50**(6), 241–250.

Yoshioka, N., Hotta, Y. and Tanaka, S. (1955) Characteristics of batch settling of homogeneous slurry. *Kagaku Kogaku* **19**, 616–625.

Yuan, Z., Bogaert, H., Vansteenkiste, G. and Verstraete, W. (1998) Sludge storage for countering nitrogen shock loads and toxicity incidents. *Water Sci. Technol.* **37**(12), 173–180.

Yuan, Z.G., Bogaert, H., Leten, J. and Verstraete, W. (2000) Reducing the size of a nitrogen removal activated sludge plant by shortening the retention time of inert solids via sludge storage. *Water Res.* **34**, 539–549.

Zeidan, A., Rohani, S., Bassi, A. and Whiting, P. (2003) Review and comparison of solids settling velocity models. *Rev. Chem. Eng.* **19**, 473–530.

Zeng, R.J., van Loosdrecht, M.C.M., Yuan, Z.G. and Keller, J. (2003a) Metabolic model for glycogen-accumulating organisms in anaerobic/aerobic activated sludge systems. *Biotechnol. Bioeng.* **81**, 92–105.

Zeng, R.J., Yuan, Z.G. and Keller, J. (2003b) Model-based analysis of anaerobic acetate uptake by a mixed culture of polyphosphate-accumulating and glycogen-accumulating organisms. *Biotechnol. Bioeng.* **83**, 293–302.

Zeng, R.J., Yuan, Z.G., van Loosdrecht, M.C.M. and Keller, J. (2002) Proposed modifications to metabolic model for glycogen-accumulating organisms under anaerobic conditions. *Biotechnol. Bioeng.* **80**, 277–279.

Zhang, D.J., Li, Z.L., Lu, P.L., Zhang, T. and Xua, D.Y. (2006) A method for characterizing the complete settling process of activated sludge. *Water Res.* **40**, 2637–2644.

Zhao, H.W., Mavinic, D.S., Oldham, W.K. and Koch, F.A. (1999) Controlling factors for simultaneous nitrification and denitrification in a two-stage intermittent aeration process treating domestic sewage. *Water Res.* **33**, 961–970.

Ziglio, G., Andreottola, G., Foladori, P. and Ragazzi, M. (2001) Experimental validation of a single-OUR method for wastewater RBCOD characterization. *Water Sci. Technol.* **43**(11), 119–126.

Zilverentant, A. (1999) Process for the treatment of wastewater containing specific components e.g. ammonia. Patent PCT/NL99/00462, WO0005177, DHV (Holland).

Zima, P., Makinia, J., Swinarski, M. and Czerwionka, K. (2008) Effects of different hydraulic models on predicting longitudinal profiles of reactive pollutants in activated sludge reactors. *Water Sci. Technol.* **58**(3), 555–561.

Zima, P., Makinia, J., Swinarski, M. and Czerwionka, K. (2009) Combining CFD with biokinetic models for predicting ammonia and phosphate behavior in aeration tanks. *Water Environ. Res.* **81**, 2353–2362.

Zwietering, M.H., de Koos, J.T., Hasenack, B.E., de Witt, J.C. and van't Riet, K. (1991) Modeling of bacterial growth as a function of temperature. *Appl Environ Microbiol.* **57**, 1094–1101.

Index

Lightning Source UK Ltd.
Milton Keynes UK
06 October 2010

160790UK00001B/27/P